WITHDRAWN
ES
D0709438

The Rho GTPases in Cancer

Kenneth van Golen
Editor

The Rho GTPases in Cancer

Editor
Kenneth van Golen, PhD
Assistant Professor of Biological Science
Head Laboratory of Cytoskeletal Physiology
320 Wolf Hall
The University of Delaware
Newark, DE 19716
USA
klvg@udel.edu

ISBN 978-1-4419-1110-0 e-ISBN 978-1-4419-1111-7
DOI 10.1007/978-1-4419-1111-7
Springer New York Dordrecht Heidelberg London

Library of Congress Control Number: 2009933113

Printed on acid-free paper

Springer is part of Springer Science+Business Media (www.springer.com)

Preface

Many molecular changes within a tumor cell occur during cancer progression. Cell growth, survival, morphology, and invasive characteristics are altered and contribute to the formation of advanced disease that eventually disseminates from the primary tumor and establishes new tumors at distant sites. Nearly all aspects of cellular behavior are controlled by small guanosine triphosphatases (GTPases). The importance of these molecules is highlighted in nature, as a majority of bacterial toxins specifically target the small GTPases and thus exert their deleterious effects by inhibiting these proteins.

In the early 1980s activating mutations in the Ras proteins were found to occur in a number of human cancers and thus became a focus of cancer research. However, not all cancers contained mutant Ras; therefore, other Ras-like proteins were thought to exist and contribute to cancer formation and progression. The discovery of the Ras homology or Rho GTPases and other small GTP-binding proteins caused quite a bit of excitement in the cancer world and much work has ensued over the years demonstrating their roles in cancer progression.

Currently, the Ras superfamily of small GTP-binding proteins has more than 150 members in five major separate subfamilies: Ras, Rho, Rab, Ran, and Arf. Each of these subfamilies controls an aspect of cellular behavior: Ras is intimately involved in growth regulation, Rho regulates the cytoskeleton, Rab and Arf are primarily involved in vesicle fusion, and Ran is involved in nuclear transport.

Because of their intimate involvement with the cytoskeleton, the Rho GTPases are primarily implicated tumor cell progression from transformation to metastasis. Since their discovery in the mid-1980s, many of the Rho proteins have been analyzed for activating mutations similar to Ras. With very few exceptions, activating mutations have not been identified. Instead, aberrant expression and activation in the absence of mutations appears to be the rule in human cancers. This volume is intended to give a concise overview of the current knowledge of Rho GTPase biology in the context of cancer progression. The book is separated into four sections, each of which explores a different aspect of Rho biology and its contribution to cancer etiology.

The first section is an overview of the Rho proteins in both normal and cancer cells. Combined with an introduction from Alan Hall, one of the pioneers of Rho biology, and his colleagues, this section provides a comprehensive overview of Rho history, regulation, signal transduction, phenotype, and its role in cancer progression.

Channing Der and colleagues provide an encyclopedic overview of the Rho GTPases, providing enough detail to make any reader well-versed in the Rho field. Finally, Sofia Merajver's laboratory provides an overview, which details the roles of the Rho proteins in cancer progression. She provides us with the history of the study of the Rho GTPases, their regulatory and effector proteins in cancer and gives us a benchmark of where the field is today.

The second section of the book details the current knowledge of the Rho regulatory proteins in cancer progression: aberrant expression and activation of these proteins leads to dysfunctional Rho signaling and a cancer phenotype.

Gary Bokoch's laboratory has provided a detailed overview of the role of Rho guanine dissociation inhibitors (GDIs) in cancer. These molecules are involved in preventing the Rho protein from associating with the inner plasma membrane and exchanging GDP for GTP, and thus becoming active. Next, Tozu Kazasa's laboratory has worked on the link between heterotrimeric G proteins and Rho activation via the RGS–RhoGEFs. This aspect of Rho activation is particularly interesting in that heterotrimeric G proteins and their associated G-protein-coupled receptors are attractive and attainable therapeutic targets. Dan Billadeau's laboratory has worked extensively on the Vav RhoGEFs, which are potent oncogenes in their own right. The chapter that he offers here covers the ever-emerging details of how these specific GEFs contribute to cancer progression. Finally, Yi Zheng and colleagues describe the role of the RhoGAPS in cancer progression. Increasingly, it is becoming apparent that RhoGAPs have other functions in addition to catalyzing the hydrolysis of Rho-bound GTP and closing of the GTPase cycle.

The third section deals with the current understanding of the Rho GTPase proteins themselves in cancer. This section goes beyond an understanding of the three most studied Rho proteins: RhoA, Rac1, and Cdc42. Instead, the chapters explore some of the lesser known, but eminently important GTPases.

Harry Mellor begins this section by describing the biology and contribution of RhoBTB, one of the least studied Rho proteins, in breast cancer. My own laboratory has studied the role of RhoC GTPase in the metastatic phenotype of several cancers and this, along with the work of other laboratories, is summarized in the second chapter of this section. In collaboration with George Prendergast and his colleagues, we have shown interactions between RhoC and RhoB GTPases. The biology and functions of RhoB are highlighted in a chapter by George Prendergast and colleagues. Finally, we focus on a short overview of another potentially important, yet understudied mechanism of Rho GTPase activation and control, the concept of phosphorylation by protein kinases. This brief chapter gives an overview of some of the salient work performed by a number of laboratories on the control of Rho activation by a variety of protein kinases.

Finally, Michael Olsen provides us with a comprehensive overview of one of the apparently most important Rho effector proteins, Rho kinase, a.k.a. ROCK. Dr. Olsen's laboratory has significantly contributed to the understanding of these proteins and their contribution to cancer progression and metastases. His discussion of this extremely important effector is exhaustive and provides a great deal of detail of its role in cancer progression.

Together, these chapters not only provide a detailed overview of essential background Rho GTPase biology, but also provide a comprehensive and detailed description of the state of knowledge of the field as it stands today. These chapters provide insight into an ever-expanding field that has become increasingly recognized over the past decade as being key to the understanding of cancer progression. Without a doubt, the Rho proteins and their related signaling molecules will provide valuable therapeutic targets in the near future.

Newark, DE Kenneth van Golen

Contents

Part I An Overview of the Rho GTPases

1 **Overview of Rho GTPase History** 3
Ellen V. Stevens and Channing J. Der

2 **Rho Proteins in Cancer** 29
Devin T. Rosenthal, John Chadwick Brenner, and Sofia D. Merajver

Part II The Rho Regulatory Proteins in Cancer

3 **RhoGDIs in Cancer** 45
Anthony N. Anselmo, Gary M. Bokoch, and Céline DerMardirossian

4 **Signaling through Galpha12/13 and RGS-RhoGEFs** 59
Nicole Hajicek, Barry Kreutz, and Tohru Kozasa

5 **Vav Proteins in Cancer** 77
Daniel D. Billadeau

6 **Rho GTPase-Activating Proteins in Cancer** 93
Matthew W. Grogg and Yi Zheng

Part III The Rho GTPase Proteins and Cancer

7 **RhoBTB Proteins in Cancer** 111
Caroline McKinnon and Harry Mellor

8 **RhoC GTPase in Cancer Progression and Metastasis** 123
Kenneth van Golen

9 **RhoB GTPase and FTIs in Cancer** 135
Minzhou Huang, Lisa D. Laury-Kleintop, and George C. Prendergast

10 Regulation of Rho GTPase Activity Through Phosphorylation Events: A Brief Overview 155
Heather Unger and Kenneth van Golen

11 The Rho-Regulated ROCK Kinases in Cancer 163
Grant R Wickman, Michael Samuel, Pamela A Lochhead, and Michael F Olson

Author Index 193

Subject Index 203

Contributors

Anthony N. Anselmo
Departments of Immunology and Cell Biology, The Scripps Research Institute,
10550 N. Torrey Pines Road, La Jolla, CA 92037, USA

Daniel D. Billadeau
Department of Immunology and Division of Oncology Research,
College of Medicine, Mayo Clinic, Rochester, MN 55905, USA

Gary M. Bockoch
Departments of Immunology and Cell Biology, The Scripps Research Institute,
10550 N. Torrey Pines Road, La Jolla, CA 92037, USA

John C. Brenner
Department of Internal Medicine and Comprehensive Cancer Center, The
University of Michigan, 1500 E. Medical Center Dr., Ann Arbor, MI 48109, USA

Channing J. Der
Department of Pharmacology, University of North Carolina at Chapel Hill,
Lineberger Comprehensive Cancer Center, Chapel Hill, NC 27599, USA

Céline DerMardirossian
Departments of Immunology and Cell Biology, The Scripps Research Institute,
10550 N. Torrey Pines Road, La Jolla, CA 92037, USA

Matthew W. Grogg
Division of Experimental Hematology and Cancer Biology, Children's Research
Foundation, Cincinnati Children's Hospital Medical Center, University
of Cincinnati, 3333 Brunet Avenue, Cincinnati, OH 45229, USA

Nicole Hajicek
Department of Pharmacology, College of Medicine,
University of Illinois at Chicago, Chicago, IL 60602, USA

Alan Hall
Cell Biology Program, Memorial Sloan-Kettering Cancer Center,
1275 York Ave., New York, NY 10065, USA

Minzhou Huang
Lankenau Institute for Medical Research, Wynnewood, PA 19096, USA

Aron Jaffe
Cell Biology Program, Memorial Sloan-Kettering Cancer Center,
1275 York Ave., New York, NY 10065, USA

Tohru Kozasa
Department of Pharmacology, College of Medicine,
University of Illinois at Chicago, Chicago, IL 60602, USA

Barry Kreutz
Department of Pharmacology, College of Medicine,
University of Illinois at Chicago, Chicago, IL 60602, USA

Lisa D. Laury-Kleintop
Lankenau Institute for Medical Research, Wynnewood, PA 19096, USA

Pamela A. Lochhead
Molecular and Cellular Biology, The Beatson Institute for Cancer Research,
Garscube Estate, Switchback Road, Glasgow G61 1BD, UK

Caroline McKinnon
Department of Biochemistry, School of Medicine, University of Bristol,
University Walk, Bristol BS8 1TD, UK

Harry Mellor
Department of Biochemistry, School of Medicine, University of Bristol,
University Walk, Bristol BS8 1TD, UK

Sofia D. Merajver
Department of Internal Medicine and Comprehensive Cancer Center, The
University of Michigan, 1500 E. Medical Center Dr., Ann Arbor, MI 48109, USA

Michael F. Olson
Molecular and Cellular Biology, The Beatson Institute for Cancer Research,
Garscube Estate, Switchback Road, Glasgow G61 1BD, UK

George C. Prengergast
Lankenau Institute for Medical Research, Wynnewood, PA 19096, USA
Department of Pathology, Anatomy and Cell Biology and Kimmel Cancer Center,
Thomas Jefferson University and Kimmel Cancer Center, Philadelphia, PA 19107,
USA

Devin T. Rosenthal
Department of Internal Medicine and Comprehensive Cancer Center, The
University of Michigan, 1500 E. Medical Center Dr., Ann Arbor, MI 48109, USA

Michael Samuel
Molecular and Cellular Biology, The Beatson Institute for Cancer Research,
Garscube Estate, Switchback Road, Glasgow G61 1BD, UK

Ellen V. Stevens
Department of Pharmacology, University of North Carolina at Chapel Hill, Lineberger Comprehensive Cancer Center, Chapel Hill, NC 27599, USA

Heather Unger
Department of Biological Sciences, The Center for Translational Cancer Research, Laboratory of Cytoskeletal Physiology, The University of Delaware, Newark, DE 19716, USA

Kenneth van Golen
Department of Biological Sciences, The Center for Translational Cancer Research, Laboratory of Cytoskeletal Physiology, The University of Delaware, Newark, DE 19716, USA

Grant R. Wickman
Molecular and Cellular Biology, The Beatson Institute for Cancer Research, Garscube Estate, Switchback Road, Glasgow G61 1BD, UK

Yi Zheng
Division of Experimental Hematology and Cancer Biology, Children's Research Foundation, Cincinnati Children's Hospital Medical Center, University of Cincinnati, 3333 Brunet Avenue, Cincinnati, OH 45229, USA

Introduction: The Rho GTPases and Cancer

Aron Jaffe and Alan Hall

The first members of the Rho family of small GTPases were discovered some 20 years ago based on their homology with the Ras oncogene. Since then, the focus of research has evolved from the identification of components of Rho GTPase-mediated signaling pathways to the analysis of how these signaling pathways control cell behavior during development in the adult, and in disease processes such as cancer.

Because of their sequence similarity to Ras, many of the early studies focused on whether Rho GTPases could also act as oncogenes and whether they were required for Ras-induced transformation. The simplest of these experiments involved expression of constitutively active or dominant negative versions of these GTPases, particularly Rho, Rac, and Cdc42, in fibroblasts, similar to the classic studies carried out with Ras after its discovery as a human oncogene in 1981. The picture that emerged from this body of work was complex. Although constitutively active Rho family GTPases are able to induce some of the phenotypes associated with oncogenic transformation (e.g., formation of foci, growth in soft agar), they are much less efficient than constitutively active Ras. However, the use of dominant negative forms of Rho GTPases strongly suggested that their activities are required for efficient transformation by Ras. We now know that all three GTPases make important contributions to the eukaryotic cell cycle, notably in promoting G1 progression and in organizing the actin and microtubule cytoskeletons during mitosis. The details of these studies and their implications are discussed in Sect. I, while the specific role of individual Rho family members in cancer is discussed in Sect. III.

GEFs, GAPs, and GDIs

Similar to Ras, Rho GTPases are highly regulated, both temporally and spatially, under normal conditions. Although the proteins responsible for controlling their activation, the guanine nucleotide exchange factors (GEFs), the GTPase activating proteins (GAPs), and the guanine nucleotide dissociation inhibitors (GDIs) were

A. Jaffe and A. Hall (✉)
Memorial Sloan-Kettering Cancer Center, 1275 York Avenue, New York, NY 10065, USA
e-mail: halla@mskcc.org

all identified around 15 years ago, they have been largely neglected until relatively recently. The first mammalian Rho family GEF was identified through sequence comparison with the yeast protein Cdc24 (Hart et al. 1991). Subsequently, several more were identified in transformation assays, since they can be readily activated by a simple truncation and then very efficiently transform immortalized fibroblasts (Whitehead et al. 1996). More recently, the human genome sequencing project has revealed 82 Rho family GEFs belonging to two distinct families (DH and CZH) (Bernards 2005). This very large number of GEFs (four times as many as the 22 Rho family members) was a surprise and likely reflects the large variety of inputs that can activate Rho family proteins, the large number of potential downstream effectors that are activated by Rho GTPases, and the fact that spatially localized activation is a key feature of their biological function. As it turns out, there are very few examples of a mutated GEF in human cancers. However, the in vitro transformation assays strongly suggest that aberrant GEF activity is likely to be a more potent mechanism for promoting the transformed phenotype than is constitutive activation of the corresponding GTPase. Accordingly, in the last few years, there has been much more interest shown in the GEF family, and not only in the cancer field.

The identification of the first GAP came through biochemical purification of an activity detected in mammalian cell extracts (Diekmann et al. 1991). Again through analysis of the human genome, around 67 Rho family GAPs have now been identified (Bernards 2005). The inactivation of a GAP could potentially have the same effect as activation of a GEF, i.e., hyperactivation of the corresponding GTPase. There is a very clear example of this in the Ras field, where the Ras GAP protein neurofibromin (encoded by the *NF1* gene) is a bona fide tumor suppressor (Cichowski and Jacks 2001). This has raised the possibility that the loss of expression of one of the 67 Rho family GAPs might make a contribution to human cancer and consequently stimulated interest in the field.

Finally, the three Rho family GDIs are also believed to act as inhibitors of Rho activity, though how they themselves are regulated is very poorly understood (Ueda et al. 1990). There is no known equivalent activity regulating Ras proteins and it is speculated that the association of Rho GTPases with membranes (which is affected by GDIs) is a more significant aspect of their life cycle than it is with Ras. Section II is dedicated to the GEFs, GAPs, and GDIs that regulate Rho family GTPases and their potential contribution to human cancer.

Downstream Effectors

The proteins that mediate signaling downstream of Rho GTPases are structurally diverse, but share the common biochemical property of binding preferentially to the active, GTP-bound form of the GTPase. This has been exploited both in yeast two-hybrid screens and in biochemical affinity purifications to identify around 100 targets or "effector" proteins (Bernards 2005). Since there is no unique sequence that defines an effector, the full extent of targets interacting with Rho family GTPases is unknown. Within this group, several harbor catalytic activities, notably protein kinases (both ser/thr and tyr), lipid kinases, and phospholipases, but the majority are scaffold proteins. While these do not encode enzymatic activities, they interact with other proteins that do, such as NADPH oxidases, protein kinases, protein phosphatases and, perhaps best studied of all, promoters of actin polymerization. There has been sustained interest in Rho effector proteins since the first one was identified in 1991 (Abo et al. 1991) and Sect. IV describes some of the work that has been done to examine their contribution to cancer.

Cell Biology

With the growing wealth of information regarding the components of Rho GTPase-mediated signaling pathways, the major challenge now is to understand how these pathways contribute to diverse cell biological processes. Advances in microscopy and in techniques for manipulating mammalian cells in both 2D and 3D culture systems have provided the foundation for studying Rho pathways in contexts closely resembling in vivo situations. In addition, Rho GTPases are being studied in a wide range of genetically tractable organisms (yeast, *C. elegans, Drosophila*, zebrafish, and mice). These complementary approaches have generated significant advances over the last 5 years, which have been reviewed extensively elsewhere (Johndrow et al. 2004; Jaffe and Hall 2005). Two aspects of cell behavior that are crucial to normal development, but which go awry during cancer progression, and in which Rho GTPases play a very prominent role are migration and morphogenesis. Since these are not addressed specifically elsewhere in this book, we will finish with a few comments on these processes.

Cell Migration

The current view of how Rho GTPases regulate cell migration has come from numerous studies carried out both in vitro and more recently in vivo. In vivo, cells can move as individuals, as groups, or as whole sheets and although the mechanisms and pathways involved are not identical, Rho GTPases play a central role in each case. There are many excellent reviews of these studies (Ridley et al. 2003; Jaffe and Hall 2005; Sahai 2005), which have led to a model in which Cdc42 and Rac orient the cell and provide protrusive forces, respectively, at the front, while Rho mediates contractile forces at the rear. However, new biosensors developed to visualize the spatial localization of active GTPases in living cells has revealed a more complex situation. For example, active Rho can be detected at the front as well as at the rear in some migrating cells, though it would seem likely that it is doing different things in the two locations (Pertz et al. 2006).

Recently, new and significant observations have been made by examining cell migration in 3D matrices to mimic better the in vivo environment. Here too, Rho GTPases play a prominent role, but there are important differences from migration on 2D tissue culture plates. In particular, isolated tumor cells appear to use two distinct modes of migration in 3D (Fig. 1a) and in some cases they can inter-convert between the two depending on the environment. Some tumor cells use an ameboid-like mode, involving localized plasma membrane blebbing, mediated by Rho/ROCK, to generate forward movement by squeezing through gaps in the extracellular matrix (ECM), while others use a mesenchymal-like mode, which does not require Rho or ROCK activity, and involves reorganization of the ECM by matrix metalloproteases (MMPs) (Sahai and Marshall 2003). Some tumor cells have been seen to migrate as groups together with stromal fibroblasts in vivo, and this has also been reproduced to some extent using a 3D co-culture system. It appears that Rho signaling in the fibroblasts is required for remodeling of the ECM, which leads to the generation of "tracks" that are then used by the carcinoma cells to migrate in a Cdc42-dependent manner (Gaggioli et al. 2007). The generation of these culture systems, together with time-lapse video microscopy, will be critical to further our understanding of the biochemical pathways through which Rho GTPases regulate migration in the complex environments found in vivo.

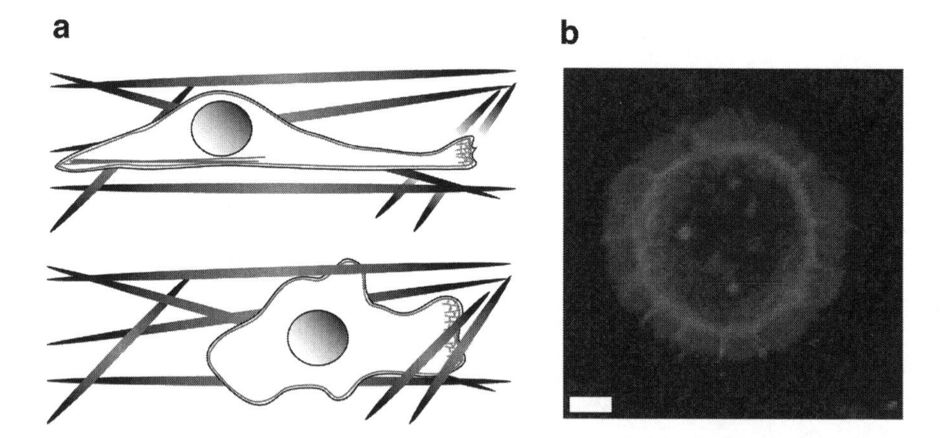

Fig. 1 (**a**) Cancer cells utilize two distinct modes of motility in 3D. *Top*, cells using mesenchymal motility have an actin-rich lamellipodium at the front of the cell, and stress fibers toward the rear of the cell to retract the cell tail. Mesenchymal motility requires MMPs to degrade the extracellular matrix, but does not require Rho-ROCK signaling. *Bottom*, cells using amoeboid motility display membrane blebbing and move by squeezing through gaps in the ECM. Amoeboid motility requires Rho-ROCK signaling but does not require MMPs. (**b**) Three-dimensional MDCK cyst. Shown is a confocal section through the middle of a cyst produced from MDCK cells grown in collagen I for 6 days. Actin (*red*) is enriched at the apical surface, facing the lumen. *Blue*, DNA. Scale bar, 10 mm

Morphogenesis

One of the hallmarks of tumor progression, particularly in epithelial cell types, is the loss of tissue architecture, which is associated with alterations in cell–cell adhesion, loss of polarity, and aberrant cell proliferation. The distinct polarized morphology of epithelial cells and the assembly and disassembly of cell–cell interactions have been studied extensively in 2D culture systems. Using such relatively simple assays, Rho and Rac were first shown to play an important role in adherens junction assembly some 10 years ago, but since then these two GTPases together with Cdc42 have been shown to play important roles in many additional aspects of epithelial morphogenesis (Braga et al. 1997; Etienne-Manneville and Hall 2002). More recently, there has been much interest taken in the ability of epithelial cells to organize themselves into in vivo-like structures (acini, tubes, and cysts) when cultured in 3D matrices: an observation that was first made over 25 years ago (Hall et al. 1982). Several groups have expressed oncogenes in these 3D culture systems to model the morphological changes associated with tumor progression (Debnath and Brugge 2005).

The best explored 3D system for analyzing the contribution of Rho GTPases to epithelial morphogenesis utilizes the canine kidney epithelial cell line MDCK. These cells proliferate and form hollow, polarized cysts when cultured for 5–7 days in collagen I or Matrigel™. The apical surface faces inside (toward the lumen) and the basal surface faces outside (toward the matrix) (Fig. 1b). When Rac is inhibited, the polarity of the cyst is inverted, with apical facing out and basal facing in (O'Brien et al. 2001). Similar work using the T4-2 transformed mammary epithelial cell line suggests that this activity of Rac in controlling polarity is regulated by PI 3-kinase (Liu et al. 2004). Together, these studies suggest that the levels of active Rac must be precisely regulated, both spatially and temporally, to achieve a polarized epithelial structure.

More recent work, again using MDCK cells, has revealed that Cdc42 is also a key player in 3D morphogenesis and it too appears to be regulated by the polarized distribution of phosphoinositides (Gassama-Diagne et al. 2006; Martin-Belmonte et al. 2007). Although these studies have not directly explored the role of Rho family GTPases in mediating effects induced by oncogenes, the tumor suppressor PTEN is required to achieve a polarized distribution of PIP2 and mislocalized PIP2 disrupts the localization of Cdc42. Specifically, loss of PTEN by siRNA produces a 3D structure with multiple lumens and this effect is phenocopied by Cdc42 siRNA (Martin-Belmonte et al. 2007). Cdc42 is a key regulator of polarity in a wide variety of cellular contexts and in many cases it acts through the evolutionarily conserved protein complex consisting of Par6 and atypical PKC (aPKC) (Etienne-Manneville and Hall 2001). It is interesting to note that, in a 3D model of mammary epithelial morphogenesis (using MCF10A cells), activation of the ErbB2 oncogene, to mimic the effects of ErbB2 overexpression in human breast tumors, promotes its binding to the Par6/aPKC complex and loss of polarity (Aranda et al. 2006). The coming years will undoubtedly see more interest in these 3D morphogenesis assays to probe oncogene function.

Animal Models

Alongside the in vitro work, there is growing interest in exploring the phenotypic consequences of modifying Rho GTPase signaling pathways in vivo. Much has been done in yeast, flies, and worms, but many tissue-specific knockouts are also being performed in the mouse. One particular avenue relevant to this book concerns the contribution of Rho pathways to tumorigenesis. Tiam-1 is a GEF for Rac and was initially identified through its ability to induce an invasive phenotype in T-lymphoma cells in an in vitro co-culture assay (Habets et al. 1994). Subsequent in vitro assays have implicated Tiam-1 in cell migration, cell polarity, morphogenesis, and proliferation. However, the role of Tiam-1 in cancer was directly assessed in vivo using a protocol for Ras-induced skin cancer (Malliri et al. 2002). There were fewer tumors in the Tiam-1 knockout mice, though interestingly a higher percentage progressed to malignancy (Malliri et al. 2002), suggesting that Tiam-1 (presumably through its effects on Rac) plays different roles during tumor initiation and tumor progression. Knockout mice for all three Rac family members have been described (Walmsley et al. 2003; Corbetta et al. 2005), and it will be interesting to see which is required downstream of Tiam-1 for tumor initiation and progression. In addition to Tiam-1, the Rho GTPase RhoC has been shown to be specifically required for tumor metastasis, but not for tumor initiation (Hakem et al. 2005) (see Chap. 8 for details).

Future Directions

It is clear that Rho GTPases play an important role at multiple steps during the onset and progression of cancer. As technology continues to advance, the precise cellular role of the numerous regulators and effectors associated with Rho pathways will emerge. One of the greatest challenges in the future will be to integrate this information with other pathway analyses, so as to promote the development of therapies that target key components required for tumor initiation or progression, without affecting normal cell behavior.

References

Abo A, Pick E, Hall A, Totty N, Teahan CG, Segal AW (1991) Activation of the NADPH oxidase involves the small GTP-binding protein p21rac1. Nature 353:668–670

Aranda V, Haire T, Nolan ME, Calarco JP, Rosenberg AZ, Fawcett JP, Pawson T, Muthuswamy SK (2006) Par6-aPKC uncouples ErbB2 induced disruption of polarized epithelial organization from proliferation control. Nat Cell Biol 8:1235–1245

Bernards A (2005) Ras superfamily and interacting proteins database. Methods Enzymol 407:1–9

Braga VM, Machesky LM, Hall A, Hotchin NA (1997) The small GTPases Rho and Rac are required for the establishment of cadherin-dependent cell–cell contacts. J Cell Biol 137:1421–1431

Cichowski K, Jacks T (2001) NF1 tumor suppressor gene function: narrowing the GAP. Cell 104:593–604

Corbetta S, Gualdoni S, Albertinazzi C, Paris S, Croci L, Consalez GG, de Curtis I (2005) Generation and characterization of Rac3 knockout mice. Mol Cell Biol 25:5763–5776

Debnath J, Brugge JS (2005) Modelling glandular epithelial cancers in three-dimensional cultures. Nat Rev Cancer 5:675–688

Diekmann D, Brill S, Garrett MD, Totty N, Hsuan J, Monfries C, Hall C, Lim L, Hall A (1991) Bcr encodes a GTPase-activating protein for p21rac. Nature 351:400–402

Etienne-Manneville S, Hall A (2001) Integrin-mediated activation of Cdc42 controls cell polarity in migrating astrocytes through PKCzeta. Cell 106:489–498

Etienne-Manneville S, Hall A (2002) Rho GTPases in cell biology. Nature 420:629–635

Gaggioli C, Hooper S, Hidalgo-Carcedo C, Grosse R, Marshall JF, Harrington K, Sahai E (2007) Fibroblast-led collective invasion of carcinoma cells with differing roles for RhoGTPases in leading and following cells. Nat Cell Biol 9:1392–1400

Gassama-Diagne A, Yu W, ter Beest M, Martin-Belmonte F, Kierbel A, Engel J, Mostov K (2006) Phosphatidylinositol-3,4,5-trisphosphate regulates the formation of the basolateral plasma membrane in epithelial cells. Nat Cell Biol 8:963–970

Habets GG, Scholtes EH, Zuydgeest D, van der Kammen RA, Stam JC, Berns A, Collard JG (1994) Identification of an invasion-inducing gene, Tiam-1, that encodes a protein with homology to GDP-GTP exchangers for Rho-like proteins. Cell 77:537–549

Hakem A, Sanchez-Sweatman O, You-Ten A, Duncan G, Wakeham A, Khokha R, Mak TW (2005) RhoC is dispensable for embryogenesis and tumor initiation but essential for metastasis. Genes Dev 19:1974–1979

Hall HG, Farson DA, Bissell MJ (1982) Lumen formation by epithelial cell lines in response to collagen overlay: a morphogenetic model in culture. Proc Natl Acad Sci USA 79:4672–4676

Hart MJ, Eva A, Evans T, Aaronson SA, Cerione RA (1991) Catalysis of guanine nucleotide exchange on the CDC42Hs protein by the dbl oncogene product. Nature 354:311–314

Jaffe AB, Hall A (2005) RHO GTPASES: biochemistry and biology. Annu Rev Cell Dev Biol 21:247–269

Johndrow JE, Magie CR, Parkhurst SM (2004) Rho GTPase function in flies: insights from a developmental and organismal perspective. Biochem Cell Biol 82:643–657

Liu H, Radisky DC, Wang F, Bissell MJ (2004) Polarity and proliferation are controlled by distinct signaling pathways downstream of PI3-kinase in breast epithelial tumor cells. J Cell Biol 164:603–612

Malliri A, van der Kammen RA, Clark K, van der Valk M, Michiels F, Collard JG (2002) Mice deficient in the Rac activator Tiam1 are resistant to Ras-induced skin tumours. Nature 417:867–871

Martin-Belmonte F, Gassama A, Datta A, Yu W, Rescher U, Gerke V, Mostov K (2007) PTEN-mediated apical segregation of phosphoinositides controls epithelial morphogenesis through Cdc42. Cell 128:383–397

O'Brien LE, Jou TS, Pollack AL, Zhang Q, Hansen SH, Yurchenco P, Mostov KE (2001) Rac1 orientates epithelial apical polarity through effects on basolateral laminin assembly. Nat Cell Biol 3:831–838

Pertz O, Hodgson L, Klemke RL, Hahn KM (2006) Spatiotemporal dynamics of RhoA activity in migrating cells. Nature 440:1069–1072

Ridley AJ, Schwartz MA, Burridge K, Firtel RA, Ginsberg MH, Borisy G, Parsons JT, Horwitz AR (2003) Cell migration: integrating signals from front to back. Science 302:1704–1709

Sahai E (2005) Mechanisms of cancer cell invasion. Curr Opin Genet Dev 15:87–96

Sahai E, Marshall CJ (2003) Differing modes of tumour cell invasion have distinct requirements for Rho/ROCK signalling and extracellular proteolysis. Nat Cell Biol 5:711–719

Ueda T, Kikuchi A, Ohga N, Yamamoto J, Takai Y (1990) Purification and characterization from bovine brain cytosol of a novel regulatory protein inhibiting the dissociation of GDP from and the subsequent binding of GTP to rhoB p20, a ras p21-like GTP-binding protein. J Biol Chem 265:9373–9380

Walmsley MJ, Ooi SK, Reynolds LF, Smith SH, Ruf S, Mathiot A, Vanes L, Williams DA, Cancro MP, Tybulewicz VL (2003) Critical roles for Rac1 and Rac2 GTPases in B cell development and signaling. Science 302:459–462
Whitehead IP, Khosravi-Far R, Kirk H, Trigo-Gonzalez G, Der CJ, Kay R (1996) Expression cloning of lsc, a novel oncogene with structural similarities to the Dbl family of guanine nucleotide exchange factors. J Biol Chem 271:18643–18650Introduction:The Rho GTPases and Cancer

Part I
An Overview of the Rho GTPases

Chapter 1
Overview of Rho GTPase History

Ellen V. Stevens and Channing J. Der

Introduction

It is estimated that approximately 1% of the human genome encodes proteins that either regulate or are regulated by direct interaction with members of the Rho (Ras-homologous) family of small GTPases (Jaffe and Hall 2005). Rho GTPases are low molecular weight proteins (21–28 kDa) that control complex biological processes. The most widely studied and best-characterized members are the three "classical" members, RhoA, Rac1, and Cdc42, and much of our current concepts of Rho GTPase function are based on their properties (Aspenstrom et al. 2007; Wennerberg and Der 2004). They act as bi-molecular switches that cycle between two conformational states: GDP-bound ("inactive" state) and GTP-bound ("active" state) and they hydrolyze GTP. In the "active" GTP-bound state, Rho is able to activate downstream effectors. There are 20 human family members (Fig. 1.1) that receive extracellular ligand-stimulated signals from upstream cell surface receptors including tyrosine kinase, cytokine, adhesion, integrin, and G-protein-coupled receptors (GPCRs). They transmit signals to a plethora of cytoplasmic effectors that regulate essentially all aspects of normal cell physiology (Etienne-Manneville and Hall 2002). These include the regulation of cell proliferation and survival, cell polarity and morphogenesis, adhesion, migration, growth, gene transcription, and vesicular trafficking. Moreover, a causal role for the aberrant regulation of Rho GTPases has been found for human disease, including cancer, and in neurological and developmental disorders (Govek et al. 2005; Sahai and Marshall 2002). Rho GTPases are also targeted by pathogenic bacteria to facilitate their infection of human cells (Finlay 2005; Munter et al. 2006). In this chapter, we provide a historical overview of some of the key discoveries that provide the foundation for our current knowledge of Rho GTPase regulation and function, with an emphasis on the importance of deregulated Rho GTPase function in human disease.

E.V. Stevens and C.J. Der (✉)
Department of Pharmacology, University of North Carolina at Chapel Hill, Lineberger Comprehensive Cancer Center, Chapel Hill, 27599, NC, USA
e-mail: channing_der@med.unc.edu

K. van Golen (ed.), *The Rho GTPases in Cancer*,
DOI 10.1007/978-1-4419-1111-7_1, © Springer Science+Business Media, LLC 2010

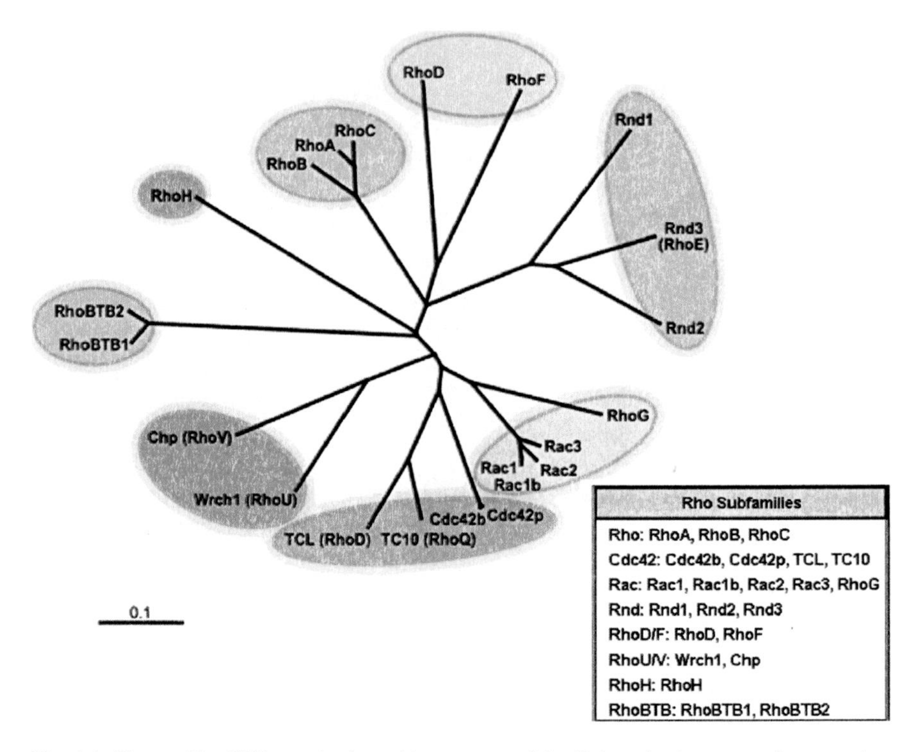

Fig. 1.1 Human Rho GTPases. Amino acid sequences of the G-domains (corresponding to amino acids 5–173 of Rac1) of human Rho GTPase family members were aligned using the ClustalX and the dendritic-tree was made with TreeViewX. The 20 members of the Rho GTPases, including two splice variants, can be subdivided into eight major branches: the Rho, Cdc42, Rac, Rnd, RhoD/F, RhoU/V, RhoH, and RhoBTB subfamilies

Evolution of Rho GTPases

Discovery

The 20 members of the human Rho GTPase family have been identified by various approaches (Table 1.1). The first *Rho* gene was identified serendipitously in 1985 through screening of a sea slug *Aplysia* cDNA library for the *Aplysia* homolog of human chorionic gonadotropin hormone (Madaule and Axel 1985) (Fig. 1.2). At that time in the early 1980s, point mutations in the Ras genes (*H-Ras*, *Ki-Ras*, and *N-Ras*) were already discovered in human cancers (Malumbres and Barbacid 2003). The mRNA of Rho encoded a 192 amino acid protein that shared 35% sequence identity with H-Ras, thereby giving the name Ras-homologous (Rho) (Madaule and Axel 1985). Cross hybridization and yeast-two hybrid screens identified three human homologues designated RhoA, RhoB, and RhoC (Chardin et al. 1988;

Table 1.1 Discovery of human Rho GTPases

Protein	Other names	Discovery	References
RhoA	ARHA, Rho H12	Identified as a gene related to Aplysia Rho in a screen of a human peripheral T cell cDNA library. Aplysia Rho was isolated inadvertently in a screen of an Aplysia abdominal ganglion cDNA library for sequences related to the alpha subunit of human chorionic gonadotropin hormone, and found to a encode *R*as *ho*mologous protein.	(Madaule and Axel 1985; Yeramian et al. 1987)
RhoB	ARHB, Rho H6	Identified as a gene related to Aplysia Rho in a screen of a human peripheral T cell cDNA library; identified as a gene that is rapidly induced in Rat-2 fibroblasts by transient activation of the retroviral tyrosine kinase oncoprotein Src.	(Chardin 1988; Jahner and Hunter 1991; Madaule and Axel 1985)
RhoC	ARHC, Rho H9	Identified as a gene related to Aplysia Rho in a screen of a human peripheral T cell cDNA library; identified as a gene overexpressed in breast, melanoma and other cancers.	(Chardin 1988; Clark et al. 2000; Madaule and Axel 1985; van Golen et al. 1999)
Rac1	TC25	Identified as a gene with sequence identity to an oligonucleotide encoding the peptide sequence FDTAGQEDYD from Cdc42/G25K purified from placenta. Analyses suggested that it was a *Ra*s-related *C*3 botulinum toxin substrate 1, although this was later found not to be the case; identified using a mixed-oligonucleotide probe corresponding to the conserved DTAGQE GTP binding motif of Ras to isolate ras-like coding sequences expressed in a human NTera2 teratocarcinoma cDNA library.	(Didsbury et al. 1989; Drivas et al. 1990)
Rac1b	Splice variant of Rac1	Identified as a larger fragment by RT-PCR amplification of human Rac1 sequence than expected and found to be a splice variant of Rac1 preferentially expressed in colorectal and breast tumors.	(Jordan et al. 1999; Schnelzer et al. 2000)
Rac2		Identified as a gene with sequence identity to an oligonucleotide encoding the peptide sequence FDTAGQEDYD from Cdc42/G25K purified from placenta. Analyses suggested that it was a *Ra*s-related *C*3 botulinum toxin substrate 2, although this was later found not to be the case.	(Didsbury et al. 1989)
Rac3		Identified in a search for Rac-related sequences by using the murine *rac*1 cDNA fragment to screen a human K562 cell line cDNA library.	(Haataja et al. 1997)

(continued)

Table 1.1 (continued)

Protein	Other names	Discovery	References
RhoG	ARHG	To isolate coding sequences specifically accumulated in late G1, a differential screening was performed on a cDNA library prepared from CCL39 hamster lung fibroblasts stimulated for 5 h with serum.	(Vincent et al. 1992)
Cdc42p	CDC42Hs, G25K	Identified as a GTP-binding protein expressed in placenta with strong sequence identity with the Saccharomyces cerevisiae gene product discovered as the 42nd mutant involved in the morphogenetic events of the *c*ell *d*ivision *c*ycle. The first Cdc42 gene isolated was from S. cerevisiae, and the placenta-derived G25K was later molecularly cloned.	(Adams et al. 1990; Polakis et al. 1989); (Johnson and Pringle 1990; Shinjo et al. 1990)
Cdc42b	G25K; splice variant of Cdc42p	Identified by screening a human fetal brain cDNA library with oligonucleotides corresponding to peptide sequences of purified G25K, and found to have strong sequence identity with Saccharomyces cerevisiae CDC42. This was later shown to be a splice variant of the human placental Cdc42.	(Munemitsu et al. 1990)
TC10	RhoQ, ARHQ, RasL7A	Identified using a mixed-oligonucleotide probe corresponding to the conserved DTAGQE GTP binding motif of Ras to isolate ras-like coding sequences expressed in a human NTera2 teratocarcinoma cDNA library.	(Drivas et al. 1990)
TCL	TC10β, RhoT, RhoJ, ARHJ, RasL7B	Identified as a TC10-like protein in a database search for Rho-related sequences in human and murine expressed sequence tag (EST) data bases TC10-like.	(Vignal et al. 2000)
Wrch-1	RhoU, ARHU, Cdc42L1	Identified by suppressive subtraction analysis of Wnt-1-regulated genes in C57MG mouse mammary epithelial cells, as a *W*nt-1 *r*esponsive *C*dc42 *h*omolog.	(Tao et al. 2001)
Chp	Wrch-2, RhoV, ARHV	Identified using a two-hybrid Ras recruitment system that detected protein-protein interactions, at the inner surface of the plasma membrane, with the Pak2 regulatory domain and proteins expressed in a myristoylated rat pituitary cDNA library.	(Aronheim et al. 1998)
Rnd1	Rho6, ARHS	Human homolog of a gene identified in a mixed oligonucleotide probe, corresponding to the Rho effector domain sequence (YVPTVFENYVADIE) and used to screen a bovine brain cDNA library. The bovine cDNA was then used to screen and isolate the human counterpart.	(Nobes et al. 1998)

Rnd2	Rho7, ARHN, RhoN	Human homolog of a bovine gene identified in a mixed oligonucleotide probe, corresponding to the Rho effector domain sequence (YVPTVFENYVADIE), used to screen a bovine brain cDNA library. The bovine cDNA was then used to screen and isolate the human counterpart.	(Nobes et al. 1998)
Rnd3	RhoE, Rho8, ARHE	Identified in a yeast interaction trap screen for p190 RhoGAP interacting proteins (RhoE); Nucleotide sequence identity with DTAGQE GTP binding motif; Human homolog of a bovine gene identified in a mixed oligonucleotide probe, corresponding to the Rho effector domain sequence (YVPTVFENYVADIE), used to screen a bovine brain cDNA library; identified using a DNA-chip expression array that identified p53-inducible genes.	(Foster et al. 1996; Nobes et al. 1998; Ongusaha et al. 2006)
RhoBTB1		Identified as a human homolog of a gene encoding the RacA Rho-related protein in Dictyostelium discoideum; identified as a gene within the10q21 deletion in head and neck squamous cell carcinomas and to define candidate tumor suppressor genes.	(Beder et al. 2006; Rivero et al. 2001)
RhoBTB2	DBC2	Identified as a human homolog of a gene encoding a Rho-related protein in Dictyostelium discoideum; identified in a representational difference analysis as a gene *d*eleted in *b*reast *c*ancer in 8p21.	(Hamaguchi et al. 2002; Rivero et al. 2001)
RhoH	TTF, ARHH	Identified as a fusion partner with LAZ3/BCL6 in a nonHodgkin's lymphoma cell line and cause by a t(3;4) translocation (*T*ranslocation *T*hree *F*our).	(Dallery et al. 1995)
Rif	ARHF, RhoF	Identified in a database search in the human Expressed Sequence Tag database, encoding a *R*ho *i*n *f*ilopodia	(Ellis and Mellor 2000)
RhoD	ARHD, RhoHP1	Isolated by PCR amplification using oligonucleotides corresponding to the GXXXXGKS/T and WDTAGQE GTP-binding motifs of Rab/Rho proteins, from mouse kidney RNA.	(Chavrier et al. 1992; Murphy et al. 1996)

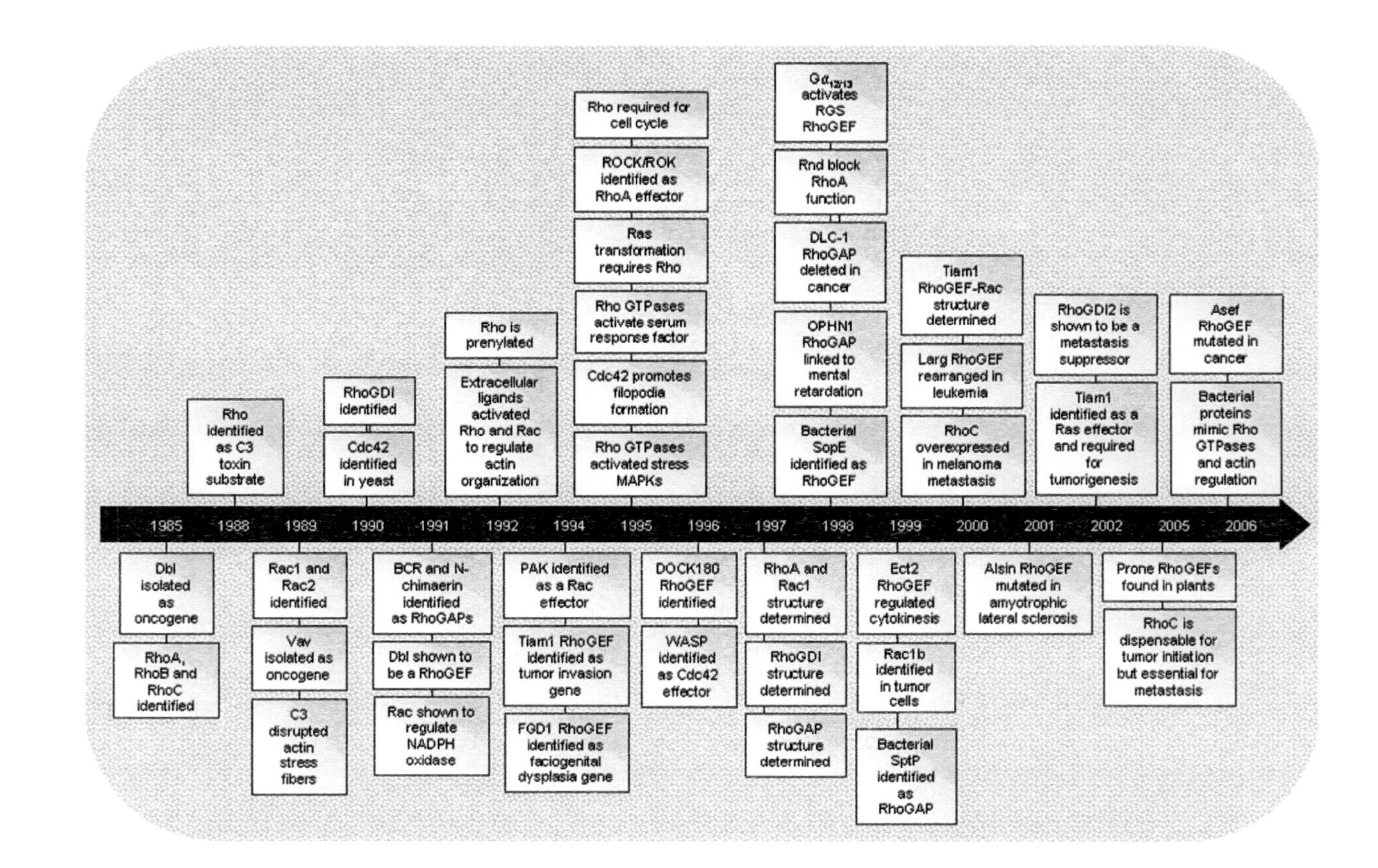

Fig. 1.2 Timeline of key discoveries in Rho research history

Yeramian et al. 1987). Also, Rho sequences were found to be evolutionarily conserved in the genomes of *Drosophila* and yeast.

The next Rho family GTPases identified were Rac1 and Rac2, which were isolated as genes with sequence identity to an oligonucleotide encoding the peptide sequence FDTAGQEDYD from Cdc42. Analyses suggested that it was a *Ra*s-related *C*3 botulinum toxin substrate, although this was later found not to be the case, due to contamination with Rho. A third member, Rac3, was later discovered in a search for Rac-related sequences by using the murine *Rac*1 cDNA fragment to screen a human K562 cell line cDNA library (Haataja et al. 1997). Finally, a splice variant of Rac1, designated Rac1b, was found to be expressed preferentially in breast and colon cancers (Jordan et al. 1999; Schnelzer et al. 2000).

Cdc42 was discovered by two independent, parallel directions of study. Cdc42 was identified initially in a mutagenesis screen in *S. cerevisiae* as a temperature-sensitive (*ts*) mutation, *cdc42–1*^{ts}, that blocked bud formation at 37°C but allowed the cell mass and volume to increase, resulting in greatly enlarged, unbudded cells (Adams et al. 1990). The predicted amino acid sequence of Cdc42 showed strong homology with Rho (Johnson and Pringle 1990). Concurrently, a 25 kDa guanine nucleotide binding protein, designated G_p or G25K, was purified from bovine brain and human placental membranes with peptide sequences showing a high sequence similarity to *S. cerevisiae* Cdc42. Subsequent isolation of two independent human cDNA isolates indicated the existence of two highly conserved (95% identical) proteins, the ubiquitously expressed Cdc42Hs (Shinjo et al. 1990) and the brain isoform G25K (Munemitsu et al. 1990), and these proteins were subsequently designated Cdc42p and Cdc42b, respectively. The two Cdc42 proteins arise from alternative splicing, are identical in amino acids 1 to 163, but diverge from residues 163 to 191 (Marks and Kwiatkowski 1996).

Additional studies identified other members of the Rho family. Although Cdc42 has no highly related isoforms, a number of Cdc42-related proteins have been identified. These include TC10 (gene expressed in teratocarcinoma cells) (Drivas et al. 1990) and a TC10-like (TCL) protein. Chp (*C*dc42 *h*omologous *p*rotein) was discovered through its ability to bind to p21 activated kinases (PAK) and Wrch-1 (Wnt-1 responsive Cdc42 homolog) as a gene upregulated by Wnt-1 signaling (Aronheim et al. 1998; Taneyhill and Pennica 2004).

RhoG was identified as a gene stimulated by growth serum (Vincent et al. 1992), whereas TTF/RhoH was identified as a gene found translocated in B cell lymphomas (Dallery-Prudhomme et al. 1997). The first member of the Rnd proteins (Rnd3/RhoE) was identified as a p190RhoGAP-interacting protein (Foster et al. 1996), and later studies identified the three Rnd proteins by low stringency hybridization screening (Nobes et al. 1998). Finally, the advancement of the genomic projects enabled screening of the whole human and murine genomes for new Rho members (e.g., Rif, RhoBTB1, RhoBTB2), which to date consists of 20 human genes encoding 23 proteins.

Rho GTPases fall into eight subfamilies based on sequence and structure homology: Rho subfamily (RhoA, RhoB, and RhoC), Rac subfamily (Rac1, Rac1b, Rac2, Rac3, and RhoG), Cdc42 subfamily (Cdc42p, Cdc42b, TC10, and TCL), Rnd subfamily

(Rnd1, Rnd2, and Rnd3), RhoBTB subfamily (RhoBTB1 and RhoBTB2), RhoD/RhoF, Wrch1/Chp, and RhoH subfamilies (Fig. 1.1) (Wennerberg and Der 2004). Out of these subfamilies, RhoH, RhoBTB, Wrch1/Chp, and Rnd have characteristics that make them atypical compared to the Rho, Rac, Cdc42, and RhoD/RhoF subfamilies (Aspenstrom et al. 2007). Rho GTPases are not found in prokaryotes, but are conserved in eukaryotic evolution, present lower eukaryotes (e.g., slime mold and yeast), plants, and mammals (Boureux et al. 2007).

Regulation and Function of Rho-Family Members

Nucleotide Binding, Lipid Modification, and Subcellular Localization

Much of our knowledge of the structure and biochemistry of Rho GTPases was first extrapolated from information derived from the study of Ras proteins. Rho GTPases share approximately 25% amino acid identity with Ras, in particular in sequences important for GDP/GTP binding and regulation, effector binding, and posttranslational lipid modifications. Like Ras, Rho GTPases possess a set of consensus GDP/GTP-binding sequence elements, an effector interaction domain, and C-terminal CAAX tetrapeptide motifs. The general structure of Rho GTPases is composed of an N-terminal G-domain involved in GTP-binding and hydrolysis, an effector binding domain, and a C-terminal membrane targeting domain (Fig. 1.3). Within the G-domain lies a 13 amino acid "insert sequence" positioned between residues analogous to Ras residues 122 and 123, and that are unique to Rho GTPases (Valencia et al. 1991). Several members of the family include additional N- and/or C-terminal sequences. In particular, RhoBTB1 and RhoBTB2 contain extensive additional C-terminal sequences that include tandem BTB (broadcomplex, tramtrack, bric a brac) domains, and lack the CAAX motif found on 16 of 20 Rho family proteins.

The regulatory cycle of GTP binding and hydrolysis is similar to Ras (Fig. 1.4). Switch I and switch II domains, corresponding to Ras residues 30–38 and 59–76, respectively, undergo a conformational change upon cycling between GDP and GTP that alters the affinity of the effector-binding domain for effectors (Vetter and Wittinghofer 2001). As described below, regulatory proteins that control Rho GTPase activity interact with switch I and/or II sequences. Tumor-associated missense mutations render Ras proteins constitutively active (primarily at residues G12 or Q61) by impairing intrinsic and GAP-stimulated GTP hydrolysis.

The Ras C-terminal CAAX motif (C = cysteine, A = aliphatic, X = terminal residue) was shown to signal for a series of posttranslational modifications: farnesyltransferase catalyzed covalent addition of a C15 farnesyl isoprenoid lipid to the cysteine residue of the CAAX tetrapeptide, Ras converting enzyme (Rce1)-catalyzed proteolytic cleavage of the AAX residues, and isoprenylcysteine

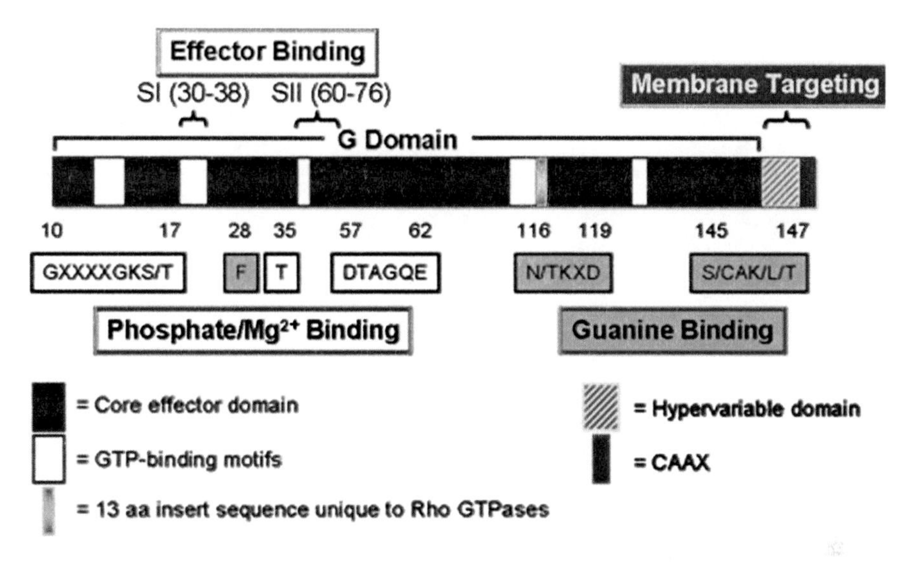

Fig. 1.3 Structural domains in Rho GTPases. Highlighted residues are important for GTPase activity: core effector domain (purple), GTP-binding motifs (white), Rho insert domain (blue), hypervariable domain (stripes), and the CAAX domain (spots). The exchange between the GDP- and the GTP-bound states is accompanied by the conformational changes in Rho in two amino terminal regions switch I (which corresponds to Ras residues 30–38) and switch II (Ras residues 60–76). These switch regions are critical for proper interaction with upstream regulators such as GEFs and GDIs and downstream effector molecules, hence termed "effector region." Another region involved in effector-mediated signaling is an alpha-helical "insert region," positioned between residues corresponding to Ras residues 122 and 123, that is present in Rho GTPases, but not other Ras superfamily members. The hypervariable domain (which exhibits the greatest sequence divergence between highly related isoforms) and the CAAX motif (cysteine-aliphatic-aliphatic-terminal amino acid) are membrane targeting signals. The CAAX motif signals for a series of posttranslational modifications to promote subcellular localization of Rho GTPases to the plasma membrane and other endomembranes

carboxylmethyltransferase (ICMT)-catalyzed carboxylmethylation of the farnesylated cysteine residue. The CAAX-signaled modifications, together with a second membrane targeting signal, either a polybasic domain, or palmitoylated-cysteine(s), immediately upstream of the terminal CAAX tetrapeptide motif facilitate Ras membrane association and subcellular localization. The majority of Rho GTPases also terminate with a C-terminal CAAX motif, and additionally contain upstream polybasic or palmitate fatty acid-modified cysteines that are required for proper membrane attachment of Rho GTPases (Cox and Der 1992; Zhang and Casey 1996). However, whereas some are modified by a C15 farnesyl group, others are modified by a related enzyme, geranylgeranyltransferase-I, to catalyze the addition of a more hydrophobic C20 geranylgeranyl isoprenoid lipid. Finally, additional C-terminal sequences adjacent to the CAAX motif (e.g., basic residues, palmitoylated cysteines) dictate the precise subcellular localization of Rho GTPases. Rho GTPases exhibit a diversity of subcellular locations, including

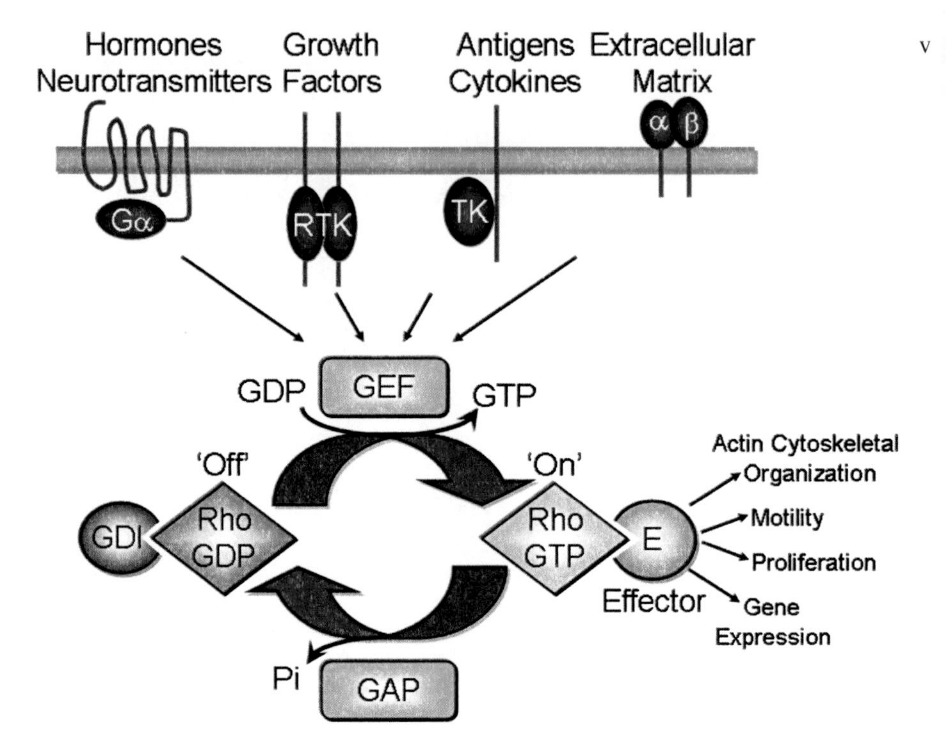

Fig. 1.4 Rho GTPase regulation and signaling. Rho GTPases associate with regulators and effectors that participate in a multitude of signaling events. Extracellular molecules stimulate upstream receptors such as G protein-coupled receptors, receptor tyrosine kinases (RTK), tyrosine kinases associated receptors (TK), and integrins that then activate Rho GTPases mostly through GEFs. The "active" GTP Rho GTPases undergo a conformation change that increases its affinity for downstream effectors important for many cellular processes including actin organization, motility, invasion, and gene expression. GAPs negatively regulate Rho GTPase function by increasing intrinsic GTP hydrolysis rate, locking them in a GDP-bound state. GDIs lock Rho GTPases in a GDP-bound state through binding and masking their C-terminal moiety thereby prohibiting membrane association, nucleotide exchange and effector binding

the plasma membrane, endosomes, the Golgi, and the endoplasmic reticulum (Vega and Ridley 2007).

Regulators of Rho GTPases

Studies in the early 1990s identified three classes of Rho-specific regulatory proteins that control the GDP/GTP Rho GTPase cycle: guanine nucleotide exchange factors (GEFs) (Rossman et al. 2005), GTPase-activating proteins (GAPs) (Moon and Zheng 2003), and GDP dissociation inhibitors (GDIs) (DerMardirossian and

Bokoch 2005) (Fig. 1.4). Since GDP and GTP are at mM levels in cells, Rho GTPases are always nucleotide-bound. GEFs catalyze the release of bound GDP, resulting in the formation of the active GTP bound state of the GTPase because GTP levels are tenfold higher than GDP in cells. GTPase inactivation involves GAP stimulating the intrinsically low GTP hydrolytic activity, converting GTPases to the inactive GDP-bound state. As reflected in their name, GDIs block the dissociation of guanine nucleotide from the Rho GTPase, but additionally possess other Rho GTPase regulatory functions (Gosser et al. 1997; Hoffman et al. 2000; Keep et al. 1997). Interestingly, there are Ras-specific GEFs and GAPs, but no GDIs for Ras.

RhoGDIs. In 1990, the first RhoGDI was identified and characterized by Takai and colleagues as the first regulator of Rho GTPases (Fukumoto et al. 1990; Ueda et al. 1990). RhoGDIs are generally considered as negative regulators of Rho GTPase activity by their ability to (1) prevent GDP dissociation and GEF-stimulated exchange, (2) prevent intrinsic and GAP-stimulated GTP hydrolysis and effector binding, and (3) membrane association. In the latter function, RhoGDI binds and masks the C-terminal isoprenyl group of GTPases via an immunoglobulin-like domain and the N-terminal regulatory portion of RhoGDI binds the switch I and switch II regions of the Rho GTPases, thereby locking Rho GTPases in an inactive complex in the cytosol (Gosser et al. 1997; Hoffman et al. 2000; Keep et al. 1997). However, RhoGDIs can also act as positive regulators via targeting Rho GTPases to subcellular membranes where they interact with different effectors or protect Rho GTPases from caspase degradation (Simpson et al. 2004).

There are three RhoGDI isoforms in humans: RhoGDI1 (GDI/GDIα), RhoGDI2 (LyGDI/D4GDI/GDIβ), and RhoGDI3 (GDIγ). RhoGDI1 is expressed ubiquitously, RhoGDI2 is normally hematopoietic cell-restricted, and RhoGDI3 is expressed preferentially in lung, testes, and brain. However, aberrant expression of RhoGDI2 has been seen in human cancers (Berman et al. 2004; Gildea et al. 2002; Theodorescu et al. 2004; Zhang and Zhang 2006). Different RhoGDI isoforms may have distinct and overlapping biological functions that are not fully understood. RhoGDI binding specificity to Rho GTPases is different for each isoforms, but still remains largely uncharacterized (DerMardirossian and Bokoch 2005), with some Rho family GTPases not regulated by RhoGDIs (Chenette et al. 2006). How RhoGDIs themselves are regulated remains largely unexplored. Recent studies have identified protein kinases that phosphorylate RhoGDIs and regulate their ability to bind Rho GTPases (DerMardirossian et al. 2004; Mehta et al. 2001), as well as RhoGDI-binding proteins that regulate the subcellular location of the RhoGDI/Rho GTPase complex (DerMardirossian et al. 2006), and caspase-mediated cleavage that create truncated, functionally divergent variants of RhoGDIs (Essmann et al. 2000; Krieser and Eastman 1999; Na et al. 1996). Since the stoichiometry of RhoGDIs and Rho GTPases is critical for dictating their degree of association, regulation of RhoGDI expression is also likely to be important. Hence, the regulation of RhoGDIs is complex and multifaceted.

RhoGEFs. There are two families of human RhoGEFs. The largest family, composed of 69 human members (Rossman et al. 2005; Schmidt and Hall 2002), was first discovered with the identification of the Dbl oncoprotein as a potent transforming gene.

Dbl was identified in a genomic DNA (derived from a diffuse B-cell lymphoma) gene transfer assay using the NIH3T3 mouse fibroblasts focus formation assay. The discovery that Dbl functions as a RhoGEF was made by Cerione and colleagues in 1991 (Hart et al. 1991). Similar genomic or cDNA expression screens for transforming genes identified additional members, including Vav (Katzav et al. 1989), Ect2 (Miki et al. 1993), Tim (Chan et al. 1994), Ost/Dbs (Horii et al. 1994; Whitehead et al. 1995a), Lbc (Toksoz and Williams 1994), Lsc (Whitehead et al. 1996), Lfc (Whitehead et al. 1995b), and Net1 (Chan et al. 1996). Tiam1 was identified in a screen for invasion inducing genes (Habets et al. 1994). The Dbl homology (DH) catalytic domain is followed by a pleckstrin-homology (PH) domain. Although PH domains are found in many other classes of signaling proteins, the invariant C-terminal association of this domain with essentially all Dbl family RhoGEFs reflects their critical role in facilitating DH domain catalytic activity. The roles of the PH domain are diverse and varied, and include interaction with phospholipids that may activate the catalytic domain (DH) of GEFs or promote localization to the plasma membrane. Dbl family RhoGEFs also possess a plethora of additional protein- and lipid-interaction domains that facilitate their regulation by upstream signaling networks.

The second family of RhoGEFs is composed of 11 DOCK proteins (also called CZH proteins) related to Dock180 (Cote and Vuori 2002; Meller et al. 2005). Dock180 (180 kDa protein) was identified initially in 1996 as a binding partner for the Crk adaptor protein (Hasegawa et al. 1996). The *Drosophila* ortholog, Myoblast city, was identified originally as a gene involved in muscle development (Rushton et al. 1995) and later, as a protein with strong homology with DOCK180 and shown essential for myoblast fusion and dorsal closure (Erickson et al. 1997). The *C. elegans* ortholog of Dock180, Ced-5, was identified as a protein required for cell migration and phagocytosis (Wu and Horvitz 1998). Although earlier studies established Dock180 and related proteins as Rac activators and can interact with nucleotide-free Rac, the first evidence that it functions as a GEF was not made until 2002 (Brugnera et al. 2002; Cote and Vuori 2002). A second family member, zizimin1, was identified as a Cdc42-interacting protein (Meller et al. 2002). Database homology searches reveal that Dock180 and zizimin1 are members of a novel family conserved in a wide variety of eukaryotes. DOCK proteins are characterized by two regions of high sequence conservation, designated Dock-homology region-1 and -2 (DHR1 and DHR2; also called CZH1 and CZH2/DOCKER, respectively) that are distinct in sequence from the DH domains. DOCK family RhoGEFs also contain additional domains, independent of the domains that promote exchange of nucleotides that are able to influence and determine their activation by upstream signals and their activation of downstream signaling to different effector molecules.

RhoGAPs. The discovery of GAPs for Ras proteins (p120 RasGAP and NF1 neurofibromin) prompted interest in searching for GAPs specific for Rho GTPases. The first identification of GAP activity for Rho GTPases was made in 1989 by Hall and colleagues (Garrett et al. 1989). They identified a 29 kDa protein with GAP activity for RhoA and subsequent limited protein sequence analyses revealed homology with the Breakpoint cluster protein (Bcr) and Bcr-related N-chimaerin protein (Diekmann et al. 1991). They then showed that Bcr and N-chimaerin possessed GAP activity for Rac1. These studies helped define a ~150 amino acid RhoGAP catalytic domain that is distinct from the catalytic domains of GAPs for other Ras family GTPases. Recent human genome analysis has unveiled up to 70 RhoGAPs in humans (Tcherkezian and Lamarche-Vane 2007). RhoGAPs regulate different Rho family members, with some very specific and others active for multiple GTPases. Thus, similar to GEFs, the number of RhoGAPs greatly outnumbers the number of Rho GTPases, and some Rho GTPase family members may not be regulated by GAPs. It is not clear why there are so many RhoGAPs and their nodes of regulation are less defined than RhoGEFs. Nevertheless, similar to GEFs, GAPs are usually multidomain proteins that suggest their regulation by distinct signaling mechanisms and their function as signaling integrators. For example, Rac-mediated phosphorylation of p190RhoGAP results in the association with p120-catenin to locally inhibit Rho activity and allow formation of adherens junctions (Nimnual et al. 2003). In human cancers, the expression of DLC-1 and related isoforms is lost in human cancers, supporting a tumor suppressor function for some RhoGAPs (Durkin et al. 2007).

Atypical Rho GTPases

The Rho GTPases subfamilies, RhoH, Rnd, Wrch1/Chp, and RhoBTB have structural and biochemical characteristics that make them atypical (Aspenstrom et al. 2004; Chardin 2006; Wennerberg and Der 2004). Some arose late in evolution and exhibit more restricted tissue expression, suggesting more specialized roles in cell physiology (Boureux et al. 2007). Some are functionally different from the classical Rho GTPases and rarely undergo regulation by GDP/GTP cycling. For example, Rnd3/RhoE possesses divergence at residues normally required for efficient intrinsic and GAP-stimulated GTPase activity (analogous to Ras residues G12 and Q61), and was shown to persist in the GTP-bound state (Foster et al. 1996). Similarly, RhoH was shown to have GTPase-inactivating amino acid substitutions resulting in the constitutively activated GTP-bound state (Li et al. 2002). Other atypical Rho GTPases, such as RhoBTB1 and RhoBTB2/DBC2, are also expected to be GTPase deficient. Therefore, instead of having GEFs and GAPs to regulate their on/off cycling, they are often regulated by gene expression levels. Posttranslational mechanisms, such as phosphorylation, or targeted destruction such as proteasomal degradation are also involved. In addition, they may be regulated by protein:protein interactions.

Wrch-1 and Chp do not have amino acid replacements at residues required for GTPase activity but do exhibit enhanced intrinsic nucleotide exchange activity (Aronheim et al. 1998; Shutes et al. 2004). Therefore, it is not clear whether their activities will be regulated by GEFs and GAPs. The Rac1 splice variant (Rac1b) also has elevated intrinsic exchange activity and resides in the GTP-bound conformation (Schnelzer et al. 2000).

The more recently identified Rho GTPase subfamilies, RhoBTB1 and RhoBTB2/DBC2, are much larger and have additional domains when compared to the classical Rho GTPases (Rivero et al. 2001). They contain two Broad Complex/Tramtrack/Bric-a-brac (BTB) domains that promote protein:protein interactions and are a Cullin 3 binding partner involved in proteosomal degradation (Wilkins et al. 2004). Finally, the dysfunctional regulation of atypical Rho GTPases (e.g., Wrch-1, RhoBTB2, and Rac1b) may be involved in the pathogenesis of diseases such as cancer.

Upstream Signaling and Downstream Effectors and Rho Biological Processes

Implicating Rho GTPases in Signal Transduction

Although the Rho gene was identified in 1985, it was not until 1992 that the Rho family was brought into the spotlight for the signal transduction field. Two seminal studies by Ridley, Hall, and colleagues established that Rho and Rac were important for extracellular stimulus-regulated actin cytoskeletal reorganization (Ridley and Hall 1992; Ridley et al. 1992). The key observations were that lysophosphatic acid (LPA) was the factor in calf serum that stimulated Rho-dependent formation of actin stress fibers, and that platelet-derived growth factor (PDGF)-stimulated membrane ruffling could be blocked by a Rac dominant negative protein. Subsequent studies also linked bradykinin-stimulated GPCR signaling to activation of Cdc42, to cause formation of actin-rich, membrane microspikes (filopodia) (Kozma et al. 1995; Nobes and Hall 1995). The ability of RhoA, Rac1, and Cdc42 to regulate diverse signaling cascades linking membrane receptors to the construction of distinct actin structures has been confirmed in mammals, worms, yeast, flies, and many other organisms.

In addition to regulation of actin cytoskeletal organization, Rho GTPases also participate in the regulation of many other signal transduction pathways beyond actin regulation. In 1991, Rac was shown to regulate the NADPH oxidase complex and superoxide production in neutrophils (Abo et al. 1991). In 1995, studies implicated Rho GTPases as regulators of the p38 and JNK stress-activated protein kinase cascades (Coso et al. 1995; Minden et al. 1995), transcription factor activity and gene transcription (Hill et al. 1995). Rho GTPase signaling was shown to be a critical regulator of cell cycle progression (Olson et al. 1995; Yamamoto et al. 1993).

Rho GTPases were also shown to comprise signaling networks involving other GTPases (Ridley and Hall 1992; Ridley et al. 1992). In particular, in 1995, Rac, Rho, and Cdc42 were found to be key mediators of Ras-induced oncogenic transformation (Khosravi-Far et al. 1995; Qiu et al. 1997; Qiu et al. 1995a; Qiu et al. 1995b). Similarly, GPCR-mediated growth transformation also involved activation of Rho GTPases (Martin et al. 2001; Zohn et al. 1998). Thus, like Ras, Rho GTPases function as signaling nodes, where divergent upstream signaling stimuli converge to activate Rho GTPase function, and where activated Rho GTPases then interact with multiple downstream effectors to regulate cytoplasmic signaling networks that control a multitude of cytoplasmic and nuclear events.

Similar to Ras, Rho effectors have been identified by a variety of experimental approaches. These include yeast two-hybrid library screens, biochemical approaches, and database searches for consensus GTP-dependent binding motifs. Below we highlight some of the key findings that have laid the foundation for the effectors that facilitate Rho GTPase signaling networks. In light of the diversity of functions regulated by Rho GTPases, it is not surprising that each interacts with a spectrum of functionally diverse effectors.

Linking Extracellular Signals with Rho GTPases

As described above, growth factor induced signaling from cell surface receptors to Rho GTPases was first described by Ridley, Hall, and colleagues (Ridley and Hall 1992, 1994; Ridley et al. 1992). At that time, the precise links that connected activated receptor tyrosine kinases or GPCRs with Rho GTPases were not known. Although the involvement of Dbl family proteins as Rho GEFs was not known, the ability to block Rho GTPase activation by dominant negative Rho GTPases that form nonproductive complexes with RhoGEFs (Feig 1999), implicated RhoGEFs as the key links (Fig. 1.4).

An additional clue that RhoGEFs may link extracellular signals with Rho GTPases came from observations that the removal of the N-terminal sequences of Dbl, Vav, Asef, Taim1, Ect2, and other Dbl family members leads to their constitutive activation when expressed in vivo (Katzav et al. 1991; Kawasaki et al. 2000; Miki et al. 1993; Ron et al. 1989; van Leeuwen et al. 1995). This suggested that their N-terminal sequences functioned as autoinhibitory regulatory domains that block RhoGEF catalytic activity. Therefore, activation of GEF is through the relief of autoinhibition via posttranslational modifications or by binding to other proteins via their N-terminal sequences. These N-terminal sequences often contain distinct domains and motifs that suggest their recognition by diverse upstream signals.

The first elucidation of how upstream signals cause Rho GTPase activation via activation of a RhoGEF came from studies of Vav. Vav has been implicated downstream of many receptors, including the receptor tyrosine kinases stimulated by the epidermal growth factor and platelet-derived growth factor (Bustelo 2000). After receptor stimulation, Vav is phosphorylated by Src and Syk tyrosine kinase families that then stimulate Vav activation of RhoA, RhoB, RhoG, Rac1, and Cdc42 (Abe et al. 2000; Crespo et al. 1997; Han et al. 1997). The Vav1 DH domain is autoinhibited by interaction with an adjacent N-terminal acidic region that includes tyrosine residues. When these residues are phosphorylated, autoinhibition is relieved and the catalytic activity of the DH domain is activated (Aghazadeh et al. 2000). A similar phosphorylation activation mechanism is also seen with Tim and other RhoGEFs (Yohe et al. 2007).

LPA binds to Edg/LPA GPCR and results in the activation of $G_{\alpha 12/13}$ subunits (Hart et al. 1998; Kozasa et al. 1998). The activated GTP-bound $G_{\alpha 12/13}$ subunits bind to the RGS (regulators of G protein signaling) domain of a family of RhoGEFs (p115RhoGEF, LARG, and PDZ-RhoGEF) resulting in their relocalization to the cell surface and subsequent activation (Fukuhara et al. 2001). These RhoGEFs are activators of RhoA and actin stress fiber formation.

Some RhoGEFs are regulated by other protein:protein interactions. Tiam1, which contains a Ras binding domain similar to Raf, is activated upon binding to GTP-bound Ras (Lambert et al. 2002). The Asef RhoGEF, an activator of Cdc42, contains an N-terminal binding domain for the APC tumor suppressor (Kawasaki et al. 2000). Asef is activated by the binding of wild-type but not tumor-associated mutant, APC tumor suppressor proteins (Mitin et al. 2007).

In addition, investigations dissecting the pathways from growth factor receptors to Rac activation uncovered growth factor induced membrane ruffling was blocked by PI 3-kinase inhibitors (Nobes and Hall 1995). Subsequent studies revealed some RhoGEFs are activated by phosphatidylinositol 3-kinase activation and production of phosphatidylinositol 3,4,5-trisphosphate (PIP3) (Schmidt and Hall 2002). For example, the P-Rex1 Rac-specific GEF is activated by PIP3 association with its PH domain (Welch et al. 2002). However, full activation involves coordinated activation by heterotrimeric G $\beta\gamma$ subunits, to promote membrane association (Barber et al. 2007). Finally, Rho proteins are not only regulated by receptor tyrosine kinases and GPCRs, but also regulated by cell adhesion molecules such as integrins, cadherins, and Ig superfamily members (Braga 2002; DeMali et al. 2003; Thompson et al. 2002). The pathways signaling from these cell surface molecules to cause Rho GTPase activation remain yet to be fully understood.

Actin Cytoskeleton Regulation

The best characterized function of Rho GTPases is their regulation of the actin cytoskeleton (Jaffe and Hall 2005). Previous findings showed that in fibroblasts the activation of RhoA, Rac1, and Cdc42 induce the assembly of stress fibers, lamellipodia, and filopodia, respectively. There is now considerable, although incomplete, knowledge of the immediate and indirect effectors that facilitate Rho GTPase regulation of actin. Both Rac and Cdc42 utilize members of the WASP family to regulate the Arp2/3 complex, which facilitates actin filament formation. Rho interaction with the mDia1 diaphanous-related formins promotes actin filament elongation. Rho activation of ROCK (also called ROK) also promotes regulation of actin stress fibers.

In 1996, WASP was discovered to interact with Cdc42 and stimulated actin polymerization (Kirchhausen and Rosen 1996). Purification of proteins required for Cdc42 actin polymerization in cell extracts revealed the Arp2/3 complex, important for nucleation during actin polymerization, and N-WASP (WASP homologue) (Bi and Zigmond 1999). SCAR/WAVE was identified as a component of the Arp2/3 complex via yeast-two hybrid screen with p21 (member of Arp2/3 complex)

(Machesky and Insall 1998). In 2000, Rac was shown to associate with WAVE and Arp2/3 through IRSp5 (Miki et al. 2000). Overall, there are a number of targets downstream of Rho and the importance of each is highly complex and variable depending on circumstances such as environment and cell type. Future studies will more clearly elucidate connections between Rho proteins and actin cytoskeletal organization.

Protein Kinase Effectors

Lim and colleagues identified in 1993 the first Rac and Cdc42 downstream target, the serine/threonine p21-activated kinase (PAK) (Manser et al. 1994). These targets were discovered by the selection of proteins that were in a complex with GTP-bound ("active") but not GDP-bound ("inactive") Rac or Cdc42. Several cytoskeletal targets have been identified downstream of PAK. These include LIM kinase, which phosphorylates the actin depolymerizing protein, cofilin, inhibiting its function (Maekawa et al. 1999). This inhibition stabilizes actin filament arrays such as stress fibers.

Following PAK, in 1996 several groups discovered RhoA-binding kinase (ROCK) downstream of Rho. A number of Rho effectors were identified via yeast two-hybrid screens, the first being rhoteckin and rhophilin (Reid et al. 1996; Watanabe et al. 1996). Rho targets have also been identified through database mining such as protein-kinase-N (PKN) which was identified through its homology to rhophilin (Watanabe et al. 1996). ROCK was the first target shown to affect actin organization (Leung et al. 1996). ROCK elevates MLC phosphorylation by inhibiting the MLC phosphatase (Kimura et al. 1996). Subsequent work established that ROCK is able to directly phosphorylate the regulatory MLC and enhance myosin activation. In addition, ROCK activates LIMK to lead to inhibition of cofilin causing actin stabilization. These studies established that Rho promotes myosin contractility and drives the formation of stress fibers and focal adhesions (Ridley 1999).

Conclusion

The functional diversity of individual Rho GTPase family members has yet to be fully elucidated. The signaling networks involving Rho GTPases have significantly increased in the past decade. Although there are 20 human family Rho GTPases, Rac1, RhoA, and Cdc42 have been the most notably studied and our knowledge about Rho GTPase function is predominantly based on these members. In the future, it should become easier to decipher the roles for other family members, with the advent of novel tools and reagents. Some of the experimental approaches that are likely to be the most instrumental for unveiling the individual roles for each Rho protein will be RNA interference and mouse knockouts. In addition, high-throughput microarray and mass-spectrometry analysis will allow analysis of global changes in gene and protein expression for a more comprehensive understanding of Rho signaling. This will also aid in the understanding of different roles for highly homologous isoforms, such as RhoA, RhoB, and RhoC, and splice variants such as Rac1b. As other family members are studied, they may have overlapping or distinct functions from the three classical Rho GTPases. They also may show importance in new biological processes, where Rho members have never been implicated before. Studies on atypical Rho GTPases have broadened the number of biological pathways involving Rho signaling. We still need to uncover the mechanisms by which atypical Rho GTPases are activated and inactivated. For instance, RhoBTB members contain sequences beyond the core GTPase domain that are most likely involved in innovative modes of regulation. In addition, determining their binding partners may give insight as to their roles in oncogenic transformation, such as Rac1b, and possibly other diseases. In addition, how GAPs, GEFs, and GDIs interact with family members and either terminate or emit signaling in physiologically relevant pathways remains another obstacle for the Rho GTPase field. Ultimately, the next step for a more comprehensive study would be to perform systematic surveys of tissue distribution, subcellular function, substrate specificity, and tissue- and species-specific roles for every Rho family member and regulator.

Acknowledgments We thank Kent Rossman for help with preparation of the dendrogram of the Rho family GTPases and Lanika DeGrafenreid for assistance in manuscript preparation. We apologize to all colleagues whose work could not be cited due to space limitations. Our studies are supported by NIH grants CA063071 and CA67771.

References

Abe, K., Rossman, K.L., Liu, B., Ritola, K.D., Chiang, D., Campbell, S.L., Burridge, K., and Der, C.J. (2000). Vav2 is an activator of Cdc42, Rac1, and RhoA. *J Biol Chem* **275**, 10141–10149.

Abo, A., Pick, E., Hall, A., Totty, N., Teahan, C.G., and Segal, A.W. (1991). Activation of the NADPH oxidase involves the small GTP-binding protein p21rac1. *Nature* **353**, 668–670.

Adams, A.E., Johnson, D.I., Longnecker, R.M., Sloat, B.F., and Pringle, J.R. (1990). CDC42 and CDC43, two additional genes involved in budding and the establishment of cell polarity in the yeast Saccharomyces cerevisiae. *J Cell Biol* **111**, 131–142.

Aghazadeh, B., Lowry, W.E., Huang, X.Y., and Rosen, M.K. (2000). Structural basis for relief of autoinhibition of the Dbl homology domain of proto-oncogene Vav by tyrosine phosphorylation. *Cell* **102**, 625–633.

Aronheim, A., Broder, Y.C., Cohen, A., Fritsch, A., Belisle, B., and Abo, A. (1998). Chp, a homologue of the GTPase Cdc42Hs, activates the JNK pathway and is implicated in reorganizing the actin cytoskeleton. *Curr Biol* **8**, 1125–1128.

Aspenstrom, P., Fransson, A., and Saras, J. (2004). Rho GTPases have diverse effects on the organization of the actin filament system. *Biochem J* **377**, 327–337.

Aspenstrom, P., Ruusala, A., and Pacholsky, D. (2007). Taking Rho GTPases to the next level: the cellular functions of atypical Rho GTPases. *Exp Cell Res* **313**, 3673–3679.

Barber, M.A., Donald, S., Thelen, S., Anderson, K.E., Thelen, M., and Welch, H.C. (2007). Membrane translocation of P-Rex1 is mediated by G protein betagamma subunits and phosphoinositide 3-kinase. *J Biol Chem* **282**, 29967–29976.

Beder, L.B., Gunduz, M., Ouchida, M., Gunduz, E., Sakai, A., Fukushima, K., Nagatsuka, H., Ito, S., Honjo, N., Nishizaki, K., et al. (2006). Identification of a candidate tumor suppressor gene RHOBTB1 located at a novel allelic loss region 10q21 in head and neck cancer. *J Cancer Res Clin Oncol* **132**, 19–27.
Berman, D.M., Shih Ie, M., Burke, L.A., Veenstra, T.D., Zhao, Y., Conrads, T.P., Kwon, S.W., Hoang, V., Yu, L.R., Zhou, M., et al. (2004). Profiling the activity of G proteins in patient-derived tissues by rapid affinity-capture of signal transduction proteins (GRASP). *Proteomics* **4**, 812–818.
Bi, E., and Zigmond, S.H. (1999). Actin polymerization: Where the WASP stings. *Curr Biol* **9**, R160–163.
Boureux, A., Vignal, E., Faure, S., and Fort, P. (2007). Evolution of the Rho family of ras-like GTPases in eukaryotes. *Mol Biol Evol* **24**, 203–216.
Braga, V.M. (2002). Cell-cell adhesion and signalling. *Curr Opin Cell Biol* **14**, 546–556.
Brugnera, E., Haney, L., Grimsley, C., Lu, M., Walk, S.F., Tosello-Trampont, A.C., Macara, I.G., Madhani, H., Fink, G.R., and Ravichandran, K.S. (2002). Unconventional Rac-GEF activity is mediated through the Dock180-ELMO complex. *Nat Cell Biol* **4**, 574–582.
Bustelo, X.R. (2000). Regulatory and signaling properties of the Vav family. *Mol Cell Biol* **20**, 1461–1477.
Chan, A.M., McGovern, E.S., Catalano, G., Fleming, T.P., and Miki, T. (1994). Expression cDNA cloning of a novel oncogene with sequence similarity to regulators of small GTP-binding proteins. *Oncogene* **9**, 1057–1063.
Chan, A.M., Takai, S., Yamada, K., and Miki, T. (1996). Isolation of a novel oncogene, NET1, from neuroepithelioma cells by expression cDNA cloning. *Oncogene* **12**, 1259–1266.
Chardin, P. (1988). The ras superfamily proteins. *Biochimie* **70**, 865–868.
Chardin, P. (2006). Function and regulation of Rnd proteins. *Nat Rev Mol Cell Biol* **7**, 54–62.
Chardin, P., Madaule, P., and Tavitian, A. (1988). Coding sequence of human rho cDNAs clone 6 and clone 9. *Nucleic Acids Res* **16**, 2717.
Chavrier, P., Simons, K., and Zerial, M. (1992). The complexity of the Rab and Rho GTP-binding protein subfamilies revealed by a PCR cloning approach. *Gene* **112**, 261–264.
Chenette, E.J., Mitin, N.Y., and Der, C.J. (2006). Multiple sequence elements facilitate Chp Rho GTPase subcellular location, membrane association, and transforming activity. *Mol Biol Cell* **17**, 3108–3121.
Clark, E.A., Golub, T.R., Lander, E.S., and Hynes, R.O. (2000). Genomic analysis of metastasis reveals an essential role for RhoC. *Nature* **406**, 532–535.
Coso, O.A., Chiariello, M., Yu, J.C., Teramoto, H., Crespo, P., Xu, N., Miki, T., and Gutkind, J.S. (1995). The small GTP-binding proteins Rac1 and Cdc42 regulate the activity of the JNK/SAPK signaling pathway. *Cell* **81**, 1137–1146.
Cote, J.F., and Vuori, K. (2002). Identification of an evolutionarily conserved superfamily of DOCK180-related proteins with guanine nucleotide exchange activity. *J Cell Sci* **115**, 4901–4913.
Cox, A.D., and Der, C.J. (1992). Protein prenylation: more than just glue? *Curr Opin Cell Biol* **4**, 1008–1016.
Crespo, P., Schuebel, K.E., Ostrom, A.A., Gutkind, J.S., and Bustelo, X.R. (1997). Phosphotyrosine-dependent activation of Rac-1 GDP/GTP exchange by the vav proto-oncogene product. *Nature* **385**, 169–172.
Dallery, E., Galiegue-Zouitina, S., Collyn-d'Hooghe, M., Quief, S., Denis, C., Hildebrand, M.P., Lantoine, D., Deweindt, C., Tilly, H., Bastard, C., et al. (1995). TTF, a gene encoding a novel small G protein, fuses to the lymphoma-associated LAZ3 gene by t(3;4) chromosomal translocation. *Oncogene* **10**, 2171–2178.
Dallery-Prudhomme, E., Roumier, C., Denis, C., Preudhomme, C., Kerckaert, J.P., and Galiegue-Zouitina, S. (1997). Genomic structure and assignment of the RhoH/TTF small GTPase gene (ARHH) to 4p13 by in situ hybridization. *Genomics* **43**, 89–94.
DeMali, K.A., Wennerberg, K., and Burridge, K. (2003). Integrin signaling to the actin cytoskeleton. *Curr Opin Cell Biol* **15**, 572–582.
DerMardirossian, C., and Bokoch, G.M. (2005). GDIs: central regulatory molecules in Rho GTPase activation. *Trends Cell Biol* **15**, 356–363.

DerMardirossian, C., Rocklin, G., Seo, J.Y., and Bokoch, G.M. (2006). Phosphorylation of RhoGDI by Src regulates Rho GTPase binding and cytosol-membrane cycling. *Mol Biol Cell* **17**, 4760–4768.

DerMardirossian, C., Schnelzer, A., and Bokoch, G.M. (2004). Phosphorylation of RhoGDI by Pak1 mediates dissociation of Rac GTPase. *Mol Cell* **15**, 117–127.

Didsbury, J., Weber, R.F., Bokoch, G.M., Evans, T., and Snyderman, R. (1989). rac, a novel ras-related family of proteins that are botulinum toxin substrates. *J Biol Chem* **264**, 16378–16382.

Diekmann, D., Brill, S., Garrett, M.D., Totty, N., Hsuan, J., Monfries, C., Hall, C., Lim, L., and Hall, A. (1991). Bcr encodes a GTPase-activating protein for p21rac. *Nature* **351**, 400–402.

Drivas, G.T., Shih, A., Coutavas, E., Rush, M.G., and D'Eustachio, P. (1990). Characterization of four novel ras-like genes expressed in a human teratocarcinoma cell line. *Mol Cell Biol* **10**, 1793–1798.

Durkin, M.E., Yuan, B.Z., Zhou, X., Zimonjic, D.B., Lowy, D.R., Thorgeirsson, S.S., and Popescu, N.C. (2007). DLC-1:a Rho GTPase-activating protein and tumour suppressor. *J Cell Mol Med* **11**, 1185–1207.

Ellis, S., and Mellor, H. (2000). The novel Rho-family GTPase rif regulates coordinated actin-based membrane rearrangements. *Curr Biol* **10**, 1387–1390.

Erickson, M.R., Galletta, B.J., and Abmayr, S.M. (1997). Drosophila myoblast city encodes a conserved protein that is essential for myoblast fusion, dorsal closure, and cytoskeletal organization. *J Cell Biol* **138**, 589–603.

Essmann, F., Wieder, T., Otto, A., Muller, E.C., Dorken, B., and Daniel, P.T. (2000). GDP dissociation inhibitor D4-GDI (Rho-GDI 2), but not the homologous rho-GDI 1, is cleaved by caspase-3 during drug-induced apoptosis. *Biochem J* **346 Pt 3**, 777–783.

Etienne-Manneville, S., and Hall, A. (2002). Rho GTPases in cell biology. *Nature* **420**, 629–635.

Feig, L.A. (1999). Tools of the trade: use of dominant-inhibitory mutants of Ras-family GTPases. *Nat Cell Biol* **1**, E25–27.

Finlay, B.B. (2005). Bacterial virulence strategies that utilize Rho GTPases. *Curr Top Microbiol Immunol* **291**, 1–10.

Foster, R., Hu, K.Q., Lu, Y., Nolan, K.M., Thissen, J., and Settleman, J. (1996). Identification of a novel human Rho protein with unusual properties: GTPase deficiency and in vivo farnesylation. *Mol Cell Biol* **16**, 2689–2699.

Fukuhara, S., Chikumi, H., and Gutkind, J.S. (2001). RGS-containing RhoGEFs: the missing link between transforming G proteins and Rho? *Oncogene* **20**, 1661–1668.

Fukumoto, Y., Kaibuchi, K., Hori, Y., Fujioka, H., Araki, S., Ueda, T., Kikuchi, A., and Takai, Y. (1990). Molecular cloning and characterization of a novel type of regulatory protein (GDI) for the rho proteins, ras p21-like small GTP-binding proteins. *Oncogene* **5**, 1321–1328.

Garrett, M.D., Self, A.J., van Oers, C., and Hall, A. (1989). Identification of distinct cytoplasmic targets for ras/R-ras and rho regulatory proteins. *J Biol Chem* **264**, 10–13.

Gildea, J.J., Seraj, M.J., Oxford, G., Harding, M.A., Hampton, G.M., Moskaluk, C.A., Frierson, H.F., Conaway, M.R., and Theodorescu, D. (2002). RhoGDI2 is an invasion and metastasis suppressor gene in human cancer. *Cancer Res* **62**, 6418–6423.

Gosser, Y.Q., Nomanbhoy, T.K., Aghazadeh, B., Manor, D., Combs, C., Cerione, R.A., and Rosen, M.K. (1997). C-terminal binding domain of Rho GDP-dissociation inhibitor directs N-terminal inhibitory peptide to GTPases. *Nature* **387**, 814–819.

Govek, E.E., Newey, S.E., and Van Aelst, L. (2005). The role of the Rho GTPases in neuronal development. *Genes Dev* **19**, 1–49.

Haataja, L., Groffen, J., and Heisterkamp, N. (1997). Characterization of RAC3, a novel member of the Rho family. *J Biol Chem* **272**, 20384–20388.

Habets, G.G., Scholtes, E.H., Zuydgeest, D., van der Kammen, R.A., Stam, J.C., Berns, A., and Collard, J.G. (1994). Identification of an invasion-inducing gene, Tiam-1, that encodes a protein with homology to GDP-GTP exchangers for Rho-like proteins. *Cell* **77**, 537–549.

Hamaguchi, M., Meth, J.L., von Klitzing, C., Wei, W., Esposito, D., Rodgers, L., Walsh, T., Welcsh, P., King, M.C., and Wigler, M.H. (2002). DBC2, a candidate for a tumor suppressor gene involved in breast cancer. *Proc Natl Acad Sci U S A* **99**, 13647–13652.

Han, J., Das, B., Wei, W., Van Aelst, L., Mosteller, R.D., Khosravi-Far, R., Westwick, J.K., Der, C.J., and Broek, D. (1997). Lck regulates Vav activation of members of the Rho family of GTPases. *Mol Cell Biol* **17**, 1346–1353.

Hart, M.J., Eva, A., Evans, T., Aaronson, S.A., and Cerione, R.A. (1991). Catalysis of guanine nucleotide exchange on the CDC42Hs protein by the dbl oncogene product. *Nature* **354**, 311–314.

Hart, M.J., Jiang, X., Kozasa, T., Roscoe, W., Singer, W.D., Gilman, A.G., Sternweis, P.C., and Bollag, G. (1998). Direct stimulation of the guanine nucleotide exchange activity of p115 RhoGEF by Galpha13. *Science* **280**, 2112–2114.

Hasegawa, H., Kiyokawa, E., Tanaka, S., Nagashima, K., Gotoh, N., Shibuya, M., Kurata, T., and Matsuda, M. (1996). DOCK180, a major CRK-binding protein, alters cell morphology upon translocation to the cell membrane. *Mol Cell Biol* **16**, 1770–1776.

Hill, C.S., Wynne, J., and Treisman, R. (1995). The Rho family GTPases RhoA, Rac1, and CDC42Hs regulate transcriptional activation by SRF. *Cell* **81**, 1159–1170.

Hoffman, G.R., Nassar, N., and Cerione, R.A. (2000). Structure of the Rho family GTP-binding protein Cdc42 in complex with the multifunctional regulator RhoGDI. *Cell* **100**, 345–356.

Horii, Y., Beeler, J.F., Sakaguchi, K., Tachibana, M., and Miki, T. (1994). A novel oncogene, ost, encodes a guanine nucleotide exchange factor that potentially links Rho and Rac signaling pathways. *EMBO J* **13**, 4776–4786.

Jaffe, A.B., and Hall, A. (2005). Rho GTPases: biochemistry and biology. *Annu Rev Cell Dev Biol* **21**, 247–269.

Jahner, D., and Hunter, T. (1991). The ras-related gene rhoB is an immediate-early gene inducible by v-Fps, epidermal growth factor, and platelet-derived growth factor in rat fibroblasts. *Mol Cell Biol* **11**, 3682–3690.

Johnson, D.I., and Pringle, J.R. (1990). Molecular characterization of CDC42, a Saccharomyces cerevisiae gene involved in the development of cell polarity. *J Cell Biol* **111**, 143–152.

Jordan, P., Brazao, R., Boavida, M.G., Gespach, C., and Chastre, E. (1999). Cloning of a novel human Rac1b splice variant with increased expression in colorectal tumors. *Oncogene* **18**, 6835–6839.

Katzav, S., Cleveland, J.L., Heslop, H.E., and Pulido, D. (1991). Loss of the amino-terminal helix-loop-helix domain of the vav proto-oncogene activates its transforming potential. *Mol Cell Biol* **11**, 1912–1920.

Katzav, S., Martin-Zanca, D., and Barbacid, M. (1989). vav, a novel human oncogene derived from a locus ubiquitously expressed in hematopoietic cells. *EMBO J* **8**, 2283–2290.

Kawasaki, Y., Senda, T., Ishidate, T., Koyama, R., Morishita, T., Iwayama, Y., Higuchi, O., and Akiyama, T. (2000). Asef, a link between the tumor suppressor APC and G-protein signaling. *Science* **289**, 1194–1197.

Keep, N.H., Barnes, M., Barsukov, I., Badii, R., Lian, L.Y., Segal, A.W., Moody, P.C., and Roberts, G.C. (1997). A modulator of rho family G proteins, rhoGDI, binds these G proteins via an immunoglobulin-like domain and a flexible N-terminal arm. *Structure* **5**, 623–633.

Khosravi-Far, R., Solski, P.A., Clark, G.J., Kinch, M.S., and Der, C.J. (1995). Activation of Rac1, RhoA, and mitogen-activated protein kinases is required for Ras transformation. *Mol Cell Biol* **15**, 6443–6453.

Kimura, K., Ito, M., Amano, M., Chihara, K., Fukata, Y., Nakafuku, M., Yamamori, B., Feng, J., Nakano, T., Okawa, K., et al. (1996). Regulation of myosin phosphatase by Rho and Rho-associated kinase (Rho-kinase). *Science* **273**, 245–248.

Kirchhausen, T., and Rosen, F.S. (1996). Disease mechanism: unravelling Wiskott-Aldrich syndrome. *Curr Biol* **6**, 676–678.

Kozasa, T., Jiang, X., Hart, M.J., Sternweis, P.M., Singer, W.D., Gilman, A.G., Bollag, G., and Sternweis, P.C. (1998). p115 RhoGEF, a GTPase activating protein for Galpha12 and Galpha13. *Science* **280**, 2109–2111.

Kozma, R., Ahmed, S., Best, A., and Lim, L. (1995). The Ras-related protein Cdc42Hs and bradykinin promote formation of peripheral actin microspikes and filopodia in Swiss 3T3 fibroblasts. *Mol Cell Biol* **15**, 1942–1952.

Krieser, R.J., and Eastman, A. (1999). Cleavage and nuclear translocation of the caspase 3 substrate Rho GDP-dissociation inhibitor, D4-GDI, during apoptosis. *Cell Death Differ* **6**, 412–419.

Lambert, J.M., Lambert, Q.T., Reuther, G.W., Malliri, A., Siderovski, D.P., Sondek, J., Collard, J.G., and Der, C.J. (2002). Tiam1 mediates Ras activation of Rac by a PI(3)K-independent mechanism. *Nat Cell Biol* **4**, 621–625.

Leung, T., Chen, X.Q., Manser, E., and Lim, L. (1996). The p160 RhoA-binding kinase ROK alpha is a member of a kinase family and is involved in the reorganization of the cytoskeleton. *Mol Cell Biol* **16**, 5313–5327.

Li, X., Bu, X., Lu, B., Avraham, H., Flavell, R.A., and Lim, B. (2002). The hematopoiesis-specific GTP-binding protein RhoH is GTPase deficient and modulates activities of other Rho GTPases by an inhibitory function. *Mol Cell Biol* **22**, 1158–1171.

Machesky, L.M., and Insall, R.H. (1998). Scar1 and the related Wiskott-Aldrich syndrome protein, WASP, regulate the actin cytoskeleton through the Arp2/3 complex. *Curr Biol* **8**, 1347–1356.

Madaule, P., and Axel, R. (1985). A novel ras-related gene family. *Cell* **41**, 31–40.

Maekawa, M., Ishizaki, T., Boku, S., Watanabe, N., Fujita, A., Iwamatsu, A., Obinata, T., Ohashi, K., Mizuno, K., and Narumiya, S. (1999). Signaling from Rho to the actin cytoskeleton through protein kinases ROCK and LIM-kinase. *Science* **285**, 895–898.

Malumbres, M., and Barbacid, M. (2003). RAS oncogenes: the first 30 years. *Nat Rev Cancer* **3**, 459–465.

Manser, E., Leung, T., Salihuddin, H., Zhao, Z.S., and Lim, L. (1994). A brain serine/threonine protein kinase activated by Cdc42 and Rac1. *Nature* **367**, 40–46.

Marks, P.W., and Kwiatkowski, D.J. (1996). Genomic organization and chromosomal location of murine Cdc42. *Genomics* **38**, 13–18.

Martin, C.B., Mahon, G.M., Klinger, M.B., Kay, R.J., Symons, M., Der, C.J., and Whitehead, I.P. (2001). The thrombin receptor, PAR-1, causes transformation by activation of Rho-mediated signaling pathways. *Oncogene* **20**, 1953–1963.

Mehta, D., Rahman, A., and Malik, A.B. (2001). Protein kinase C-alpha signals rho-guanine nucleotide dissociation inhibitor phosphorylation and rho activation and regulates the endothelial cell barrier function. *J Biol Chem* **276**, 22614–22620.

Meller, N., Irani-Tehrani, M., Kiosses, W.B., Del Pozo, M.A., and Schwartz, M.A. (2002). Zizimin1, a novel Cdc42 activator, reveals a new GEF domain for Rho proteins. *Nat Cell Biol* **4**, 639–647.

Meller, N., Merlot, S., and Guda, C. (2005). CZH proteins: a new family of Rho-GEFs. *J Cell Sci* **118**, 4937–4946.

Miki, H., Yamaguchi, H., Suetsugu, S., and Takenawa, T. (2000). IRSp53 is an essential intermediate between Rac and WAVE in the regulation of membrane ruffling. *Nature* **408**, 732–735.

Miki, T., Smith, C.L., Long, J.E., Eva, A., and Fleming, T.P. (1993). Oncogene ect2 is related to regulators of small GTP-binding proteins. *Nature* **362**, 462–465.

Minden, A., Lin, A., Claret, F.X., Abo, A., and Karin, M. (1995). Selective activation of the JNK signaling cascade and c-Jun transcriptional activity by the small GTPases Rac and Cdc42Hs. *Cell* **81**, 1147–1157.

Mitin, N., Betts, L., Yohe, M.E., Der, C.J., Sondek, J., and Rossman, K.L. (2007). Release of autoinhibition of ASEF by APC leads to CDC42 activation and tumor suppression. *Nat Struct Mol Biol* **14**, 814–823.

Moon, S.Y., and Zheng, Y. (2003). Rho GTPase-activating proteins in cell regulation. *Trends Cell Biol* **13**, 13–22.

Munemitsu, S., Innis, M.A., Clark, R., McCormick, F., Ullrich, A., and Polakis, P. (1990). Molecular cloning and expression of a G25K cDNA, the human homolog of the yeast cell cycle gene CDC42. *Mol Cell Biol* **10**, 5977–5982.

Munter, S., Way, M., and Frischknecht, F. (2006). Signaling during pathogen infection. *Sci STKE* **2006**, re5.

Murphy, C., Saffrich, R., Grummt, M., Gournier, H., Rybin, V., Rubino, M., Auvinen, P., Lutcke, A., Parton, R.G., and Zerial, M. (1996). Endosome dynamics regulated by a Rho protein. *Nature* **384**, 427–432.

Na, S., Chuang, T.H., Cunningham, A., Turi, T.G., Hanke, J.H., Bokoch, G.M., and Danley, D.E. (1996). D4-GDI, a substrate of CPP32, is proteolyzed during Fas-induced apoptosis. *J Biol Chem* **271**, 11209–11213.
Nimnual, A.S., Taylor, L.J., and Bar-Sagi, D. (2003). Redox-dependent downregulation of Rho by Rac. *Nat Cell Biol* **5**, 236–241.
Nobes, C.D., and Hall, A. (1995). Rho, rac, and cdc42 GTPases regulate the assembly of multimolecular focal complexes associated with actin stress fibers, lamellipodia, and filopodia. *Cell* **81**, 53–62.
Nobes, C.D., Lauritzen, I., Mattei, M.G., Paris, S., Hall, A., and Chardin, P. (1998). A new member of the Rho family, Rnd1, promotes disassembly of actin filament structures and loss of cell adhesion. *J Cell Biol* **141**, 187–197.
Olson, M.F., Ashworth, A., and Hall, A. (1995). An essential role for Rho, Rac, and Cdc42 GTPases in cell cycle progression through G1. *Science* **269**, 1270–1272.
Ongusaha, P.P., Kim, H.G., Boswell, S.A., Ridley, A.J., Der, C.J., Dotto, G.P., Kim, Y.B., Aaronson, S.A., and Lee, S.W. (2006). RhoE is a pro-survival p53 target gene that inhibits ROCK I-mediated apoptosis in response to genotoxic stress. *Curr Biol* **16**, 2466–2472.
Polakis, P.G., Snyderman, R., and Evans, T. (1989). Characterization of G25K, a GTP-binding protein containing a novel putative nucleotide binding domain. *Biochem Biophys Res Commun* **160**, 25–32.
Qiu, R.G., Abo, A., McCormick, F., and Symons, M. (1997). Cdc42 regulates anchorage-independent growth and is necessary for Ras transformation. *Mol Cell Biol* **17**, 3449–3458.
Qiu, R.G., Chen, J., Kirn, D., McCormick, F., and Symons, M. (1995a). An essential role for Rac in Ras transformation. *Nature* **374**, 457–459.
Qiu, R.G., Chen, J., McCormick, F., and Symons, M. (1995b). A role for Rho in Ras transformation. *Proc Natl Acad Sci U S A* **92**, 11781–11785.
Reid, T., Furuyashiki, T., Ishizaki, T., Watanabe, G., Watanabe, N., Fujisawa, K., Morii, N., Madaule, P., and Narumiya, S. (1996). Rhotekin, a new putative target for Rho bearing homology to a serine/threonine kinase, PKN, and rhophilin in the rho-binding domain. *J Biol Chem* **271**, 13556–13560.
Ridley, A.J. (1999). Stress fibres take shape. *Nat Cell Biol* **1**, E64–66.
Ridley, A.J., and Hall, A. (1992). The small GTP-binding protein rho regulates the assembly of focal adhesions and actin stress fibers in response to growth factors. *Cell* **70**, 389–399.
Ridley, A.J., and Hall, A. (1994). Signal transduction pathways regulating Rho-mediated stress fibre formation: requirement for a tyrosine kinase. *EMBO J* **13**, 2600–2610.
Ridley, A.J., Paterson, H.F., Johnston, C.L., Diekmann, D., and Hall, A. (1992). The small GTP-binding protein rac regulates growth factor-induced membrane ruffling. *Cell* **70**, 401–410.
Rivero, F., Dislich, H., Glockner, G., and Noegel, A.A. (2001). The Dictyostelium discoideum family of Rho-related proteins. *Nucleic Acids Res* **29**, 1068–1079.
Ron, D., Graziani, G., Aaronson, S.A., and Eva, A. (1989). The N-terminal region of proto-dbl down regulates its transforming activity. *Oncogene* **4**, 1067–1072.
Rossman, K.L., Der, C.J., and Sondek, J. (2005). GEF means go: turning on RHO GTPases with guanine nucleotide-exchange factors. *Nat Rev Mol Cell Biol* **6**, 167–180.
Rushton, E., Drysdale, R., Abmayr, S.M., Michelson, A.M., and Bate, M. (1995). Mutations in a novel gene, myoblast city, provide evidence in support of the founder cell hypothesis for Drosophila muscle development. *Development* **121**, 1979–1988.
Sahai, E., and Marshall, C.J. (2002). RHO-GTPases and cancer. *Nat Rev Cancer* **2**, 133–142.
Schmidt, A., and Hall, A. (2002). Guanine nucleotide exchange factors for Rho GTPases: turning on the switch. *Genes Dev* **16**, 1587–1609.
Schnelzer, A., Prechtel, D., Knaus, U., Dehne, K., Gerhard, M., Graeff, H., Harbeck, N., Schmitt, M., and Lengyel, E. (2000). Rac1 in human breast cancer: overexpression, mutation analysis, and characterization of a new isoform, Rac1b. *Oncogene* **19**, 3013–3020.
Shinjo, K., Koland, J.G., Hart, M.J., Narasimhan, V., Johnson, D.I., Evans, T., and Cerione, R.A. (1990). Molecular cloning of the gene for the human placental GTP-binding protein Gp (G25K): identification of this GTP-binding protein as the human homolog of the yeast cell-division-cycle protein CDC42. *Proc Natl Acad Sci U S A* **87**, 9853–9857.

Shutes, A., Berzat, A.C., Cox, A.D., and Der, C.J. (2004). Atypical mechanism of regulation of the Wrch-1 Rho family small GTPase. *Curr Biol* **14**, 2052–2056.

Simpson, K.J., Dugan, A.S., and Mercurio, A.M. (2004). Functional analysis of the contribution of RhoA and RhoC GTPases to invasive breast carcinoma. *Cancer Res* **64**, 8694–8701.

Taneyhill, L., and Pennica, D. (2004). Identification of Wnt responsive genes using a murine mammary epithelial cell line model system. *BMC Dev Biol* **4**, 6.

Tao, W., Pennica, D., Xu, L., Kalejta, R.F., and Levine, A.J. (2001). Wrch-1, a novel member of the Rho gene family that is regulated by Wnt-1. *Genes Dev* **15**, 1796–1807.

Tcherkezian, J., and Lamarche-Vane, N. (2007). Current knowledge of the large RhoGAP family of proteins. *Biol Cell* **99**, 67–86.

Theodorescu, D., Sapinoso, L.M., Conaway, M.R., Oxford, G., Hampton, G.M., and Frierson, H.F., Jr. (2004). Reduced expression of metastasis suppressor RhoGDI2 is associated with decreased survival for patients with bladder cancer. *Clin Cancer Res* **10**, 3800–3806.

Thompson, P.W., Randi, A.M., and Ridley, A.J. (2002). Intercellular adhesion molecule (ICAM)-1, but not ICAM-2, activates RhoA and stimulates c-fos and rhoA transcription in endothelial cells. *J Immunol* **169**, 1007–1013.

Toksoz, D., and Williams, D.A. (1994). Novel human oncogene lbc detected by transfection with distinct homology regions to signal transduction products. *Oncogene* **9**, 621–628.

Ueda, T., Kikuchi, A., Ohga, N., Yamamoto, J., and Takai, Y. (1990). Purification and characterization from bovine brain cytosol of a novel regulatory protein inhibiting the dissociation of GDP from and the subsequent binding of GTP to rhoB p20, a ras p21-like GTP-binding protein. *J Biol Chem* **265**, 9373–9380.

Valencia, A., Chardin, P., Wittinghofer, A., and Sander, C. (1991). The ras protein family: evolutionary tree and role of conserved amino acids. *Biochemistry* **30**, 4637–4648.

van Golen, K.L., Davies, S., Wu, Z.F., Wang, Y., Bucana, C.D., Root, H., Chandrasekharappa, S., Strawderman, M., Ethier, S.P., and Merajver, S.D. (1999). A novel putative low-affinity insulin-like growth factor-binding protein, LIBC (lost in inflammatory breast cancer), and RhoC GTPase correlate with the inflammatory breast cancer phenotype. *Clin Cancer Res* **5**, 2511–2519.

van Leeuwen, F.N., van der Kammen, R.A., Habets, G.G., and Collard, J.G. (1995). Oncogenic activity of Tiam1 and Rac1 in NIH3T3 cells. *Oncogene* **11**, 2215–2221.

Vega, F.M., and Ridley, A.J. (2007). SnapShot: Rho family GTPases. *Cell* **129**, 1430.

Vetter, I.R., and Wittinghofer, A. (2001). The guanine nucleotide-binding switch in three dimensions. *Science* **294**, 1299–1304.

Vignal, E., De Toledo, M., Comunale, F., Ladopoulou, A., Gauthier-Rouviere, C., Blangy, A., and Fort, P. (2000). Characterization of TCL, a new GTPase of the rho family related to TC10 andCcdc42. *J Biol Chem* **275**, 36457–36464.

Vincent, S., Jeanteur, P., and Fort, P. (1992). Growth-regulated expression of rhoG, a new member of the ras homolog gene family. *Mol Cell Biol* **12**, 3138–3148.

Watanabe, G., Saito, Y., Madaule, P., Ishizaki, T., Fujisawa, K., Morii, N., Mukai, H., Ono, Y., Kakizuka, A., and Narumiya, S. (1996). Protein kinase N (PKN) and PKN-related protein rhophilin as targets of small GTPase Rho. *Science* **271**, 645–648.

Welch, H.C., Coadwell, W.J., Ellson, C.D., Ferguson, G.J., Andrews, S.R., Erdjument-Bromage, H., Tempst, P., Hawkins, P.T., and Stephens, L.R. (2002). P-Rex1, a PtdIns(3,4,5)P3- and Gbetagamma-regulated guanine-nucleotide exchange factor for Rac. *Cell* **108**, 809–821.

Wennerberg, K., and Der, C.J. (2004). Rho-family GTPases: it's not only Rac and Rho (and I like it). *J Cell Sci* **117**, 1301–1312.

Whitehead, I., Kirk, H., and Kay, R. (1995a). Retroviral transduction and oncogenic selection of a cDNA encoding Dbs, a homolog of the Dbl guanine nucleotide exchange factor. *Oncogene* **10**, 713–721.

Whitehead, I., Kirk, H., Tognon, C., Trigo-Gonzalez, G., and Kay, R. (1995b). Expression cloning of lfc, a novel oncogene with structural similarities to guanine nucleotide exchange factors and to the regulatory region of protein kinase C. *J Biol Chem* **270**, 18388–18395.

Whitehead, I.P., Khosravi-Far, R., Kirk, H., Trigo-Gonzalez, G., Der, C.J., and Kay, R. (1996). Expression cloning of lsc, a novel oncogene with structural similarities to the Dbl family of guanine nucleotide exchange factors. *J Biol Chem* **271**, 18643–18650.

Wilkins, A., Ping, Q., and Carpenter, C.L. (2004). RhoBTB2 is a substrate of the mammalian Cul3 ubiquitin ligase complex. *Genes Dev* **18**, 856–861.

Wu, Y.C., and Horvitz, H.R. (1998). C. elegans phagocytosis and cell-migration protein CED-5 is similar to human DOCK180. *Nature* **392**, 501–504.

Yamamoto, M., Marui, N., Sakai, T., Morii, N., Kozaki, S., Ikai, K., Imamura, S., and Narumiya, S. (1993). ADP-ribosylation of the rhoA gene product by botulinum C3 exoenzyme causes Swiss 3T3 cells to accumulate in the G1 phase of the cell cycle. *Oncogene* **8**, 1449–1455.

Yeramian, P., Chardin, P., Madaule, P., and Tavitian, A. (1987). Nucleotide sequence of human rho cDNA clone 12. *Nucleic Acids Res* **15**, 1869.

Yohe, M.E., Rossman, K.L., Gardner, O.S., Karnoub, A.E., Snyder, J.T., Gershburg, S., Graves, L.M., Der, C.J., and Sondek, J. (2007). Auto-inhibition of the Dbl family protein Tim by an N-terminal helical motif. *J Biol Chem* **282**, 13813–13823.

Zhang, F.L., and Casey, P.J. (1996). Protein prenylation: molecular mechanisms and functional consequences. *Annu Rev Biochem* **65**, 241–269.

Zhang, Y., and Zhang, B. (2006). D4-GDI, a Rho GTPase regulator, promotes breast cancer cell invasiveness. *Cancer Res* **66**, 5592–5598.

Zohn, I.E., Symons, M., Chrzanowska-Wodnicka, M., Westwick, J.K., and Der, C.J. (1998). Mas oncogene signaling and transformation require the small GTP-binding protein Rac. *Mol Cell Biol* **18**, 1225–1235.

Chapter 2
Rho Proteins in Cancer

Devin T. Rosenthal, John Chadwick Brenner, and Sofia D. Merajver

Introduction

The control of coordinated cell motility and the plasticity of cell polarity lie at the center of tissue development in embryonic life. Proteins homologous to those present in bacteria and yeast evolved in multicellular organisms to control these functions. In recent years, we have come to understand that when proteins which govern motility are dysregulated in adult tissues, major derangements of growth and of interactions between tissues ensue. In particular, the Ras homology proteins, beginning with Ras itself, play major and diverse roles in cancer initiation and progression. In this chapter, we review with broad brushstrokes the role of Rho proteins in cancer.

Traditionally, the Ras superfamily is divided into five different branches: Ras, Rho, Rab, Ran, and ARF families. Recent in silico analysis of the human genome has revealed the presence of several new Ras-superfamily pseudogenes for which transcriptional evidence does not currently exist, and the function of these pseudogenes is unknown.

Ras homologous (Rho)-family GTPases comprise the largest subfamily cluster of the Ras-homology superfamily of small GTPases (~21 kDa). Currently, 22 Rho family entries are annotated in the ENTREZ database, which can be divided into 10 different groups on the basis of their sequence homology to either Cdc42, Rac1, RhoA, RhoD, Rif/RhoF, Rnd3/RhoE, TTF/RhoH, Chp/RhoV, or RhoBTB.

Like most Ras homology proteins, Rho GTPases function as molecular switches where GTP-bound Rho proteins are active and GDP-bound Rho proteins are inactive. Thus, Rho-GTP binds and activates downstream effectors leading to a variety of signaling cascades, while Rho-GDP does not. Although the involvement of a guanine nucleotide is a prerequisite for members of the Rho family, GTPase activity itself is not, as members of the Rnd and RhoH subfamilies are GTPase deficient (Nobes et al. 1998; Li et al. 2002) or in the case of RhoE/Rnd3, constitutively GTP-bound (Foster

D.T. Rosenthal, J.C. Brenner and S.D. Merajver (✉)
Department of Internal Medicine and Comprehensive Cancer Center,
University of Michigan, Ann Arbor, MI, USA
e-mail: smerajve@umich.edu

K. van Golen (ed.), *The Rho GTPases in Cancer*,
DOI 10.1007/978-1-4419-1111-7_2, © Springer Science+Business Media, LLC 2010

et al.1996; Wennerberg et al. 2003). This is just one of several examples of the diversity, and consequently controversy, of the proteins classified as Rho GTPases.

Cycling between GTP- and GDP-bound states is tightly controlled by three different classes of Rho regulatory proteins: GTPase-activating proteins (GAPs), guanine nucleotide dissociation inhibitors (GDIs), and guanine nucleotide exchange factors (GEFs). GAPs and GDIs inhibit Rho activity; GAPs by accelerating Rho protein hydrolysis of GTP into GDP and GDIs by sequestering Rho proteins to the cytoplasm and blocking GDP dissociation. GEFs activate Rho proteins by triggering the release of GDP from Rho-GDP, thereby permitting the Rho protein to bind GTP.

The activation of Rho-family proteins is mediated through a wide array of mechanisms including interactions with G-protein-coupled receptors (GPCRs) and other cell surface receptors, such as cytokine, tyrosine kinase, and adhesion receptors. Once initiated, Rho signaling leads to the activation of several different pathways depending on the precursor signal, the cellular context, and the crosstalk with other activated or repressed pathways. In fact, when Rho-family proteins were first brought into the limelight following landmark publications by Ridley, Hall, and their coworkers in 1992, the number of publications on the diversity of functions of Rho proteins rose exponentially (Ridley and Hall 1992; Ridley et al. 1992). In these papers, Ridley and Hall related the assembly of focal adhesions and actin stress fibers to growth factor signaling through Rho GTPase. Since then, Rho proteins have also been shown to play roles in adhesion, migration, phagocytosis, cytokinesis, neurite extension and retraction, cell morphogenesis and polarization, growth and cell survival (Chimini and Chavrier 2000; Etienne-Manneville and Hall 2002; Evers et al. 2000; Raftopoulou and Hall 2004). Additionally, as summarized in Table 2.1, aberrant overexpression of some Rho-family proteins has been shown to promote malignant transformation and metastasis [See chart refs].

Rho-Family Structure

Ras-superfamily proteins have a highly conserved G box GDP/GTP-binding motif element at the start of the N-terminus, allowing this family of proteins to have a very high affinity for both GDP and GTP nucleotides. Specifically, the conserved domain structure is G1, GXXXXGKS/T; G2, T; G3, DXXGQ/H/T; G4, T/NKXD; and G5, C/SAK/L/T (Bourne et al. 1991). These five conserved elements compose the G domain, which has conserved structure and function shared by all Ras-superfamily proteins. Although Ras-superfamily proteins retain an overall similar conformation in the GDP or GTP bound states, two regions - Switch 1 and Switch 2 - undergo a significant shift relative to the GTPase core, causing a change in affinity for GAP or GEF regulatory proteins and downstream effectors (Bishop and Hall 2000; Repasky et al. 2004). Consequently, regulatory proteins and effectors most frequently bind to the Switch 1 and Switch 2 regions of Ras-superfamily GTPases.

A second important feature of Ras-superfamily and Rho-family proteins is the inclusion of a C-terminal CAAX amino acid motif, which undergoes a post-translational addition of a prenyl group through the transfer of either a farnesyl or a geranylgeranyl moiety to the cysteine in this motif, thereby anchoring the GTPase to the membrane. This feature is essential for Rho protein function. Of the Ras-superfamily, only two subgroups, the Rho-family and Rab-family proteins, are known to be regulated by GDIs that conceal the C-terminal prenyl modification to promote cytosolic sequestration of these GTPases (Olofsson 1999; Seabra and Wasmeier 2004). In the cases of RhoB, TC10, and TCL, however, the cysteine residue of the CAAX motif can be palmitoylated, thus preventing recognition by RhoGDI and promoting membrane localization (Michaelson et al. 2001).

In support of the structural importance of the Switch domains and the CAAX motif to Rho protein function, Dvorsky and Ahmadian calculated the relative number of interactions of all residues of Rho GTPases with GDIs, GEFs, GAPs, and effectors using available structures for the most well-studied members of the Rho-subfamily, RhoA, Rac1, and Cdc42, and concluded that most interactions occur at Switch domain 1, 2 or at the extreme C-terminus (Dvorsky and Ahmadian 2004). The relevance of the CAAX motif brings into question the inclusion of the "atypical" Rho GTPases in the Rho family, as most of the atypical Rho GTPases (except for Wrch-1) lack a CAAX motif and (with the exception of Wrch-2) are not membrane associated (Ridley 2006).

The translocation of Rho proteins from the membrane fraction into the cytosolic fraction is a key step in the inhibition of Rho GTPase proteins. Analysis of inactive forms of Rho proteins have also revealed that some classes of Rho proteins may become inactivated via phosphorylation. RhoA, for example, is phosphorylated by PKA at serine 188, and this phosphorylation has been shown to increase the apparent strength of RhoA interactions with GDIs by inhibiting binding of RhoA to other downstream targets independent of RhoA's GDP/GTP state (Dong et al. 1998; Forget et al. 2002).

Rho Regulatory Proteins

Guanine Nucleotide Exchange Factors (GEFs)

The first identified Rho GEF was cloned from a screen for transforming genes in diffuse B-cell lymphoma cells, hence the name Dbl, and has been demonstrated to function as a GEF for the Rho-family member Cdc42 (Eva et al. 1988; Hart et al. 1991). Dbl represented the first member of a newly recognized family of GEFs that specifically regulate Rho GTPases (Schmidt and Hall 2002). Dbl contains two regions of homology with the yeast GEF Cdc24; the Dbl-homology (DH) domain, which catalyzes the release of GDP from Rho family proteins by stabilizing GTPase intermediates that are devoid of nucleotide and Mg^{2+}, and an adjacent C-terminal pleckstrin homology (PH) domain, which is thought to promote Dbl

Table 2.1

	Rho subfamily																					
	Rho			Rac				Cdc42					RhoBTB		Miro		Rnd			RhoH	RhoD	
Cancer type	Rho A	Rho B	Rho C	Rac 1	Rac 2	Rac 3	RhoG	Cdc 42	TC10	TCL	Wrch-2*	Wrch-1*	RhoBTB-1*	RhoBTB-2*	Miro-1*	Miro-2*	Rnd 1	Rnd 2	Rnd 3	Rho/TTF	RhoD	Rif
Bladder	X	S	X	X				X						X								
Breast	X	X	X	X		X	i	X				X		X								
Cervical	X	S		S							i											
Colon	X		X	X				X				X							X			
Gastric	X	X	X	X				X			i											
Head and Neck	X	S	X	X	X			X					S									
Hepatocellular	X		X	i				i														
Leukemia	X			X	X	X		X			i									X		
Lung	X	S	X	X				X			i			X					X			
Lympoma	X	X	X	i				X												X		
Melanoma	X	S	X	X				X											S			
Osteosarcoma	X	S		i																		
Ovarian	X		X					X														
Pancreatic	X	S	X	i				X			i											
Prostate	X		X	X		X		X											S			

X = oncogene
S = tumor suppressor
i = indirect/observational cancer involvement
* = atypical

localization to the plasma membrane via binding to phosphoinositides, leading to subsequent interaction with Rho GTPases (Rossman et al. 2005). A misnomer in the field is that GEFs recruit GTP to the GDP/GTP-binding motif of GTPases. In actuality, the concentration of GTP is significantly higher in vivo than that of GDP; thus, a concentration gradient drives the "preferential" loading of GTP into the GTPase.

Through analysis of several structures, interactions between the DH domain and Rho GTPases have been well characterized. DH domains have three conserved regions (CR1-CR3), which form the core of this domain's three-dimensional structure (Rossman et al. 2005). The most significant difference in the structure of DH domains occurs at the helix, which is C-terminal to CR1-CR3 and contains the α6-helix. Several groups have demonstrated that the DH domain interacts extensively with the Switch 1 region of Rho GTPases, and that CR3 and the α6-helix contact Switch 2, anchoring the two proteins through a hydrophobic cleft. Likewise, a conserved basic residue and semi-conserved Asn in CR3 make important contacts with Switch 2 that contributes to the DH domain's exchange potential (Rossman et al. 2005).

The functions and mechanisms of the PH domain appear to be more diverse than those of the DH domain, and at this point are less well understood. Structural data from numerous Rho GEFs indicate that positioning of the PH domain relative to the DH domain is highly variable, likely contributing to the observation that PH domain interaction with Rho GTPases also varies greatly between Rho GEFs (Rossman et al. 2005).

Although PH domains are characterized by their ability to bind phosphoinositides, the importance of this binding to Rho GEF activity remains hotly contested. Early studies demonstrated that loss of function due to deletion of the PH domain in Dbl-family members could be restored by replacing the PH domain with another membrane localization signal (Rossman et al. 2005). Subsequent studies, however, have shown that Dbl-family PH domain binding to phospholipids occurs with low affinity and specificity and is dispensable, in some cases, for proper GEF localization (Rossman et al. 2005).

GTPase-Activating Proteins (GAPs)

GAPs exhibit a wide array of specificity and activity. Recently, experiments introducing the RhoGAP domains of either p122RhoGAP or GRAF into fibroblasts blocked the formation of RhoA-mediated actin stress fiber formation, despite the indiscriminate in vitro activity of these proteins toward Rac1 or Cdc42 (Sekimata et al. 1999; Taylor et al. 1999). In addition to the fact that many identified RhoGAPs contain several domains other than just a RhoGAP region, the aforementioned data add support to the underappreciated hypothesis that various signaling pathways converge on RhoGAPs by interacting with their regulatory motifs to control RhoGAP specificity.

Protein kinases can regulate RhoGAP activity, the best example being the regulation of p190 RhoGAP by Src family tyrosine kinases. Observationally, Src activation leads cells to exhibit characteristics commonly associated with inactivated Rho proteins, namely disruption of actin stress fibers and reduction of focal contacts. As such, it was not surprising that Src activation was shown to correlate with the phosphorylation of two Tyr residues on p190 RhoGAP proximal to the RhoGAP domain, and that this double phosphorylation is critical for RhoGAP activity (Roof et al. 1998). Phosphorylation control of the p190 RhoGAP is not limited to tyrosine phosphorylation. In fact, protein kinase C activity, which leads to phosphorylation of Ser/Thr residues, can modulate the spatial distribution of p190 RhoGAP in the cell (Brouns et al. 2000).

In addition to p190 RhoGAP, several other GTPase-activating proteins have been identified, the most famous of which is BCR, the 5' partner of an oncogenic and reciprocal fusion cloned from the breakpoint region in Philadelphia-positive chronic myeloid leukemia, BCR-ABL (Diekmann et al. 1991; Rowley 1973). The 3' end of BCR actually contains the RhoGAP domain, suggesting that the RhoGAP activity of BCR may be maintained because the translocation and fusion are reciprocal (producing both BCR-ABL and ABL-BCR chimeras). Despite this notion, however, ABL-BCR actually loses its ability to function as a RhoGAP and has negative consequences on cell adhesion, in particular adherence to endothelial cells (Zheng et al. 2006). The regulation of Rho activity as mediated by RhoGAPs is crucial to an integrated understanding of Rho function in cancer.

Rho Effector Interactions: General Considerations

Rho proteins regulate cellular structure by controlling microtubules and actin cytoskeleton structure. To accomplish this, Rho proteins regulate the function of the Diaphanous-related formins, such as mDia1. mDia1 contains a Rho-binding domain, a polyproline FH1 domain that regulates F-actin, and an FH2 domain, which regulates microtubule structure. Interestingly, the introduction of mDia1 mutants lacking the Rho-binding domain into HeLa cells resulted in the bipolar elongation as microtubules aligned parallel to F-actin bundles (Ishizaki et al. 2001). These data demonstrated the essential role of Rho signaling through mDia1 in the maintenance of normal cell morphology.

In an effort to determine whether RhoA's abilities to induce nuclear signaling and oncogenic effects were due to the effect on the cytoskeleton or through other mechanisms, transfection of several different RhoA mutant expression vectors into NIH3T3 cells followed by a colony formation assay revealed that substitutions of Leu-39, Glu-39, or Cys-42 failed to stimulate serum response factor-mediated gene expression or induce neoplastic transformation. In transient assays, however, the introduction of these three different mutants did induce significant changes in actin stress fiber formation, providing some of the first evidence that in certain cases the Rho-effector proteins regulating the structure of actin filaments are distinct from those signaling to the nucleus and hijacking growth control (Zohar et al. 1998).

Following the realization that Rho signaled through a variety of effectors to induce different effects, crystal structure data helped to clarify classes of Rho-interacting proteins. In fact, the crystal structure of RhoA bound to the effector domain of protein kinase PKN/PRK1 revealed that an antiparallel coiled-coil finger (ACC finger) fold of the effector domain bound to the Switch 1, 2, and adjacent regions by specific hydrogen bonds in RhoA. Subsequent in silico sequence analysis demonstrated that the structure of the ACC finger domain is highly conserved in Rho effector proteins (Maesaki et al. 1999).

Rho GTPases Drive Cell Migration

The migration of cells in multicellular organisms is an essential part of development, wound repair, and immune surveillance that is controlled by extracellular signals which elicit intracellular changes in the organization of actin and microtubule cytoskeletons. As a cell begins its migratory journey to a new destination, leading edge protrusions or lamellipodium extend from the surface of the cell toward the attractive extracellular signal (or away from the repulsive signal) and establish new adhesion sites along this leading edge. The cell body then begins to contract and adhesions along the lagging edge are broken in a highly coordinated process leading to efficient, directional movement.

While the Rho family members Cdc42, Rac, and Rho form integrin-based matrix adhesion complexes at the cell surface, Cdc42 and Rac also regulate the polymerization of actin to form filopodial and lamellipodial protrusions, respectively, and Rho proteins regulate the assembly of contractile actin-myosin filaments (Nobes and Hall 1995; Ridley and Hall 1992; Ridley et al. 1992).

Although Rac and Cdc42 are known to catalyze a protrusive force by inducing localized actin polymerization at the leading edge of a cell during migration, studies in *Drosophila melanogaster* demonstrated that Cdc42 loss of function does not affect the rate of migration of peripheral glial cells, but does alter the direction of migration (Sepp and Auld 2003). Experiments using high-speed supernatants from *Xenopus* egg extracts demonstrated that constitutively active Cdc42 (generated by loading Cdc42 with GTPγS - a non-hydrolysable GTP analogue) can induce F-actin polymerization in wild-type extracts, but not N-WASP (a downstream effector of Cdc42) immunodepleted extracts (Rohatgi et al. 1999). N-WASP belongs to the WASP/SCAR/WAVE family of scaffold proteins. Ensuing experiments in the Rohatgi et al. paper and several others demonstrated that members of this family stimulate actin polymerization either de novo or at the barbed edge of preexisting filaments through stimulation of the Arp2/3 complex (Rohtgi et al. 1999; Weaver et al. 2003). Interestingly, WASP and WAVE have also been shown to interact with profilin, a protein that increases the rate of actin polymerization by acting synergistically with the Arp2/3 complex (Blanchoin et al. 2000; Yang et al. 2000).

Rac activation occurs primarily at the leading edge of a migrating cell, thereby preferentially driving the formation of lamellipodium along this edge. In 2005, the mechanism regulating Rac activation to the leading edge was finally elucidated (Nishiya et al. 2005). Normally, the α_4 integrin–paxillin complex inhibits stable lamellipodia by recruiting Arf-GAP, which in turn inhibits Arf, thus blocking Rac activation. However, phosphorylation of α_4 integrin, which only occurs at the leading edge (Lim et al. 2007), prevents its interaction with paxillin and subsequently Arf-GAP, thereby releasing repression of Arf and allowing it to activate Rac.

Rho activity facilitates cell migration by orchestrating focal adhesion assembly, cell body contraction, and lagging edge retraction. The protein p160ROCK is known to work specifically downstream of Rho, not Rac, to mediate focal adhesions and stress fibers (Ishizaju et al. 1997). After becoming activated, p160ROCK stabilizes actin filaments in actin-myosin filaments by phosphorylating and activating LIMK, which then inactivates cofilin, an actin depolymerizer, through a subsequent phosphorylation (Maekawa et al. 1999; Sumi et al. 2001). As the enemy of my enemy is my friend, by inhibiting a protein that prevents actin reassembly, this p160ROCK phosphorylation cascade stabilizes actin filaments along the lagging edge of a migratory cell. Interestingly, p160ROCK has also been shown to phosphorylate and inactivate myosin light chain phosphatase, leading to increased myosin phosphorylation and the generation of a contractile force, which produces the desired movement (Mitchison and Cramer 1996).

Conclusions

It is clear that many of the Rho family members act in concert and share multiple signaling pathways to affect their functions under different stimuli. Future efforts to understand their biology should be, at least in part, focused on understanding the dynamics of the on–off switch that is at the heart of Rho protein function. We believe the era of using quantitative, sophisticated methods of modeling and imaging live cells will carry this field to the next level, where we may enhance targeted drug design by a deeper insight into how these proteins transmit information in the cell.

Acknowledgments This work was supported by the Burroughs Wellcome Fund, the Breast Cancer Research Foundation, the Department of Defense Breast Cancer Research Program (BC083217 and BC083262), and the National Institutes of Health (CA-77612).

References

Nobes CD, Lauritzen I, Mattei MG, Paris S, Hall A, Chardin P. A new member of the Rho family, Rnd1, promotes disassembly of actin filament structures and loss of cell adhesion. J Cell Biol 1998 Apr 6;141(1):187–97.

Li X, Bu X, Lu B, Avraham H, Flavell RA, Lim B. The hematopoiesis-specific GTP-binding protein RhoH is GTPase deficient and modulates activities of other Rho GTPases by an inhibitory function. Mol Cell Biol 2002 Feb;22(4):1158–71.

Foster R, Hu KQ, Lu Y, Nolan KM, Thissen J, Settleman J. Identification of a novel human Rho protein with unusual properties: GTPase deficiency and in vivo farnesylation. Mol Cell Biol 1996 Jun;16(6):2689–99.

Wennerberg K, Forget MA, Ellerbroek SM, Arthur WT, Burridge K, Settleman J, et al. Rnd proteins function as RhoA antagonists by activating p190 RhoGAP. Curr Biol 2003 Jul 1;13(13):1106–15.

Ridley AJ, Hall A. The small GTP-binding protein rho regulates the assembly of focal adhesions and actin stress fibers in response to growth factors. Cell 1992;70(3):389–99.

Ridley AJ, Paterson H, Johnston C, Diekman D., Hall A. The small GTP-binding protein rac regulates growth-factor induced membrane ruffling. Cell 1992;70(3):401–10.

Chimini G, Chavrier P. Function of Rho family proteins in actin dynamics during phagocytosis and engulfment. Nat Cell Biol 2000 Oct;2(10):E191-E196.

Etienne-Manneville S, Hall A. Rho GTPases in cell biology. Nature 2002 Dec 12;420(6916):629–35.

Evers EE, Zondag GC, Malliri A, Price LS, ten Klooster JP, van der Kammen RA, et al. Rho family proteins in cell adhesion and cell migration. Eur J Cancer 2000 Jun;36(10):1269–74.

Raftopoulou M, Hall A. Cell migration: Rho GTPases lead the way. Dev Biol 2004 Jan 1;265(1):23–32.

Bourne HR, Sanders DA, McCormick F. The GTPase superfamily: conserved structure and molecular mechanism. Nature 1991 Jan 10;349(6305):117–27.

Bishop AL, Hall A. Rho GTPases and their effector proteins. Biochem J 2000 Jun 1;348 Pt 2:241–55.

Repasky GA, Chenette EJ, Der CJ. Renewing the conspiracy theory debate: does Raf function alone to mediate Ras oncogenesis? Trends Cell Biol 2004 Nov;14(11):639–47.

Olofsson B. Rho guanine dissociation inhibitors: Pivotal molecules in cellular signalling. Cell Signal 1999 Aug;11(8):545–54.

Seabra MC, Wasmeier C. Controlling the location and activation of Rab GTPases. Curr Opin Cell Biol 2004 Aug;16(4):451–7.

Michaelson D, Silletti J, Murphy G, D'Eustachio P, Rush M, Philips MR. Differential localization of Rho GTPases in live cells: regulation by hypervariable regions and RhoGDI binding. J Cell Biol 2001 Jan 8;152(1):111–26.

Dvorsky R, Ahmadian MR. Always look on the bright site of Rho: structural implications for a conserved intermolecular interface. EMBO Rep 2004 Dec;5(12):1130–6.

Ridley AJ. Rho GTPases and actin dynamics in membrane protrusions and vesicle trafficking. Trends Cell Biol 2006 Oct;16(10):522–9.

Dong JM, Leung T, Manser E, Lim L. cAMP-induced morphological changes are counteracted by the activated RhoA small GTPase and the Rho kinase ROKalpha. J Biol Chem 1998 Aug 28;273(35):22554–62.

Forget MA, Desrosiers RR, Gingras D, Beliveau R. Phosphorylation states of Cdc42 and RhoA regulate their interactions with Rho GDP dissociation inhibitor and their extraction from biological membranes. Biochem J 2002 Jan 15;361(Pt 2):243–54.

Eva A, Vecchio G, Rao CD, Tronick SR, Aaronson SA. The predicted DBL oncogene product defines a distinct class of transforming proteins. Proc Natl Acad Sci U S A 1988 Apr;85(7):2061–5.

Hart MJ, Eva A, Evans T, Aaronson SA, Cerione RA. Catalysis of guanine nucleotide exchange on the CDC42Hs protein by the dbl oncogene product. Nature 1991 Nov 28;354(6351):311–4.

Schmidt A, Hall A. Guanine nucleotide exchange factors for Rho GTPases: turning on the switch. Genes Dev 2002 Jul 1;16(13):1587–609.

Rossman KL, Der CJ, Sondek J. GEF means go: turning on RHO GTPases with guanine nucleotide-exchange factors. Nat Rev Mol Cell Biol 2005 Feb;6(2):167–80.

Sekimata M, Kabuyama Y, Emori Y, Homma Y. Morphological changes and detachment of adherent cells induced by p122, a GTPase-activating protein for Rho. J Biol Chem 1999 Jun 18;274(25):17757–62.

Taylor JM, Macklem MM, Parsons JT. Cytoskeletal changes induced by GRAF, the GTPase regulator associated with focal adhesion kinase, are mediated by Rho. J Cell Sci 1999 Jan;112 (Pt 2):231–42.
Roof RW, Haskell MD, Dukes BD, Sherman N, Kinter M, Parsons SJ. Phosphotyrosine (p-Tyr)-dependent and -independent mechanisms of p190 RhoGAP-p120 RasGAP interaction: Tyr 1105 of p190, a substrate for c- Src, is the sole p-Tyr mediator of complex formation. Mol Cell Biol 1998 Dec;18(12):7052–63.
Brouns MR, Matheson SF, Hu KQ, Delalle I, Caviness VS, Silver J, et al. The adhesion signaling molecule p190 RhoGAP is required for morphogenetic processes in neural development. Development 2000 Nov;127(22):4891–903.
Diekmann D, Brill S, Garrett MD, Totty N, Hsuan J, Monfries C, et al. Bcr encodes a GTPase-activating protein for p21rac. Nature 1991 May 30;351(6325):400–2.
Rowley JD. Letter: A new consistent chromosomal abnormality in chronic myelogenous leukaemia identified by quinacrine fluorescence and Giemsa staining. Nature 1973 Jun 1;243(5405):290–3.
Zheng X, Guller S, Beissert T, Puccetti E, Ruthardt M. BCR and its mutants, the reciprocal t(9;22)-associated ABL/BCR fusion proteins, differentially regulate the cytoskeleton and cell motility. BMC Cancer 2006;6:262.
Ishizaki T, Morishima Y, Okamoto M, Furuyashiki T, Kato T, Narumiya S. Coordination of microtubules and the actin cytoskeleton by the Rho effector mDia1. Nat Cell Biol 2001 Jan;3(1):8–14.
Zohar M, Teramoto H, Katz BZ, Yamada KM, Gutkind JS. Effector domain mutants of Rho dissociate cytoskeletal changes from nuclear signaling and cellular transformation. Oncogene 1998 Aug 27;17(8):991–8.
Maesaki R, Ihara K, Shimizu T, Kuroda S, Kaibuchi K, Hakoshima T. The structural basis of Rho effector recognition revealed by the crystal structure of human RhoA complexed with the effector domain of PKN/PRK1. Mol Cell 1999 Nov;4(5):793–803.
Nobes CD, Hall A. Rho, rac, and cdc42 GTPases regulate the assembly of multimolecular focal complexes associated with actin stress fibers, lamellipodia and filopodia. Cell 1995;81(1):53–62.
Sepp KJ, Auld VJ. RhoA and Rac1 GTPases mediate the dynamic rearrangement of actin in peripheral glia. Development 2003 May;130(9):1825–35.
Rohatgi R, Ma L, Miki H, Lopez M, Kirchhausen T, Takenawa T, et al. The interaction between N-WASP and the Arp2/3 complex links Cdc42-dependent signals to actin assembly. Cell 1999 Apr 16;97(2):221–31.
Weaver AM, Young ME, Lee WL, Cooper JA. Integration of signals to the Arp2/3 complex. Curr Opin Cell Biol 2003 Feb;15(1):23–30.
Blanchoin L, Pollard TD, Mullins RD. Interactions of ADF/cofilin, Arp2/3 complex, capping protein and profilin in remodeling of branched actin filament networks. Curr Biol 2000 Oct 19;10(20):1273–82.
Yang C, Huang M, DeBiasio J, Pring M, Joyce M, Miki H, et al. Profilin enhances Cdc42-induced nucleation of actin polymerization. J Cell Biol 2000 Sep 4;150(5):1001–12.
Nishiya N, Kiosses WB, Han J, Ginsberg MH. An alpha4 integrin-paxillin-Arf-GAP complex restricts Rac activation to the leading edge of migrating cells. Nat Cell Biol 2005 Apr;7(4):343–52.
Lim CJ, Han J, Yousefi N, Ma Y, Amieux PS, McKnight GS, et al. Alpha4 integrins are type I cAMP-dependent protein kinase-anchoring proteins. Nat Cell Biol 2007 Apr;9(4):415–21.
Ishizaki T, Naito M, Fujisawa K, Maekawa M, Watanabe N, Saito Y, et al. p160ROCK, a Rho-associated coiled-coil forming protein kinase, works downstream of Rho and induces focal adhesions. FEBS Lett 1997 Mar 10;404(2–3):118–24.
Maekawa M, Ishizaki T, Boku S, Watanabe N, Fujita A, Iwamatsu A, et al. Signaling from Rho to the actin cytoskeleton through protein kinases and LIM-kinase. Science 1999;285(5429):895–8.
Sumi T, Matsumoto K, Nakamura T. Specific activation of LIM kinase 2 via phosphorylation of threonine 505 by ROCK, a Rho-dependent protein kinase. J Biol Chem 2001 Jan 5;276(1):670–6.
Mitchison TJ, Cramer LP. Actin-based cell motility and cell locomotion. Cell 1996 Feb 9;84(3):371–9.
Kamai T, Tsujii T, Arai K, Takagi K, Asami H, Ito Y, et al. Significant association of Rho/ROCK pathway with invasion and metastasis of bladder cancer. Clin Cancer Res 2003 Jul;9(7):2632–41.

Pervaiz S, Cao J, Chao OS, Chin YY, Clement MV. Activation of the RacGTPase inhibits apoptosis in human tumor cells. Oncogene 2001 Sep 27;20(43):6263–8.

Wu F, Chen Y, Li Y, Ju J, Wang Z, Yan D. RNA-interference-mediated Cdc42 silencing down-regulates phosphorylation of STAT3 and suppresses growth in human bladder-cancer cells. Biotechnol Appl Biochem 2008 Feb;49(Pt 2):121–8.

Knowles MA, Aveyard JS, Taylor CF, Harnden P, Bass S. Mutation analysis of the 8p candidate tumour suppressor genes DBC2 (RHOBTB2) and LZTS1 in bladder cancer. Cancer Lett 2005 Jul 8;225(1):121–30.

Bourguignon LY, Zhu H, Shao L, Zhu D, Chen YW. Rho-kinase (ROK) promotes CD44v(3,8–10)-ankyrin interaction and tumor cell migration in metastatic breast cancer cells. Cell Motil Cytoskeleton 1999;43(4):269–87.

de CP, Gauville C, Closson V, Linares G, Calvo F, Tavitian A, et al. EGF modulation of the ras-related rhoB gene expression in human breast-cancer cell lines. Int J Cancer 1994 Nov 1;59(3):408–15.

van Golen KL, Davies S, Wu ZF, Wang Y, Bucana CD, Root H, et al. A novel putative low-affinity insulin-like growth factor-binding protein, LIBC (lost in inflammatory breast cancer), and RhoC GTPase correlate with the inflammatory breast cancer phenotype. Clin Cancer Res 1999 Sep;5(9):2511–9.

Debily MA, Camarca A, Ciullo M, Mayer C, El MS, Ba I, et al. Expression and molecular characterization of alternative transcripts of the ARHGEF5/TIM oncogene specific for human breast cancer. Hum Mol Genet 2004 Feb 1;13(3):323–34.

Bourguignon LY, Zhu H, Shao L, Chen YW. Ankyrin-Tiam1 interaction promotes Rac1 signaling and metastatic breast tumor cell invasion and migration. J Cell Biol 2000 Jul 10;150(1):177–91.

Mira JP, Benard V, Groffen J, Sanders LC, Knaus UG. Endogenous, hyperactive Rac3 controls proliferation of breast cancer cells by a p21-activated kinase-dependent pathway. Proc Natl Acad Sci U S A 2000 Jan 4;97(1):185–9.

Hirsch DS, Shen Y, Wu WJ. Growth and motility inhibition of breast cancer cells by epidermal growth factor receptor degradation is correlated with inactivation of Cdc42. Cancer Res 2006 Apr 1;66(7):3523–30.

Hamaguchi M, Meth JL, von KC, Wei W, Esposito D, Rodgers L, et al. DBC2, a candidate for a tumor suppressor gene involved in breast cancer. Proc Natl Acad Sci U S A 2002 Oct 15;99(21):13647–52.

Kirikoshi H, Katoh M. Expression of WRCH1 in human cancer and down-regulation of WRCH1 by beta-estradiol in MCF-7 cells. Int J Oncol 2002 Apr;20(4):777–83.

Hamadmad SN, Hohl RJ. Erythropoietin stimulates cancer cell migration and activates RhoA protein through a mitogen-activated protein kinase/extracellular signal-regulated kinase-dependent mechanism. J Pharmacol Exp Ther 2008 Mar;324(3):1227–33.

Chen Z, Sun J, Pradines A, Favre G, Adnane J, Sebti SM. Both farnesylated and geranylgeranylated RhoB inhibit malignant transformation and suppress human tumor growth in nude mice. J Biol Chem 2000 Jun 16;275(24):17974–8.

Geiger T, Sabanay H, Kravchenko-Balasha N, Geiger B, Levitzki A. Anomalous features of EMT during keratinocyte transformation. PLoS ONE 2008;3(2):e1574.

Katoh M. Molecular cloning and characterization of WRCH2 on human chromosome 15q15. Int J Oncol 2002 May;20(5):977–82.

Fritz G, Just I, Kaina B. Rho GTPases are over-expressed in human tumors. Int J Cancer 1999 May 31;81(5):682–7.

Bellovin DI, Simpson KJ, Danilov T, Maynard E, Rimm DL, Oettgen P, et al. Reciprocal regulation of RhoA and RhoC characterizes the EMT and identifies RhoC as a prognostic marker of colon carcinoma. Oncogene 2006 Nov 2;25(52):6959–67.

Akashi H, Han HJ, Iizaka M, Nakamura Y. Growth-suppressive effect of non-steroidal anti-inflammatory drugs on 11 colon-cancer cell lines and fluorescence differential display of genes whose expression is influenced by sulindac. Int J Cancer 2000 Dec 15;88(6):873–80.

Kim MH, Park JS, Chang HJ, Baek MK, Kim HR, Shin BA, et al. Lysophosphatidic acid promotes cell invasion by up-regulating the urokinase-type plasminogen activator receptor in human gastric cancer cells. J Cell Biochem 2008 Jun 1;104(3):1102–12.

Nishigaki M, Aoyagi K, Danjoh I, Fukaya M, Yanagihara K, Sakamoto H, et al. Discovery of aberrant expression of R-RAS by cancer-linked DNA hypomethylation in gastric cancer using microarrays. Cancer Res 2005 Mar 15;65(6):2115–24.

Liu N, Zhang G, Bi F, Pan Y, Xue Y, Shi Y, et al. RhoC is essential for the metastasis of gastric cancer. J Mol Med 2007 Oct;85(10):1149–56.

Xue Y, Bi F, Zhang X, Zhang S, Pan Y, Liu N, et al. Role of Rac1 and Cdc42 in hypoxia induced p53 and von Hippel-Lindau suppression and HIF1alpha activation. Int J Cancer 2006 Jun 15;118(12):2965–72.

Adnane J, Muro-Cacho C, Mathews L, Sebti SM, Munoz-Antonia T. Suppression of rho B expression in invasive carcinoma from head and neck cancer patients. Clin Cancer Res 2002 Jul;8(7):2225–32.

Pan Q, Bao LW, Teknos TN, Merajver SD. Targeted Disruption of Protein Kinase C{varepsilon} Reduces Cell Invasion and Motility through Inactivation of RhoA and RhoC GTPases in Head and Neck Squamous Cell Carcinoma. Cancer Res 2006 Oct 1;66(19):9379–84.

Patel V, Rosenfeldt HM, Lyons R, Servitja JM, Bustelo XR, Siroff M, et al. Persistent activation of Rac1 in squamous carcinomas of the head and neck: evidence for an EGFR/Vav2 signaling axis involved in cell invasion. Carcinogenesis 2007 Jun;28(6):1145–52.

Abraham MT, Kuriakose MA, Sacks PG, Yee H, Chiriboga L, Bearer EL, et al. Motility-related proteins as markers for head and neck squamous cell cancer. Laryngoscope 2001 Jul;111(7):1285–9.

Beder LB, Gunduz M, Ouchida M, Gunduz E, Sakai A, Fukushima K, et al. Identification of a candidate tumor suppressor gene RHOBTB1 located at a novel allelic loss region 10q21 in head and neck cancer. J Cancer Res Clin Oncol 2006 Jan;132(1):19–27.

Wang D, Dou K, Xiang H, Song Z, Zhao Q, Chen Y, et al. Involvement of RhoA in progression of human hepatocellular carcinoma. J Gastroenterol Hepatol 2007 Nov;22(11):1916–20.

Wang W, Yang LY, Huang GW, Lu WQ. Expression and significance of RhoC gene in hepatocellular carcinoma. World Journal of Gastroenterology 2003;9(9):1950–3.

Ching YP, Leong VY, Lee MF, Xu HT, Jin DY, Ng IO. P21-activated protein kinase is overexpressed in hepatocellular carcinoma and enhances cancer metastasis involving c-Jun NH2-terminal kinase activation and paxillin phosphorylation. Cancer Res 2007 Apr 15;67(8):3601–8.

Chang CS, Huang SM, Lin HH, Wu CC, Wang CJ. Different expression of apoptotic proteins between HBV-infected and non-HBV-infected hepatocellular carcinoma. Hepatogastroenterology 2007 Oct;54(79):2061–8.

Reuther GW, Lambert QT, Booden MA, Wennerberg K, Becknell B, Marcucci G, et al. Leukemia-associated Rho guanine nucleotide exchange factor, a Dbl family protein found mutated in leukemia, causes transformation by activation of RhoA. J Biol Chem 2001 Jul 20;276(29):27145–51.

Nishihara H, Maeda M, Oda A, Tsuda M, Sawa H, Nagashima K, et al. DOCK2 associates with CrkL and regulates Rac1 in human leukemia cell lines. Blood 2002 Dec 1;100(12):3968–74.

Munugalavadla V, Sims EC, Borneo J, Chan RJ, Kapur R. Genetic and pharmacologic evidence implicating the p85 alpha, but not p85 beta, regulatory subunit of PI3K and Rac2 GTPase in regulating oncogenic KIT-induced transformation in acute myeloid leukemia and systemic mastocytosis. Blood 2007 Sep 1;110(5):1612–20.

Cho YJ, Zhang B, Kaartinen V, Haataja L, de C, I, Groffen J, et al. Generation of rac3 null mutant mice: role of Rac3 in Bcr/Abl-caused lymphoblastic leukemia. Mol Cell Biol 2005 Jul;25(13):5777–85.

Harnois T, Constantin B, Rioux A, Grenioux E, Kitzis A, Bourmeyster N. Differential interaction and activation of Rho family GTPases by $p210^{bcr\text{-}abl\ and\ p190bcr\text{-}abl}$. Oncogene 2003;22(41):6445–54.

Rossi D, Berra E, Cerri M, Deambrogi C, Barbieri C, Franceschetti S, et al. Aberrant somatic hypermutation in transformation of follicular lymphoma and chronic lymphocytic leukemia to diffuse large B-cell lymphoma. Haematologica 2006 Oct;91(10):1405–9.

Wang S, Yan-Neale Y, Fischer D, Zeremski M, Cai R, Zhu J, et al. Histone deacetylase 1 represses the small GTPase RhoB expression in human nonsmall lung carcinoma cell line. Oncogene 2003 Sep 18;22(40):6204–13.

Cuiyan Z, Jie H, Fang Z, Kezhi Z, Junting W, Susheng S, et al. Overexpression of RhoE in Non-small Cell Lung Cancer (NSCLC) is associated with smoking and correlates with DNA copy number changes. Cancer Biol Ther 2007 Mar;6(3):335–42.
Wilkins A, Ping Q, Carpenter CL. RhoBTB2 is a substrate of the mammalian Cul3 ubiquitin ligase complex. Genes Dev 2004 Apr 15;18(8):856–61.
Stam JC, Michiels F, van der Kammen RA, Moolenaar WH, Collard JG. Invasion of T-lymphoma cells: cooperation between Rho family GTPases and lysophospholipid receptor signaling. EMBO J 1998 Jul 15;17(14):4066–74.
Kari L, Loboda A, Nebozhyn M, Rook AH, Vonderheid EC, Nichols C, et al. Classification and prediction of survival in patients with the leukemic phase of cutaneous T cell lymphoma. J Exp Med 2003 Jun 2;197(11):1477–88.
Schwering I, Brauninger A, Distler V, Jesdinsky J, Diehl V, Hansmann ML, et al. Profiling of Hodgkin's lymphoma cell line L1236 and germinal center B cells: identification of Hodgkin's lymphoma-specific genes. Mol Med 2003 Mar;9(3–4):85–95.
Zhang B, Zhang Y, Shacter E. Rac1 inhibits apoptosis in human lymphoma cells by stimulating Bad phosphorylation on Ser-75. Mol Cell Biol 2004 Jul;24(14):6205–14.
Dallery E, Galiegue-Zouitina S, Collyn-d'Hooghe M, Quief S, Denis C, Hildebrand MP, et al. TTF, a gene encoding a novel small G protein, fuses to the lymphoma-associated LAZ3 gene by t(3;4) chromosomal translocation. Oncogene 1995 Jun 1;10(11):2171–8.
Robledo MM, Bartolome RA, Longo N, Rodriguez-Frade JM, Mellado M, Longo I, et al. Expression of functional chemokine receptors CXCR3 and CXCR4 on human melanoma cells. J Biol Chem 2001 Nov 30;276(48):45098–105.
Jiang K, Sun J, Cheng J, Djeu JY, Wei S, Sebti S. Akt mediates Ras downregulation of RhoB, a suppressor of transformation, invasion, and metastasis. Mol Cell Biol 2004 Jun;24(12):5565–76.
Clark EA, Golub TR, Lander ES, Hynes RO. Genomic analysis of metastasis reveals an essential role for RhoC. Nature 2000 Aug 3;406(6795):532–5.
Eisenmann KM, McCarthy JB, Simpson MA, Keely PJ, Guan JL, Tachibana K, et al. Melanoma chondroitin sulphate proteoglycan regulates cell spreading through Cdc42, Ack-1 and p130cas. Nat Cell Biol 1999 Dec;1(8):507–13.
Pinner S, Sahai E. PDK1 regulates cancer cell motility by antagonising inhibition of ROCK1 by RhoE. Nat Cell Biol 2008 Feb;10(2):127–37.
Khyrul WA, LaLonde DP, Brown MC, Levinson H, Turner CE. The integrin-linked kinase regulates cell morphology and motility in a rho-associated kinase-dependent manner. J Biol Chem 2004 Dec 24;279(52):54131–9.
Horii Y, Beeler JF, Sakaguchi K, Tachibana M, Miki T. A novel oncogene, ost, encodes a guanine nucleotide exchange factor that potentially links Rho and Rac signaling pathways. EMBO J 1994 Oct 17;13(20):4776–86.
Horiuchi A, Imai T, Wang C, Ohira S, Feng Y, Nikaido T, et al. Up-regulation of small GTPases, RhoA and RhoC, is associated with tumor progression in ovarian carcinoma. Lab Invest 2003 Jun;83(6):861–70.
Bourguignon LY, Gilad E, Rothman K, Peyrollier K. Hyaluronan-CD44 interaction with IQGAP1 promotes Cdc42 and ERK signaling, leading to actin binding, Elk-1/estrogen receptor transcriptional activation, and ovarian cancer progression. J Biol Chem 2005 Mar 25;280(12):11961–72.
Kusama T, Mukai M, Iwasaki T, Tatsuta M, Matsumoto Y, Akedo H, et al. Inhibition of epidermal growth factor-induced RhoA translocation and invasion of human pancreatic cancer cells by 3-hydroxy-3-methylglutaryl-coenzyme a reductase inhibitors. Cancer Res 2001 Jun 15;61(12):4885–91.
Suwa H, Ohshio G, Imamura T, Watanabe G, Arii S, Imamura M, et al. Overexpression of the rhoC gene correlates with progression of ductal adenocarcinoma of the pancreas. Br J Cancer 1998;77(1):147–52.
Fernandez-Zapico ME, Gonzalez-Paz NC, Weiss E, Savoy DN, Molina JR, Fonseca R, et al. Ectopic expression of VAV1 reveals an unexpected role in pancreatic cancer tumorigenesis. Cancer Cell 2005 Jan;7(1):39–49.

Taniuchi K, Nakagawa H, Hosokawa M, Nakamura T, Eguchi H, Ohigashi H, et al. Overexpressed P-cadherin/CDH3 promotes motility of pancreatic cancer cells by interacting with p120ctn and activating rho-family GTPases. Cancer Res 2005 Apr 15;65(8):3092–9.

Nie D, Guo Y, Yang D, Tang Y, Chen Y, Wang MT, et al. Thromboxane A2 receptors in prostate carcinoma: expression and its role in regulating cell motility via small GTPase Rho. Cancer Res 2008 Jan 1;68(1):115–21.

Yao H, Dashner EJ, van Golen CM, van Golen KL. RhoC GTPase is required for PC-3 prostate cancer cell invasion but not motility. Oncogene 2005 Nov 28.

Knight-Krajewski S, Welsh CF, Liu Y, Lyons LS, Faysal JM, Yang ES, et al. Deregulation of the Rho GTPase, Rac1, suppresses cyclin-dependent kinase inhibitor p21(CIP1) levels in androgen-independent human prostate cancer cells. Oncogene 2004 Jul 15;23(32):5513–22.

Gnanapragasam VJ, Leung HY, Pulimood AS, Neal DE, Robson CN. Expression of RAC 3, a steroid hormone receptor co-activator in prostate cancer. Br J Cancer 2001 Dec 14;85(12):1928–36.

Black PC, Mize GJ, Karlin P, Greenberg DL, Hawley SJ, True LD, et al. Overexpression of protease-activated receptors-1,-2, and-4 (PAR-1, -2, and -4) in prostate cancer. Prostate 2007 May 15;67(7):743–56.

Bektic J, Pfeil K, Berger AP, Ramoner R, Pelzer A, Schafer G, et al. Small G-protein RhoE is underexpressed in prostate cancer and induces cell cycle arrest and apoptosis. Prostate 2005 Sep 1;64(4):332–40.

Part II
The Rho Regulatory Proteins in Cancer

Chapter 3
RhoGDIs in Cancer

Anthony N. Anselmo, Gary M. Bokoch, and Céline DerMardirossian

GDIs Function

The dynamics of Rho GTPase action are regulated by both an activity cycle and a cytosol-to-membrane cycle (Fig. 3.1). Rho GTPases are activated by the exchange of bound GDP for ambient GTP, which is stimulated by guanine nucleotide exchange factors (GEFs) and are inactivated by hydrolysis of bound GTP to GDP catalyzed by GTPase-activating proteins (GAPs). This activity cycle is regulated by guanine nucleotide dissociation inhibitors (GDIs), which act to sterically shield the Rho GTPases from the action of GEFs and GAPs. Thus for Rho GTPases to become active, it is thought that they must first be released from GDIs in order for membrane-associated GEFs to catalyze activation. Superimposed on this activity cycle is a cytosol/membrane cycle that is directly controlled by Rho GDIs (see below).

Three human GDIs have been identified: RhoGDI (or RhoGDIα) is the most abundant and ubiquitously expressed (Fukumoto et al. 1990; Leonard et al. 1992). It forms cytosolic complexes with most Rho GTPases. D4GDI (or RhoGDIβ) is abundantly expressed only in hematopoietic tissues, particularly in B- and T-lymphocytes (Gorvel et al. 1998; Lelias et al. 1993). Both RhoGDI and D4GDI localize to the cytosol and form 1:1 complexes with multiple Rho GTPases. Finally, RhoGDIγ (or RhoGDI3) is specifically expressed in lung, brain, and testis (Adra et al. 1997; Zalcman et al. 1996). Unlike the two other GDIs, RhoGDIγ is associated with vesicular membranes, and exhibits specificity for interactions with RhoB and RhoG.

In general, the cytosolic GDIs (RhoGDI and D4GDI) exhibit several similar activities (see Fig. 3.1) (reviewed in DerMardirossian and Bokoch 2005; Dovas and Couchman 2005; Dransart et al. 2005): (1) They inhibit the dissociation of GDP from Rho GTPases, maintaining the Rho GTPases in an inactive form and sterically preventing GTPase activation by GEFs. (2) They interact (with lower affinity) with

A.N. Anselmo, G.M. Bokoch (✉), and C. DerMardirossian
Departments of Immunology and Cell Biology, The Scripps Research Institute,
10550 N. Torrey Pines Road, La Jolla, CA 92037, USA
e-mail: bokoch@scripps.edu

K. van Golen (ed.), *The Rho GTPases in Cancer*,
DOI 10.1007/978-1-4419-1111-7_3, © Springer Science+Business Media, LLC 2010

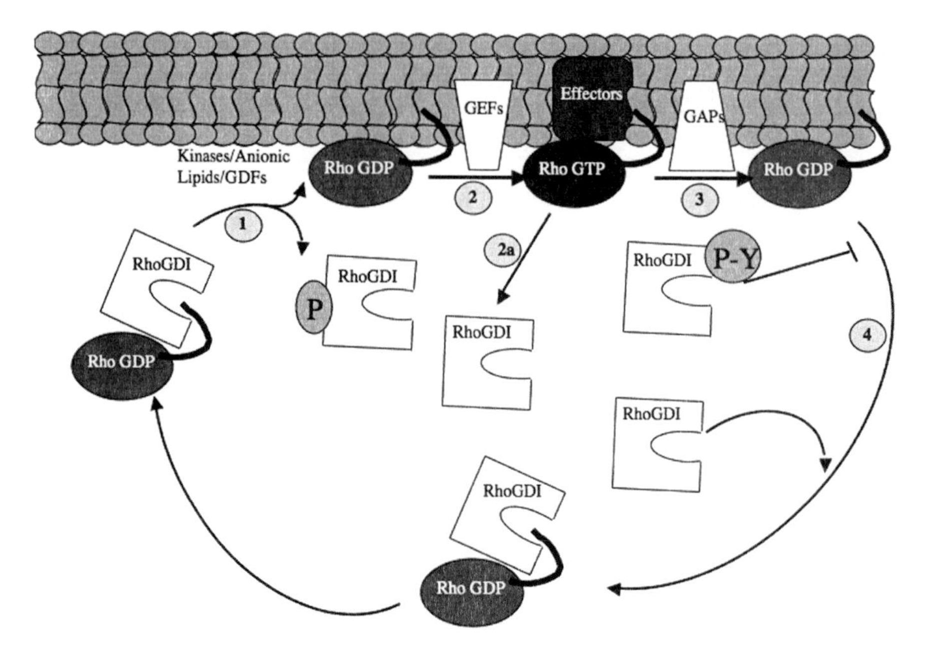

Fig. 3.1 A model of the Rho GTPase-GDI cycle. In resting cells, GDP-bound Rho GTPases are complexed to RhoGDI in the cytosol. In response to cell stimulation, Rho GTPases are released from RhoGDI through the actions of GDI displacement factors (GDFs), anionic lipids with complex dissociative activity, and/or through kinase-mediated phosphorylation of RhoGDI (1). Exchange of GDP to GTP is catalyzed by guanine nucleotide exchange factors (GEFs) (2). Active Rho GTPases can then bind and regulate various effectors. At this step, RhoGDI might extract Rho GTPases from the membrane in its GTP-bound form (2a) to terminate the signal or to relocalize Rho GTPases to a distinct membrane compartment within the cell. GTPase-activating protein (GAP) mediates the conversion of Rho GTPases to the GDP form (3) and allows their re-association with RhoGDI (4). However, when RhoGDI is tyrosine phosphorylated by Src, it has less affinity for Rho GTPases and remains membrane-localized, thereby prolonging Rho GTPase activity P: Phosphorylation on Ser/Thr P-Y: Phosphorylation on Tyrosine

the GTP-bound form of the Rho GTPases to inhibit GTP hydrolysis, thereby blocking both intrinsic and GAP-catalyzed GTPase activity and preventing interactions with effector targets. (3) They modulate the cytosol/membrane cycle of Rho GTPases. These proteins maintain individual Rho GTPases as soluble cytosolic proteins by forming high affinity (1.6–30 nM) (Gosser et al. 1997; Nomanbhoy and Cerione 1996) complexes in which the geranylgeranyl C-terminal membrane-targeting moiety of the Rho GTPases is masked by insertion into a hydrophobic pocket within the immunoglobulin-like domain of RhoGDI (Gosser et al. 1997; Grizot et al. 2001; Hoffman et al. 2000; Keep et al. 1997; Longenecker et al. 1999; Scheffzek et al. 2000). When Rho proteins are released from GDIs, they are able to insert into the lipid bilayer of the plasma membrane through their isoprenylated C-terminus. The released Rho GTPases interact with, and are activated by, membrane-associated GEFs, which initiate their association with effector targets at the membrane. A re-complexation with RhoGDI and D4GDI, possibly resulting from GTP hydrolysis, is postulated to promote inactivation by recycling of the Rho GTPase back into the cytosol. Trafficking of Rho GTPases to the cell membrane and their activation/deactivation is therefore a highly complex process.

It is evident that GDIs act to limit Rho GTPase activity by "isolating" Rho proteins at three distinct points: (1) by inhibiting the dissociation of GDP from Rho proteins and preventing the exchange of GDP for GTP by GEFs, they maintain the Rho GTPase in an inactive form; (2) by interacting with the GTP-bound form of the Rho GTPase, they block interaction with effector targets; and (3) by regulating Rho GTPase cycling, they inhibit intrinsic and RhoGAP-mediated GTP hydrolysis, and indirectly regulate Rho GTPase extraction from the plasma membrane. The release of Rho proteins from GDIs and their re-association are therefore two critical points in the regulation of Rho GTPase signaling dynamics, yet the mechanisms and protein components that modulate this basic Rho GTPase-GDI cycle remain poorly defined. Increasing our understanding of this regulatory cycle is important, as there is substantial evidence that the misregulation of GDIs contributes to developmental defects, cancer, and infectious disease. Nowhere is this clearer than in the RhoGDI-knockout mouse, in which the loss of RhoGDI function in the kidney results in dramatic defects in renal cell function (Togawa et al. 1999).

General Mechanism of Regulation of GDIs

The GDIs use the repertoire of mechanisms described above to regulate Rho GTPase activities. In general, Rho GTPase release from GDIs is thought to be regulated via three nonexclusive mechanisms that are described below. GDIs, by directly interacting with receptors/displacement factors or by phospho-regulation of their interaction with Rho GTPases, may also determine exactly *where* localized Rho GTPase activation takes place.

GDI displacement factors. Protein-protein interactions are one mechanism for release of Rho GTPases from GDIs. Proteins that interact with GDIs to disrupt complexes with Rho GTPases are termed "displacement factors" (GDFs) and act to stimulate their delivery to membranes and subsequent activation. The ERM (Erzin, Radixin, Moesin) family of proteins (Takahashi et al. 1997) that are involved in cortical actin reorganization events, the tyrosine kinase Etk (Kim et al. 2002), and the p75 neurotropin receptor (Yamashita and Tohyama 2003) have all been described to induce the release of RhoA from RhoGDI.

Interestingly, the ERM proteins may act to sequester RhoGDI after cell activation. In T cells, CD43-Erzin complexes appear to sequester and inhibit RhoGDI from extracting Rho GTPases after membrane insertion, thereby prolonging positive signaling (Allenspach et al. 2001). Etk protein (a nonreceptor tyrosine kinase) also releases RhoA from RhoGDI complexes. In this case, the pleckstrin-homology domain of Etk is sufficient to bind RhoA and induce its release from RhoGDI, indicating that this is not a phosphorylation-mediated effect. Finally, the neurotrophin receptor p75 interacts directly via its cytoplasmic domain with RhoGDI, resulting in RhoA activation in axons (Yamashita and Tohyama 2003). A synthetic peptide that specifically binds to the RhoGDI binding site on p75 blocked neuronal RhoA activation and reversed axon retraction. It is not clear how many different GDFs exist or how important they are in general for regulating Rho-GDI function.

Biologically active lipids. In conjunction with the GDFs, lipids regulate RhoGTPase-GDI association and dissociation. Several biologically relevant lipids have been reported to have the ability to decrease the affinity of RhoGDI for Rho and Rac. Cell-free assays showed that arachidonic acid, phosphatidic acid, and PtdIns(4,5) P2 could disrupt the Rac1-RhoGDI complex (Bourmeyster et al. 1992; Chuang et al. 1993). Moreover, phosphatidic acid and inositol phospholipids inhibited RhoGDI function at physiologically relevant concentrations *in vitro.* A detailed investigation of the effects of anionic lipids at physiological membrane lipid ratios revealed that phosphatidylglycerol and phosphatidic acid were particularly effective, and could be directly coupled to Rac activation (Faure et al. 1999; Ugolev et al. 2006).

Regulation by phosphorylation. Phosphorylation of GDIs is emerging as an important mechanism for modulating the dynamics of Rho GTPase-Rho GDI cycling. Differentially charged isoforms of RhoGDI and D4GDI have been detected in cells by two-dimensional gel analysis, consistent with the existence of phosphorylated forms of these proteins (Bourmeyster and Vignais 1996; Gorvel et al. 1998). RhoA activation initiated by the thrombin receptor in endothelial cells required the protein kinase C-alpha (PKCα)-mediated phosphorylation of RhoGDI (Mehta et al. 2001). RhoGDI is also phosphorylated, both *in vitro* and in vivo, by p21-activated kinase 1 (Pak1), a downstream effector of Rac and Cdc42 (DerMardirossian et al. 2004). Phosphorylation by Pak1 occurs on two sites (Ser101 and Ser174) in RhoGDI on the external surface of the hydrophobic cleft in which the Rho GTPase prenyl group binds. Surprisingly, the result of phosphorylation at Ser101 and Ser174 is the selective release of Rac1, but not RhoA, from cytosolic RhoGDI complexes, leading to its subsequent activation by exchange factors. The structural basis for such selective Rho GTPase release remains undefined. This mechanism was implicated in Rac1 activation by the growth factors EGF and PDGF.

More recently, RhoGDI has been shown to be phosphorylated by Src tyrosine kinase on Tyr156 both *in vitro* and in vivo (DerMardirossian et al. 2006). As the Tyr156 phosphorylation site lies at the RhoGDI-Rho GTPase binding interface, phosphorylation dramatically decreases the binding of RhoGDI to Rho, Rac, and Cdc42. Pre-formed RhoGDI-Rho GTPase complexes cannot be phosphorylated by Src, suggesting that it is only the free form of RhoGDI, subsequent to the release from Rho GTPase partners that is acted upon by Src. Indeed, Src only binds to free RhoGDI, and Rho GTPase activities after membrane association and GTP-loading are prolonged in Src-transformed fibroblasts presumably due to the inability of P-Tyr156 RhoGDI to bind and recycle the active Rho GTPases.

Interestingly, P-Tyr156 RhoGDI itself becomes membrane localized, apparently at sites of Rho GTPase activity. Such tight membrane binding is distinct from the transient membrane interaction observed with wild-type RhoGDI but is reminiscent, however, of the reported behavior of RhoGDI mutants in which several nearby amino acids (D45, D185) were mutated (Dransart et al. 2004). This result suggests that phosphorylation at Tyr156 (1) prevents the interaction and re-binding of membrane-associated Rho GTPases with RhoGDI, thereby prolonging the period of Rho GTPase activation, and (2) interrupts the steady state recycling function of RhoGDI, thereby "delaying" the normally transient localization of RhoGDI at the membrane and revealing these transient steps as they become rate limiting. The regulation of GDIs and Rho GTPase activity by Src-mediated phosphorylation may play significant roles in cellular regulation and disease, including Src-dependent cancers. A recent study has shown that the phosphorylation of D4GDI at Tyr 153 by Src enhances its metastasis suppressor functions in association with increased membrane localization (Wu et al. 2009).

GDIs in Cancer

The past few years have seen an increase in the amount of research related to the role of the GDIs in cancer biology. Much of this recent work explores the importance of the GDIs in tumor survival, as well as cancer invasion and metastasis. Not unexpectedly, these studies suggest the possibility of the GDIs as effective anti-cancer drug targets.

Tumor Apoptosis: The ability to evade apoptosis is one of the hallmarks of cancers (Hanahan and Weinberg 2000). Increasing evidence suggests that RhoGDI protects at least some cancer cell types from apoptotic death (see Table 3.1). Ectopic overexpression of RhoGDI protects the MDA-MB-231 human breast cancer cell line from apoptosis induced by chemotherapeutic agents such as etoposide (a topoisomerase II inhibitor) and doxorubicin (a DNA intercalator) (Zhang et al. 2005). In agreement with this, depletion of RhoGDI by RNAi sensitizes the MDA-MB-231 cells to etoposide treatment. Further, a modest increase in apoptosis has been observed in this tumor cell line upon depletion of RhoGDI even in the absence of any chemotherapeutic insult (Zhang et al. 2005).

Table 3.1

RhoGDI		D4-GDI	
Apoptosis ↑	Metastasis ↑	Apoptosis ↑	Metastasis ↑
Breast (↓) (Zhang et al. 2005) # **Kidney** (↓) (Reimer et al. 2007) #	**Breast** (**NC**) (Jiang et al. 2003) * Ovarian (↑) (Jones et al. 2002) *	**Ovarian** (↓) (Goto et al. 2006) */#	**Breast** (↑) (Zhang and Zhang 2006) # Breast (NC) (Jiang et al. 2003) * Bladder (↓) (Gildea et al. 2002) */#, (Theodorescu et al. 2004) * Leukemia (↓) (Nakata et al. 2008) */#

References categorized by RhoGDIs isoform, tumor type, and the cellular process studied (apoptosis or metastasis). "↑" represents a positive correlation between GDIs expression levels and the extent of apoptosis or metastasis. "↓" represents a negative correlation. NC means that no correlation (positive or negative) exists. "*" represents that the study was performed in vivo, while "#" studies were performed *in vitro*

In a like fashion, RhoGDI has also been implicated in regulating the sensitivity of immortalized human embryonic kidney cells (HEKs) to busulfan, a chemotherapeutic alkylating agent used historically in the treatment of chronic myeloid leukemia (previous to the development of imatinib). Busulfan inhibits progression of tumors through the cell cycle and results in their subsequent apoptosis (Reimer et al. 2007). In a similar study, depletion of RhoGDI sensitized HEKs to apoptosis by paclitaxel, a microtubule stabilizing agent that acts as a mitotic inhibitor (Reimer et al. 2007).

Some circumstantial evidence also exists for a role for D4GDI in modulating apoptotic responsiveness. In a survey of ovarian cancer tissues, higher levels of aberrantly-expressed D4GDI correlated with resistance to treatment with paclitaxel (Goto et al. 2006). D4GDI is known to be a substrate for cleavage by caspase-3 during apoptosis, resulting in an intact 22 kDa cleavage product which translocates to the nucleus (Krieser and Eastman 1999; Kwon et al. 2002; Na et al. 1996). Cleavage of D4GDI is not required for cell death since mutation of the two aspartic acid residues critical for cleavage did not inhibit apoptosis (Krieser and Eastman 1999; Kwon et al. 2002). The exact consequence of the nuclear translocation of the D4GDI cleavage product still remains to be determined.

The mechanism by which RhoGDI and D4GDI protect cells from apoptosis is not fully understood; however, as Rho GTPases can regulate the activity of proapoptotic signaling pathways (e.g., JNK, p38), increases in GDIs levels may act by suppressing Rho GTPase-dependent apoptotic signals. An additional hypothesis is that increased RhoGDI may block caspase-3-mediated cleavage of Rac1, presumably required in this case for cell survival (Zhang et al. 2005).

Tumor invasion and metastasis. A positive correlation exists between RhoGDI and D4GDI expression and increased tumor invasiveness in ovarian and breast cancers (Table 3.1). In a study of microdissected ovarian tumor samples, RhoGDI was expressed at almost sixfold higher levels in highly invasive metastatic ovarian tumors relative to noninvasive tumors exhibiting low malignant potential (Jones et al. 2002). D4GDI expression is similarly correlative with respect to breast cancer cell motility and invasiveness. D4GDI, which exhibits a primarily hematopoietic expression profile, is aberrantly expressed in a number of breast cancer cell lines (MDA-MB-231, MDA-MB-468, T47D, BT549, SKBR3, BT474) (Zhang and Zhang 2006). Depletion of D4GDI by RNAi in the breast cancer cell lines MDA-MB-231 and MCF-12A grown on a thick layer of Matrigel™ induced the reversion of cells from a mesenchymal to an epithelial phenotype. Further, the targeted reduction of D4GDI in breast cancer cell lines resulted in the inhibition of cell motility and invasiveness as determined in standard *in vitro* transwell migration assays (Schunke et al. 2007; Zhang and Zhang 2006). All of these phenotypes were rescued by restorative expression of a mutated, nontargetable full-length cDNA encoding for the wild-type D4GDI (Zhang and Zhang 2006).

A distinct relationship exists between D4GDI expression levels and bladder tumor cell metastasis. Gildea et al. (2002) observed that D4GDI is expressed at a level more than 64-fold higher in the nonmetastatic human bladder cancer cell line T24 than in the much more aggressive derivative T24T cell line. When T24T cells were stably transfected with a plasmid driving constitutive expression of D4GDI from a CMV promoter, the metastatic potential of these tumor cells, as determined by injection into immunodeficient mice, was severely compromised. Consistent with this result, D4GDI blocked cell migration of these cells in a standard wound healing assay, and also blocked cell invasion in an organotypic *in vitro* invasion assay (Gildea et al. 2002).

Likewise, in certain leukemias and lymphomas, evidence of an inverse relationship between D4GDI expression and cell invasion exists. Two point mutations (V68L,V69A) of the D4GDI gene were found in the leukemic cell line, Reh, originally derived from a patient with acute lymphoblastic leukemia (Nakata et al. 2008). Ectopic expression of this D4GDI mutant, which exhibited dominant inhibitory effects on GDP dissociation *in vitro*, increased the invasiveness of human leukemic Raji cells in immune-compromised SCID mice. Conversely, wild-type D4GDI overexpression inhibited Raji cell infiltration of systemic organs in these same mice (Nakata et al. 2008). Similarly, while leukemic cell motility and adhesion were enhanced by dominant negative D4GDI expression *in vitro*, these cellular activities were suppressed by wild-type D4GDI. This suggests that the spontaneous mutation of D4GDI may be an important contributing factor in the development and progression of malignant leukemia.

Angiogenesis. After their initial formation, solid tumors invariably sustain growth by promoting aberrant angiogenesis. RhoGDI may play an inhibitory role in tumor-promoted angiogenesis. In a recent study, the drug cerivastatin, known to block angiogenesis in human microvascular endothelial cells (HMECs) *in vitro* through the inhibition of the pro-angiogenic RhoA/FAK/PI3K/AKT pathway, induced a correlative upregulation in RhoGDI expression (Vincent et al. 2003). Similarly, the anti-angiogenic effects of vinblastine and rapamycin on an endothelial cell line (EA.hy926) correlated with the upregulation of RhoGDI *in vitro* (Campostrini et al. 2006). This suggests that the targeted upregulation of vascular endothelial RhoGDI serves to directly inhibit RhoA and thus the downstream activation of a proangiogenic pathway. Still, the precise mechanism that underlies the correlative upregulation of RhoGDI and the inhibition of the pro-angiogenic RhoA pathway has yet to be definitively established.

Mechanism for the Biological Effects of Changing GDI Protein Levels in Cancer

How might the positive and negative effects of RhoGDI and D4GDI on various aspects of tumor progression and metastasis relate to regulatory effects on Rho GTPase cycling? Certainly the increase in sensitivity to chemotherapeutic agents and the increase in metastatic capability observed in some cancers with reduced RhoGDI or D4GDI levels may be suggestive of increased Rho GTPase activity. Rho GTPases can both positively regulate cell death signaling pathways and increase cell motility and invasiveness (see Chap. 2, this book). The reduction of RhoGDI expression would tend to free up Rho GTPases for activation by membrane-localized GEFs, and the activated GTPases would consequently be able to insert into local membranes, where they could interact with and stimulate proapoptotic and metastatic effectors.

The positive effects of RhoGDI overexpression on metastasis are somewhat more difficult to explain, given the necessary roles of Rho family GTPases in regulating the cytoskeletal processes involved in cell motility. However, it is clear that the relative levels of RhoA versus Rac/Cdc42 GTPases may be more critical than their absolute levels. RhoA and Rac/Cdc42 often act antagonistically during cell migration, and overactivity of one Rho GTPase can disrupt the coordinated functioning of the motile cell as a whole. Thus, an increase in GDIs levels may differentially affect individual Rho GTPases, thereby changing the ratios of active Rho GTPases. There are some data to indicate that there are differential affinities of RhoGDI versus D4GDI for individual Rho GTPases (Gorvel et al. 1998), although most members of the broader Rho family have not been analyzed (DerMardirossian and Bokoch 2005). Certainly other mechanisms are conceivable, including the possibility that GDIs play positive roles in directing individual Rho GTPases to membranes where they will become locally activated for a specified cell function. As noted previously, GDIs have been shown to bind to Rho GTPases in their GTP states, protecting them from the action of GAPs. Additionally, the formation of Rho GTPase-RhoGDI complexes acts to shield the GTPases from proteolytic digestion by caspases (Zhang et al. 2005). Thus, increases in GDIs might maintain a pool of activatable GTPases that would otherwise have been lost. Finally, the transcriptional regulation of the expression of other genes, such as cyclooxygenase-2, by D4GDI in breast cancer cells may also be relevant (Schunke et al. 2007).

RhoGDIs as Prognostic Indicators in Cancer

RhoGDI, D4GDI, and RhoGDIγ expression levels may be valuable in the prognosis of certain cancer patients, depending on the type of cancer (Table 3.1). In breast cancer, for example, a reduction in RhoGDIγ expression levels relative to normal breast tissues correlates with poor patient prognosis. Indeed, higher levels of metastasis, local recurrence, and death from breast cancer were observed in patients with drastically reduced levels of RhoGDIγ. No such correlation was observed for RhoGDI or D4GDI in this study (Jiang et al. 2003). Similarly, there was no correlation between D4GDI expression in tissue samples from two cohorts of breast cancer patients (Schunke et al. 2007).

However, increased RhoGDI expression levels negatively correlated with disease-free survival for patients with nonsmall cell lung carcinomas (NSCLCs) (Blackhall et al. 2004). Similarly, increased expression of D4GDI correlated with the resistance of ovarian tumors to treatment with paclitaxel *in vitro*. Not surprisingly, increased morbidity and mortality was observed in those ovarian cancer patients with elevated D4GDI levels (Goto et al. 2006). In this cancer, D4GDI serves then as an excellent indicator of the patient populations most likely to respond to paclitaxel therapy. The same was not true in bladder cancer, however, since it was the reduced expression of D4GDI that correlated with poor prognosis (Theodorescu et al. 2004).

This might have been expected, since in bladder cancer, D4GDI actually acts as a tumor metastasis suppressor gene (Gildea et al. 2002; Theodorescu et al. 2004).

Tumor Suppression by HMG-CoA Inhibitors (Statins) and RhoGDI Function

To date, no current cancer chemotherapeutics are known to directly target RhoGDI activity. However, the 3-hydroxy-3-methylglutaryl coenzyme A (HMG-CoA) reductase inhibitors (also known as statins) may indirectly act as RhoGDI functional inhibitors.

HMG-CoA reductase is the rate-limiting enzyme in the synthesis of mevalonate, an important intermediate in the production of cholesterol and other isoprenoids, including trans-farnesyl and trans-geranylgeranyl pyrophosphates. HMG-CoA reductase inhibitors (statins), widely used in the treatment of hypercholesterolemia and prevention of cardiovascular disease, may exhibit an added benefit in tumor suppression (Sassano and Platanias 2008). Indeed, a number of statins have been shown to have therapeutic effects in acute myeloid leukemia, chronic lymphocytic leukemia, and colorectal cancers, among others (breast cancer, lung cancer, prostate cancer, pancreatic cancer, and multiple myelomas) (Sassano and Platanias 2008). Among those statins included in these studies were lovastatin, atorvastatin, simvastatin, and cerivastatin.

In human acute myeloid leukemias, the HMG-CoA reductase inhibitor lovastatin induces neoplasm-specific cell death in a manner that is dependent on its ability to inhibit protein geranylgeranylation (Xia et al. 2001). Similarly, simvastatin has been shown to cause apoptosis in chronic lymphocytic leukemia cells *in vitro* (Chapman-Shimshoni et al. 2003). While lovastatin induced only a modest degree of apoptosis in colorectal cancer cell lines, it did sensitize the lines to cell death *in vitro* in response to chemotherapeutic agents such as the thymidylate synthase inhibitor 5-fluorouracil (5-FU) and the platinum-based agent cisplatin (Agarwal et al. 1999).

While many of the cancer-related statin studies have been concerned with their therapeutic effects, statins may also possibly be used in cancer prevention. Indeed, in a retrospective study of over 480,000 veterans, the use of statins was associated with a decrease in the incidence of pancreatic cancer (Khurana et al. 2007). An increase in the prophylactic effect of the statins in regards to pancreatic cancer was found to correlate with the duration of treatment (Khurana et al. 2007). The results of clinical trials to assess the prophylactic potential of statins in cancer biology will be available in the near future.

How might statins exert their anti-neoplastic effects? By inhibiting the HMG-CoA-dependent synthesis of isoprenoids such as trans-farnesyl-pyrophosphates and trans-geranylgeranyl-pyrophosphates, statins affect the posttranslational isoprenylation of Rho GTPases and, hence, their ability to interact effectively with GDIs and to localize properly to cell membranes (Hori et al. 1991). Seemingly counterintuitively, statins can increase the total cellular levels and cellular GTP loading of at least a subset of Rho GTPases, including RhoA, RhoB, RhoC, and Rac (Cordle et al. 2005; Turner et al. 2008). The latter may be the indirect result of the decreased affinity of nonprenylated Rho GTPases for GDIs, which normally insulate the Rho

GTPases from contact with membrane-localized RhoGEFs (see Fig. 3.1). Indeed, a marked reduction in the interaction of RhoGDI with Rac has been directly measured in THP-1 monocytes and human erythroleukemia (HEL) cells upon treatment with statins (Cordle et al. 2005; Turner et al. 2008).

Given that statins increase RhoA and RhoC levels, as well as enhance their GTP loading in tumors, this would seem to be at odds with the fact that activated RhoA and RhoC promote cellular transformation and oncogenesis (see Chap. 2, this book). However, the correct membrane localization of activated Rho GTPases may be critically required for their action in tumor formation and survival. Hence, the anti-neoplastic effects of statins are likely dependent on the proper membrane-cytosol cycling of RhoA, RhoC, and Rac1. The cytotoxic effects of statins may be at least partially dependent on their inhibition of the interactions between GDIs and their substrate Rho GTPases (see Fig. 3.2). In support of this hypothesis, an increase in the GTP loading of RhoA and C that was similar to that observed in statin-treated HEL cells was also observed in those cells depleted of RhoGDI by targeted siRNA treatment (Turner et al. 2008). The relative contribution of changes in Rho GTPase-GDI complexation and cycling to the action of statins in mediating tumor suppression remains to be investigated.

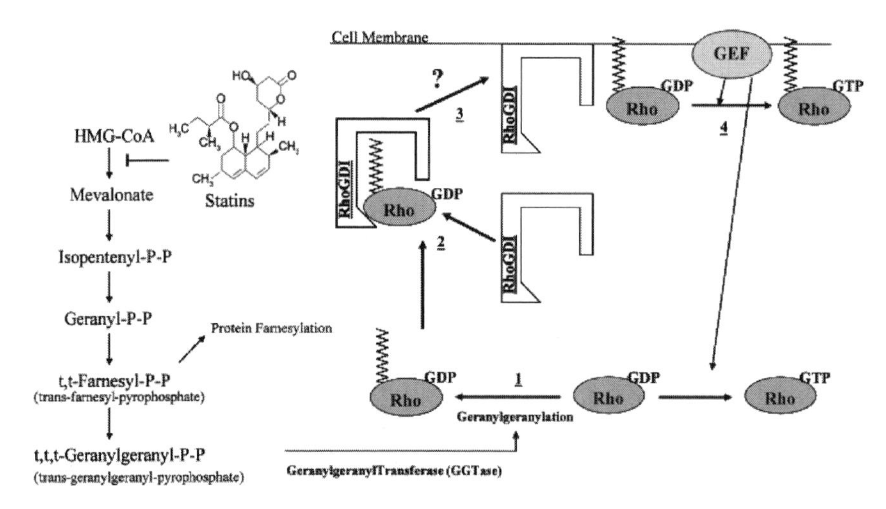

Fig. 3.2 Statins inhibit the geranylgeranylation of Rho GTPases and their interactions with Rho GDI in this Rho. *(Left)* Statins (lovastatin depicted here) block synthesis of mevalonate, a critical intermediate in the synthesis of cholesterol, as well as the isoprenoids trans-farnesyl-pyrophosphate and trans-geranylgeranyl-pyrophosphate. *(Right, model)* Geranylgeranyl transferase (GGTase) catalyzes the addition of a geranylgeranyl prenyl group to individual Rho GTPases (1). The geranylgeranylation of the Rho GTPase is required for its interaction with Rho GDIs to form a cytosolic complex (2). The membrane translocation of the Rho GDI-Rho GTPase complex and the subsequent membrane release of the Rho GTPase occur by poorly defined mechanisms (3) (see Fig. 3.1). In some cell types, guanine nucleotide exchange factors (GEF) catalyze the GTP loading (activation) of nonprenylated cytoplasmic Rho GTPases that (presumably) transiently interact at the membrane by mass action (4). Thus statins, by inhibiting Rho GTPase geranylgeranylation, block stable membrane-localized function of Rho GTPases while promoting a loss-of-function Rho GDI phenotype

GDIs as Drug Targets?

The correlation and functional data implicating GDIs in tumor apoptosis and metastasis suggest that they may serve both as prognostic indicators and as molecular target(s) for cancer therapy. Tumor-restricted inhibition of GDI-RhoGTPase interactions would likely be one mechanism to modulate responsiveness to chemotherapy and/or to reduce tumor aggressiveness. In this regard, antisense and/or RNAi-mediated approaches to targeting the gene expression of GDIs might also be effective. The development of rational therapeutic strategies will depend on a greater understanding of how individual Rho GTPases contribute to specific signaling events at various times in the pathophysiological process of cancer. More research remains to be done in order to validate the GDIs as viable drug targets in cancer.

References

Adra, C.N., D. Manor, J.L. Ko, S. Zhu, T. Horiuchi, L. Van Aelst, R.A. Cerione, and B. Lim. 1997. RhoGDIgamma: a GDP-dissociation inhibitor for Rho proteins with preferential expression in brain and pancreas. *Proc Natl Acad Sci U S A*. 94:4279–84.

Agarwal, B., S. Bhendwal, B. Halmos, S.F. Moss, W.G. Ramey, and P.R. Holt. 1999. Lovastatin augments apoptosis induced by chemotherapeutic agents in colon cancer cells. *Clin Cancer Res*. 5:2223–9.

Allenspach, E.J., P. Cullinan, J. Tong, Q. Tang, A.G. Tesciuba, J.L. Cannon, S.M. Takahashi, R. Morgan, J.K. Burkhardt, and A.I. Sperling. 2001. ERM-dependent movement of CD43 defines a novel protein complex distal to the immunological synapse. *Immunity*. 15:739–50.

Blackhall, F.H., D.A. Wigle, I. Jurisica, M. Pintilie, N. Liu, G. Darling, M.R. Johnston, S. Keshavjee, T. Waddell, T. Winton, F.A. Shepherd, and M.S. Tsao. 2004. Validating the prognostic value of marker genes derived from a non-small cell lung cancer microarray study. *Lung Cancer*. 46:197–204.

Bourmeyster, N., M.J. Stasia, J. Garin, J. Gagnon, P. Boquet, and P.V. Vignais. 1992. Copurification of rho protein and the rho-GDP dissociation inhibitor from bovine neutrophil cytosol. Effect of phosphoinositides on rho ADP-ribosylation by the C3 exoenzyme of Clostridium botulinum. *Biochemistry*. 31:12863–9.

Bourmeyster, N., and P.V. Vignais. 1996. Phosphorylation of Rho GDI stabilizes the Rho A-Rho GDI complex in neutrophil cytosol. *Biochem Biophys Res Commun*. 218:54–60.

Campostrini, N., D. Marimpietri, A. Totolo, C. Mancone, G.M. Fimia, M. Ponzoni, and P.G. Righetti. 2006. Proteomic analysis of anti-angiogenic effects by a combined treatment with vinblastine and rapamycin in an endothelial cell line. *Proteomics*. 6:4420–31.

Chapman-Shimshoni, D., M. Yuklea, J. Radnay, H. Shapiro, and M. Lishner. 2003. Simvastatin induces apoptosis of B-CLL cells by activation of mitochondrial caspase 9. *Exp Hematol*. 31:779–83.

Chuang, T.H., B.P. Bohl, and G.M. Bokoch. 1993. Biologically-active lipids are regulators of Rac-GDI complexation. *J.Biol.Chem*. 268:26206–26211.

Cordle, A., J. Koenigsknecht-Talboo, B. Wilkinson, A. Limpert, and G. Landreth. 2005. Mechanisms of statin-mediated inhibition of small G-protein function. *J Biol Chem*. 280:34202–9.

DerMardirossian, C., and G.M. Bokoch. 2005. GDIs: central regulatory molecules in Rho GTPase activation. *Trends Cell Biol*. 15:356–63.

DerMardirossian, C., G. Rocklin, J.Y. Seo, and G.M. Bokoch. 2006. Phosphorylation of RhoGDI by Src regulates Rho GTPase binding and cytosol-membrane cycling. *Mol Biol Cell*. 17:4760–8.

DerMardirossian, C., A. Schnelzer, and G.M. Bokoch. 2004. Phosphorylation of RhoGDI by Pak1 mediates dissociation of Rac GTPase. *Mol Cell*. 15:117–27.

Dovas, A., and J.R. Couchman. 2005. RhoGDI: multiple functions in the regulation of Rho family GTPase activities. *Biochem J*. 390:1–9.

Dransart, E., A. Morin, J. Cherfils, and B. Olofsson. 2004. Uncoupling of inhibitory and shuttling functions of rhoGDIs. *J.Biol Chem*. 280:4674–4683.

Dransart, E., B. Olofsson, and J. Cherfils. 2005. RhoGDIs revisited: novel roles in Rho regulation. *Traffic*. 6:957–66.

Faure, J., P.V. Vignais, and M.C. Dagher. 1999. Phosphoinositide-dependent activation of Rho A involves partial opening of the RhoA/Rho-GDI complex. *Eur J Biochem*. 262:879–89.

Fukumoto, Y., K. Kaibuchi, Y. Hori, H. Fujioka, S. Araki, T. Ueda, A. Kikuchi, and Y. Takai. 1990. Molecular-cloning and characterization of a novel type of regulatory protein (GDI) for the Rho proteins, Ras p21-like small GTP-binding proteins. *Oncogene*. 5:1321–1328.

Gildea, J.J., M.J. Seraj, G. Oxford, M.A. Harding, G.M. Hampton, C.A. Moskaluk, H.F. Frierson, M.R. Conaway, and D. Theodorescu. 2002. RhoGDI2 is an invasion and metastasis suppressor gene in human cancer. *Cancer Res*. 62:6418–23.

Gorvel, J.P., T.C. Chang, J. Boretto, T. Azuma, and P. Chavrier. 1998. Differential properties of D4/LyGDI versus RhoGDI: phosphorylation and Rho GTPase selectivity. *FEBS Lett*. 422:269–273.

Gosser, Y.Q., T.K. Nomanbhoy, B. Aghazadeh, D. Manor, C. Combs, R.A. Cerione, and M.K. Rosen. 1997. C-terminal binding domain of Rho GDP-dissociation inhibitor directs N-terminal inhibitory peptide to GTPases. *Nature*. 387:814–819.

Goto, T., M. Takano, M. Sakamoto, A. Kondo, J. Hirata, T. Kita, H. Tsuda, Y. Tenjin, and Y. Kikuchi. 2006. Gene expression profiles with cDNA microarray reveal RhoGDI as a predictive marker for paclitaxel resistance in ovarian cancers. *Oncol Rep*. 15:1265–71.

Grizot, S., J. Faure, F. Fieschi, P.V. Vignais, M.C. Dagher, and E. Pebay-Peyroula. 2001. Crystal structure of the Rac1-RhoGDI complex involved in NADPH oxidase activation. *Biochemistry*. 40:10007–10013.

Hanahan, D., and R.A. Weinberg. 2000. The hallmarks of cancer. *Cell*. 100:57–70.

Hoffman, G.R., N. Nassar, and R.A. Cerione. 2000. Structure of the Rho family GTP-binding protein Cdc42 in complex with the multifunctional regulator RhoGDI. *Cell*. 100:345–56.

Hori, Y., A. Kikuchi, M. Isomura, M. Katayama, Y. Miura, H. Fujioka, K. Kaibuchi, and Y. Takai. 1991. Post-translational modifications of the C-terminal region of the rho protein are important for its interaction with membranes and the stimulatory and inhibitory GDP/GTP exchange proteins. *Oncogene*. 6:515–22.

Jiang, W.G., G. Watkins, J. Lane, G.H. Cunnick, A. Douglas-Jones, K. Mokbel, and R.E. Mansel. 2003. Prognostic value of rho GTPases and rho guanine nucleotide dissociation inhibitors in human breast cancers. *Clin Cancer Res*. 9:6432–40.

Jones, M.B., H. Krutzsch, H. Shu, Y. Zhao, L.A. Liotta, E.C. Kohn, and E.F. Petricoin, 3rd. 2002. Proteomic analysis and identification of new biomarkers and therapeutic targets for invasive ovarian cancer. *Proteomics*. 2:76–84.

Keep, N.H., M. Barnes, I. Barsukov, R. Badii, L.Y. Lian, A.W. Segal, P.C. Moody, and G.C. Roberts. 1997. A modulator of rho family G proteins, rhoGDI, binds these G proteins via an immunoglobulin-like domain and a flexible N-terminal arm. *Structure*. 5:623–33.

Khurana, V., A. Sheth, G. Caldito, and J.S. Barkin. 2007. Statins reduce the risk of pancreatic cancer in humans: a case-control study of half a million veterans. *Pancreas*. 34:260–5.

Kim, O., J. Yang, and Y. Qiu. 2002. Selective activation of small GTPase RhoA by tyrosine kinase Etk through its pleckstrin homology domain. *J.Biol Chem*. 277:30066–30071.

Krieser, R.J., and A. Eastman. 1999. Cleavage and nuclear translocation of the caspase 3 substrate Rho GDP-dissociation inhibitor, D4-GDI, during apoptosis. *Cell Death Differ*. 6:412–9.

Kwon, K.B., E.K. Park, D.G. Ryu, and B.H. Park. 2002. D4-GDI is cleaved by caspase-3 during daunorubicin-induced apoptosis in HL-60 cells. *Exp Mol Med*. 34:32–7.

Lelias, J.M., C.N. Adra, G.M. Wulf, J.C. Guillemot, M. Khagad, D. Caput, and B. Lim. 1993. cDNA cloning of a human mRNA preferentially expressed in hematopoietic cells and with homology to a GDP-dissociation inhibitor for the rho GTP-binding proteins. *Proc Natl Acad Sci U S A*. 90:1479–83.

Leonard, D., M.J. Hart, J.V. Platko, A. Eva, W. Henzel, T. Evans, and R.A. Cerione. 1992. The identification and characterization of a GDP-dissociation inhibitor (GDI) for the CDC42Hs protein. *J Biol Chem*. 267:22860–8.

Longenecker, K., P. Read, U. Derewenda, Z. Dauter, X. Liu, S. Garrard, L. Walker, A.V. Somlyo, R.K. Nakamoto, A.P. Somlyo, and Z.S. Derewenda. 1999. How RhoGDI binds Rho. *Acta Crystallogr D Biol Crystallogr*. 55 (Pt 9):1503–15.

Mehta, D., A. Rahman, and A.B. Malik. 2001. Protein kinase C-alpha signals Rho-guanine nucleotide dissociation inhibitor phosphorylation and Rho activation and regulates the endothelial cell barrier function. *J.Biol.Chem*. 276:22614–22620.

Na, S., T.H. Chuang, A. Cunningham, T.G. Turi, J.H. Hanke, G.M. Bokoch, and D.E. Danley. 1996. D4-GDI, a substrate of CPP32, is proteolyzed during Fas-induced apoptosis. *J Biol Chem*. 271:11209–13.

Nakata, Y., K. Kondoh, S. Fukushima, A. Hashiguchi, W. Du, M. Hayashi, J. Fujimoto, J. Hata, and T. Yamada. 2008. Mutated D4-guanine diphosphate-dissociation inhibitor is found in human leukemic cells and promotes leukemic cell invasion. *Exp Hematol*. 36:37–50.

Nomanbhoy, T.K., and R.A. Cerione. 1996. Characterization of the interaction between RhoGDI and Cdc42Hs using fluorescence spectroscopy. *J.Biol Chem*. 271:10004–10009.

Reimer, J., S. Bien, J. Sonnemann, J.F. Beck, T. Wieland, H.K. Kroemer, and C.A. Ritter. 2007. Reduced expression of Rho guanine nucleotide dissociation inhibitor-alpha modulates the cytotoxic effect of busulfan in HEK293 cells. *Anticancer Drugs*. 18:333–40.

Sassano, A., and L.C. Platanias. 2008. Statins in tumor suppression. *Cancer Lett*. 260:11–9.

Scheffzek, K., I. Stephan, O.N. Jensen, D. Illenberger, and P. Gierschik. 2000. The Rac-RhoGDI complex and the structural basis for the regulation of Rho proteins by RhoGDI. *Nat.Struct. Biol*. 7:122–126.

Schunke, D., P. Span, H. Ronneburg, A. Dittmer, M. Vetter, H.J. Holzhausen, E. Kantelhardt, S. Krenkel, V. Muller, F.C. Sweep, C. Thomssen, and J. Dittmer. 2007. Cyclooxygenase-2 is a target gene of rho GDP dissociation inhibitor beta in breast cancer cells. *Cancer Res*. 67:10694–702.

Takahashi, K., T. Sasaki, A. Mammoto, K. Takaishi, T. Kameyama, S. Tsukita, S. Tsukita, and Y. Takai. 1997. Direct interaction of the Rho GDP dissociation inhibitor with ezrin/radixin/moesin initiates the activation of the Rho small G protein. *J.Biol.Chem*. 272:23371–23375.

Theodorescu, D., L.M. Sapinoso, M.R. Conaway, G. Oxford, G.M. Hampton, and H.F. Frierson, Jr. 2004. Reduced expression of metastasis suppressor RhoGDI2 is associated with decreased survival for patients with bladder cancer. *Clin Cancer Res*. 10:3800–6.

Togawa, A., J. Miyoshi, H. Ishizaki, M. Tanaka, A. Takakura, H. Nishioka, H. Yoshida, T. Doi, A. Mizoguchi, N. Matsuura, Y. Niho, Y. Nishimune, S. Nishikawa, and Y. Takai. 1999. Progressive impairment of kidneys and reproductive organs in mice lacking Rho GDIalpha. *Oncogene*. 18:5373–80.

Turner, S.J., S. Zhuang, T. Zhang, G.R. Boss, and R.B. Pilz. 2008. Effects of lovastatin on Rho isoform expression, activity, and association with guanine nucleotide dissociation inhibitors. *Biochem Pharmacol*. 75:405–13.

Ugolev, Y., S. Molshanski-Mor, C. Weinbaum, and E. Pick. 2006. Liposomes comprising anionic but not neutral phospholipids cause dissociation of [Rac(1 or 2)-RhoGDI] complexes and support amphiphile-independent NADPH oxidase activation by such complexes. *J Biol Chem*. May 15, 2006, e pub ahead of print; URL: PM16702219.

Vincent, L., P. Albanese, H. Bompais, G. Uzan, J.P. Vannier, P.G. Steg, J. Soria, and C. Soria. 2003. Insights in the molecular mechanisms of the anti-angiogenic effect of an inhibitor of 3-hydroxy-3-methylglutaryl coenzyme A reductase. *Thromb Haemost*. 89:530–7.

Wu, Y., Moissoglu, K., Wang, H., Wang, X., Frierson, H.F., Schwartz, M.A., Theodorescu, D. 2009. Src phosphorylation of RhoGDI2 regulates its metastasis suppressor function. Proc Natl Acad Sci U S A. 7:106(14):5807-12.

Xia, Z., M.M. Tan, W.W. Wong, J. Dimitroulakos, M.D. Minden, and L.Z. Penn. 2001. Blocking protein geranylgeranylation is essential for lovastatin-induced apoptosis of human acute myeloid leukemia cells. *Leukemia*. 15:1398–407.

Yamashita, T., and M. Tohyama. 2003. The p75 receptor acts as a displacement factor that releases Rho from Rho-GDI. *Nat.Neurosci.* 6:461–467.

Zalcman, G., V. Closson, J. Camonis, N. Honore, M.F. Rousseau-Merck, A. Tavitian, and B. Olofsson. 1996. RhoGDI-3 is a new GDP dissociation inhibitor (GDI). Identification of a non-cytosolic GDI protein interacting with the small GTP-binding proteins RhoB and RhoG. *J.Biol.Chem.* 271 30366–30374.

Zhang, B., Y. Zhang, M.C. Dagher, and E. Shacter. 2005. Rho GDP dissociation inhibitor protects cancer cells against drug-induced apoptosis. *Cancer Res*. 65:6054–62.

Zhang, Y., and B. Zhang. 2006. D4-GDI, a Rho GTPase regulator, promotes breast cancer cell invasiveness. *Cancer Res*. 66:5592–8.

Chapter 4
Signaling through Galpha12/13 and RGS-RhoGEFs

Nicole Hajicek, Barry Kreutz, and Tohru Kozasa

Introduction

The monomeric GTPase Rho is a well-established regulator of a variety of intracellular processes including formation of actin stress fibers and assembly of focal adhesions, gene transcription, and control of cell growth (Van Aelst and D'Souza-Schorey 1997; Hall 1998). With the discovery that another monomeric GTPase, Ras, behaves as a potent oncogene, intensive attention was focused on this protein family as potential regulators of cellular transformation. Early studies indicated that overexpression of wild-type Rho or a constitutively activated mutant can induce a transformed phenotype in murine fibroblasts and established that Rho activation is an important component in Ras-mediated transformation (Avraham and Weinberg 1989; Khosravi-Far et al. 1995; Qiu et al. 1995; Sahai, Olson and Marshall 2001). The precise mechanisms by which Rho exerts its transforming potential are not yet fully understood. Unlike Ras, which is frequently mutated in human cancers, activating mutations in Rho have not been found (Bos 1988; Boettner and Van Aelst 2002; Sahai and Marshall 2002). However, analysis of several cancer types has revealed that some isoforms of Rho are overexpressed in these tissues, and that higher expression levels tend to correspond to a more invasive phenotype (Fritz, Just, and Kaina 1999; Clark et al. 2000; Fritz et al. 2002). Another potential link between Rho and transformation is the Dbl family of RhoGEFs (*g*uanine nucleotide *e*xchange *f*actors), which activate monomeric GTPases by facilitating the exchange of GDP for GTP. Many of these proteins were originally identified in screens for putative oncogenes capable of transforming NIH3T3 fibroblasts. One subfamily of RhoGEFs in particular, which contains *r*egulator of *G* protein *s*ignaling (RGS)-like domains in their N-termini, have been shown to act as direct links between Rho and the heterotrimeric G proteins Gα12

N. Hajicek, B. Kreutz and T. Kozasa (✉)
Department of Pharmacology, College of Medicine,
University of Illinois at Chicago, Chicago, IL, 60602, USA
e-mail: tkozas@uic.edu

K. van Golen (ed.), *The Rho GTPases in Cancer*,
DOI 10.1007/978-1-4419-1111-7_4, © Springer Science+Business Media, LLC 2010

and Gα13, which themselves have potent transforming capabilities (Xu et al. 1993; Xu, Voyno-Yasenetskaya and Gutkind 1994). This property makes Gα12 and Gα13 unique among other Gα subunits. However, GTPase deficient mutants of Gαs and Gαi have been found in human tumors of endocrine origin (Radhika and Dhanasekaran 2001). For a review of G protein-mediated signaling pathways and transformation, see Fukuhara, Chikumi, and Gutkind (2001) and Radhika and Dhanasekaran (2001).

RGS-RhoGEFs define a unique subclass of GEFs, which represent a point of integration between monomeric and heterotrimeric G protein signaling networks. The molecular mechanisms that underlie and regulate signaling from Gα12 and Gα13 through RGS-RhoGEFs to RhoA and the associated physiological functions constitute the scope of this chapter.

For the purposes of this chapter, unless stated otherwise, the term "Rho" refers specifically to the RhoA protein, and not to the Rho family of monomeric GTPases, which consists of Rho, Rac, Cdc42, and their associated isoforms. Although a detailed characterization of the similarities and differences in signaling and regulation among the Rho isoforms (Rho A-C) is of fundamental importance, it is beyond the scope of this chapter.

RGS-RhoGEFs: Form and Function

p115RhoGEF. p115RhoGEF was initially identified as a ubiquitously expressed 115 kD protein, which bound preferentially to the nucleotide-free form of RhoA (Hart et al. 1996). In vitro, p115RhoGEF stimulated the release of GDP from RhoA, but not Rac1, Cdc42, or K-Ras, demonstrating that this protein functions as a Rho-specific GEF. Subsequent analysis of p115RhoGEF revealed that its N-terminus contained a domain with weak sequence identity to the RGS domain and which specifically stimulated the intrinsic GTPase activity of Gα13 and to a lesser extent Gα12 (Kozasa et al. 1998) (Fig. 4.1). This report was the first to identify a GAP (*G*TPase *a*ctivating *p*rotein) for α subunits of the G12 family. Furthermore, Gα13 but not Gα12 was shown to stimulate the GEF activity of p115RhoGEF in vitro (Hart et al. 1998) (Fig. 4.2). However, Gα12 can block Gα13-mediated stimulation of p115RhoGEF activity, suggesting these α subunits compete for binding to p115RhoGEF, perhaps in the RGS domain. Complementary to these biochemical studies, p115RhoGEF was also shown to activate gene transcription from the SRE (*s*erum *r*esponse *e*lement) in a Rho-dependent manner, and importantly, Gα13 but not other heterotrimeric Gα subunits synergistically enhanced p115Rho-GEF-mediated SRE activation (Mao et al. 1998b). A functional link between p115RhoGEF and Rho activation downstream from the Gα12/13-coupled thrombin receptor has also been demonstrated (Majumdar et al. 1999). Taken together, these pieces of evidence strongly support a novel, direct mechanism linking Gα13 to RhoA activation via p115RhoGEF.

PDZ-RhoGEF. A search for other proteins with sequence homology to the RGS-like domain of p115RhoGEF identified another putative member of the RGS-RhoGEF

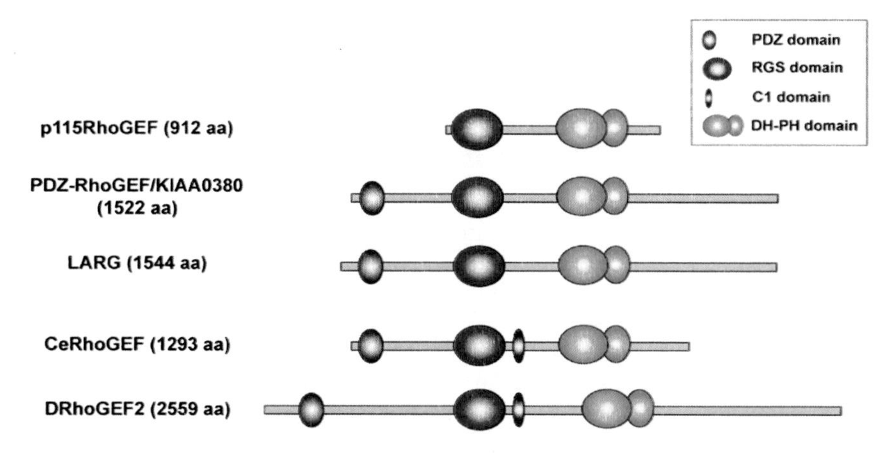

Fig. 4.1 Domain structure of RGS-RhoGEFs. The three known human RGS RhoGEFs are depicted as linear cartoons. All three proteins contain an RGS domain in their N-terminus. These domains specifically stimulate the intrinsic GTPase activity of the G12 family of heterotrimeric G proteins. Additionally, the N-terminal region of PDZ-RhoGEF and LARG each contain a PDZ domain. Like virtually all other GEFs for monomeric GTPases, p115RhoGEF, PDZ-RhoGEF, and LARG contain tandem DH and PH domains, which catalyze exchange of GDP for GTP on the monomeric GTPase RhoA. For comparison, CeRhoGEF and DRhoGEF2, RGS-RhoGEFs from *C. elegans* and *D. melanogaster* respectively, are depicted as well. Unlike human RGS-RhoGEFs, CeRhoGEF and DRhoGEF2 each contain one C1 domain.

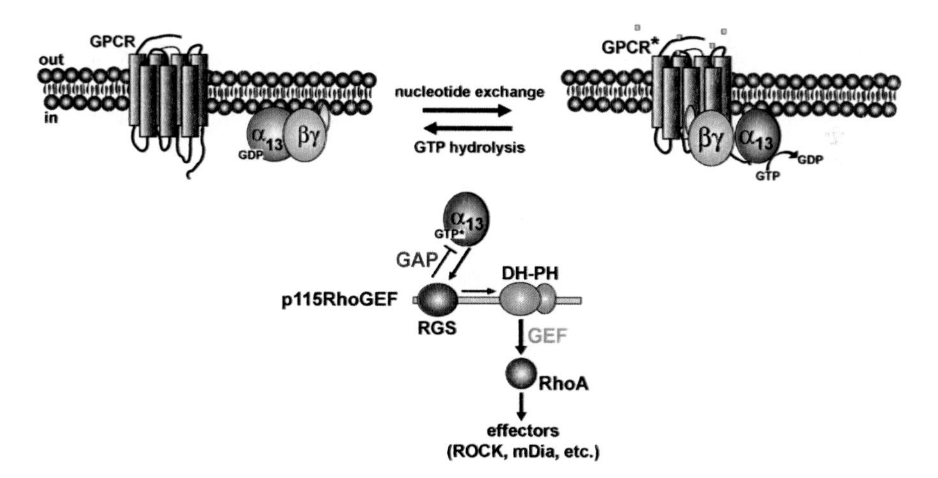

Fig. 4.2 G13-mediated signaling. In the resting state, Gα13 is bound to GDP and the Gβγ dimer. Upon stimulation with ligand, GPCRs coupled to G13 facilitate the exchange of GDP for GTP resulting in activation of the Gα subunit and release of Gβγ. Both Gα*GTP and free Gβγ have the capacity to regulate signaling through downstream effectors. One class of effectors for Gα13 is GEFs for the monomeric GTPase RhoA, such as p115RhoGEF. Upon binding of Gα13*GTP to p115RhoGEF, the DH/PH domains activate RhoA by facilitating the exchange of GDP for GTP. p115RhoGEF also contains a RGS domain in its N-terminus which specifically stimulates the intrinsic GTPase (GAP) activity of Gα13.

protein family, PDZ-RhoGEF, (*P*SD95/*D*iscs large/*Z*O-1) also known as KIAA0380 (Kozasa et al. 1998). The murine ortholog of PDZ-RhoGEF, GTRAP48, was first identified by yeast two-hybrid screening as a protein which interacts with the glutamate transporter EAAT4 and modulates its activity (Jackson et al. 2001). Northern blot analysis revealed that GTRAP48 and PDZ-RhoGEF mRNA is enriched in the central nervous system, but can also be found in other tissues as well (Fukuhara et al. 1999; Jackson et al. 2001). Jackson et al. (2001) also demonstrated that GTRAP48 was able to stimulate the binding of GTPγS to RhoA but not Rac or Cdc42 in vitro, demonstrating that this protein functions as a Rho-specific GEF. Subsequent biochemical analysis of PDZ-RhoGEF confirmed that the isolated DH and PH domains (*D*bl *h*omology/*p*leckstrin *h*omology) are sufficient to stimulate specific nucleotide exchange on Rho in vitro (Rumenapp et al. 1999; Togashi et al. 2000; Derewenda et al. 2004). Togashi et al. (2000) further demonstrated that overexpression of full-length PDZ-RhoGEF in COS7 cells increased the amount of GTP bound to RhoA, but not Rac1 or Cdc42, suggesting that it functions as a Rho-specific GEF in a cellular context as well. In support of this notion, PDZ-RhoGEF has been linked to several Rho-dependent cellular responses such as SRE-mediated gene transcription and neurite retraction (Fukuhara et al. 1999; Togashi et al. 2000). Interestingly, GTRAP48 and PDZ-RhoGEF are able to co-immunoprecipitate with activated forms of Gα13, suggesting these proteins act as direct links between heterotrimeric G proteins and RhoA (Fukuhara et al. 1999; Jackson et al. 2001). In support of this model, deletion mutants of PDZ-RhoGEF lacking its DH/PH domains abrogate Rho-dependent SRE activation by constitutively active mutants of Gα12/13 (Fukuhara et al. 1999) or LPA-induced neurite retraction (Togashi et al. 2000) demonstrating indirectly that PDZ-RhoGEF is regulated by G12 family members in vivo. In vitro, however, neither PDZ-RhoGEF nor GTRAP48 has detectable GAP activity toward Gα12 or Gα13 (Wells et al. 2002). Thus, the precise mechanism by which G12 family members regulate the RhoGEF activity of PDZ-RhoGEF and GTRAP48 remains uncharacterized.

LARG. LARG (*l*eukemia *a*ssociated *R*ho*G*EF) was originally identified as a novel protein fused to the MLL (*m*ixed *l*ineage *l*eukemia) gene in a patient with acute myeloid leukemia (Kourlas et al. 2000). Several domains characteristic of a signal transduction molecule were identified in the full-length LARG protein including PDZ, DH, PH, and a domain similar to the RGS-like domain in PDZ-RhoGEF which was shown to bind to activated forms of Gα12/13 (Fukuhara et al. 1999). In particular, the presence of tandem DH/PH domains strongly suggested that this protein was a GEF for Rho family GTPases. Consistent with this model, overexpression of full-length LARG is able to stimulate SRE-dependent gene transcription (Fukuhara, Chikumi, and Gutkind 2000; Reuther et al. 2001). In vitro analysis revealed that the isolated DH/PH domains of LARG bound to the nucleotide-free form of RhoA but not Rac1 or Cdc42 (Reuther et al. 2001). The same study demonstrated that in vitro the DH/PH domains of LARG could incorporate fluorescently labeled GDP onto RhoA but not Rac1 or Cdc42. Overexpression of full-length LARG can increase levels of RhoA·GTP in cells but not Rac1·GTP or Cdc42·GTP, demonstrating that LARG functions as a RhoA-specific GEF (Fukuhara,

Chikumi and Gutkind 2000; Reuther et al. 2001). LARG's RGS domain has also been shown to bind to activated forms of Gα12 and Gα13 (Fukuhara, Chikumi, and Gutkind 2000; Booden, Siderovski, and Der 2002). Furthermore, in vitro reconstitution assays directly demonstrated LARG's specific GTPase activity toward Gα12 and Gα13, as well as stimulation of its RhoGEF activity by the same Gα subunits (Suzuki et al. 2003). LARG thus represents the first RGS-RhoGEF whose activity is directly stimulated in vitro by Gα12, although as discussed below, Gα12 can only stimulate the RhoGEF activity of tyrosine-phosphorylated LARG (Suzuki et al. 2003).

RGS-RhoGEF Interaction with Other Proteins

In addition to interacting with their effector Rho and upstream activators Gα12 and Gα13, RGS-RhoGEFs are also capable of binding to various other cellular proteins. In particular, PDZ-RhoGEF and LARG are well known to interact with multiple cell surface receptors by virtue of their PDZ domain. Their ability to bind to and regulate the activity of plexins is one such example. In mammals, plexins are transmembrane receptors that mediate the repulsive cues of semaphorins to initiate collapse of neurite growth cones. Multiple studies have demonstrated that plexin B interacts with both PDZ-RhoGEF and LARG to activate RhoA and promote growth cone collapse (Aurandt et al. 2002; Hirotani et al. 2002; Perrot, Vazquez-Prado, and Gutkind 2002; Swiercz et al. 2002). Additionally, interaction between LARG and insulin-like growth factor-1 receptor (Taya et al. 2001) and CD44 (Bourguignon et al. 2006) has also been reported. The LPA receptor also interacts with LARG as well as PDZ-RhoGEF (Yamada et al. 2005). LARG's GEF activity has been reported to be regulated by interaction with the Rho effector mDia1, suggesting a positive feedback loop between these proteins regulates Rho activation (Kitzing et al. 2007). p115RhoGEF has been reported to interact with the C-terminus of HIV-1's transmembrane protein gp41 (Zhang et al. 1999). This interaction inhibits the ability of p115RhoGEF to initiate Rho-dependent stress fiber formation and gene transcription. Interestingly, mutations in gp41 that block the interaction with p115RhoGEF inhibit the ability of the HIV-1 virus to produce infectious particles in certain cell types. This finding suggests that the interaction between gp41 and p115RhoGEF modulates the ability of the virus to replicate.

Mechanisms of Regulation

Regulation by Gα. Biochemical evidence has clearly demonstrated that the GEF activity of p115RhoGEF and LARG can be directly stimulated by Gα12 and Gα13 (Hart et al. 1998; Suzuki et al. 2003). However, the precise mechanism by which these α subunits stimulate the exchange activity of the DH/PH domains of RGS-RhoGEFs remains unclear. One report demonstrated that upon activation with AlF_4^-, Gα13 was able to bind directly to a truncated version of p115RhoGEF consisting of the DH and PH domains, although it was unable to stimulate the guanine nucleotide exchange activity of this p115RhoGEF fragment (Wells et al. 2002). This finding suggests that Gα13 may be able to bind to additional regions of p115RhoGEF outside of the RGS domain, although the role of these site(s) in stimulating GEF activity remains uncharacterized. The notion that Gα13 has multiple interaction interfaces with RGS-RhoGEFs is also supported by kinetic and thermodynamic analysis of the binding between Gα13 and LARG. Using surface plasmon resonance to measure the affinity of the interaction between Gα13 and fragments of LARG, the presence of additional binding sites in the region of the DH/PH domains of LARG has been demonstrated (N. Suzuki and T. Kozasa, unpublished observations; Suzuki et al. 2009). In addition to analysis of the Gα13 binding surfaces on RGS-RhoGEFs, preliminary characterization of the surfaces of Gα13 necessary for binding to RGS-RhoGEFs has also been reported. One study, utilizing chimeras of Gα13 and Gαi2, revealed that the switch regions of Gα13 are necessary but not sufficient for activation of RGS-RhoGEFs in cell-based assays (Vazquez-Prado et al. 2004). Rather, a large portion of the Ras-like domain of Gα13 appears to be necessary for efficient Rho activation. Subsequent studies utilizing chimeras of Gα12 and Gα13 further identified that the last 100 amino acid residues of Gα13 are required for RhoA activation in cells and in vitro (Kreutz et al. 2007). Interestingly, this study also demonstrated that the surfaces of Gα13 required for GAP activity and activation of GEF activity are distinct. Specifically, the amino terminus of Gα13 is responsible for mediating the GAP response to RGS-RhoGEFs, while the carboxyl terminus mediates activation of the RGS-RhosGEFs p115RhoGEF and LARG.

RGS-RhoGEFs combine GTPase activity and effector into a single molecule. At first glance, this integration seems counterintuitive: Interaction between activated Gα12/13 and the RGS domain should efficiently terminate signaling from the α subunit before it is able to initiate downstream signaling. However, this model assumes that RGS domains behave as dedicated Gα antagonists, and emerging evidence suggests that this is not the case. The first studies to examine the effects of RGS proteins on Gα-mediated signaling analyzed the kinetics of GIRK (*G* protein coupled *i*nward *r*ectifier K^+) channel activation and deactivation in heterologous expression systems (Doupnik et al. 1997; Saitoh et al. 1997). These studies demonstrated that expression of some RGS proteins could profoundly increase the rate of channel recovery after removal of agonist, suggesting that the lifetime of GTP on the Gα subunit and not the inherent kinetic properties of the channel itself dictate the speed of channel deactivation. Surprisingly, however, expression of these RGS proteins also enhanced the rate of channel activation. Importantly, although the activation kinetics increased, there was no decrease in steady-state channel current, suggesting that in this situation, RGS proteins may actually enhance the rate of Gα activation. Subsequent biochemical analysis and computer modeling support the notion that RGS proteins may in fact increase the kinetics of Gα activation by ensuring a constant supply of Gα·GDP which can be rapidly activated by GPCRs (Zhong et al. 2003).

Oligomerization. Another possible mechanism regulating RGS-RhoGEF activity is oligomerization. Initial characterization of p115RhoGEF revealed that a construct which lacked the first 248 amino acids and was truncated after the PH domain induced significantly higher focus formation in NIH 3T3 fibroblasts than a construct lacking the first 82 amino acids (Hart et al. 1996). Consistent with these data, truncation of p115RhoGEF after the PH domain results in a construct with an increased ability to stimulate gene expression from the SRE promoter compared to the full-length protein (Wells et al. 2001; Chikumi et al. 2004). This C-terminal truncation alone is sufficient to increase p115RhoGEF's transforming potential (Chikumi et al. 2004). Interestingly, the in vitro exchange activity of p115RhoGEF is not significantly altered, or in some instances is even decreased by this C-terminal deletion (Wells et al. 2001; Chikumi et al. 2004). These results suggest that p115RhoGEF activity may be negatively regulated in vivo through its C-terminus by as yet unknown cellular components. In order to identify potential regulatory proteins, yeast two-hybrid screening has been utilized to identify proteins associated with the murine ortholog of p115RhoGEF, Lsc. However, the only Lsc interacting protein identified in this screen was a C-terminal fragment of Lsc itself, suggesting that it homo-oligomerizes via its C-terminal region (Eisenhaure et al. 2003). This homo-oligomerization was subsequently confirmed using co-immunoprecipitation (Chikumi et al. 2004). Importantly, Eisenhaure et al. (2003) demonstrated that oligomerization and negative regulation of Lsc activity are distinct functions of its C-terminus, as disruption of a putative coiled-coil domain impairs oligomerization, but does not result in a gain of function in reporter gene assays. LARG and PDZ-RhoGEF have also been shown to oligomerize via their C-terminal domains and are able to homo-oligomerize and possibly form hetero-oligomers with each other, but not with p115RhoGEF (Chikumi et al. 2004). Like p115RhoGEF though, deletion of their C-termini has little effect on their RhoGEF activity in vitro, while dramatically enhancing their ability to both activate gene transcription and induce transformation (Chikumi et al. 2004). While less is known about the function and regulation of oligomerization of LARG and PDZ-RhoGEF, recent evidence has suggested that oligomerization of LARG may regulate its subcellular localization (Grabocka and Wedegaertner 2007). Further study is required to fully understand the role that oligomerization plays in regulating RGS-RhoGEF activity.

Phosphorylation. Another mechanism regulating RGS-RhoGEF activity is post-translational modification in the form of phosphorylation. p115RhoGEF has been shown to be phosphorylated in a PKCα-dependent manner in human umbilical vein endothelial cells in response to thrombin stimulation (Holinstat et al. 2003). This phosphorylation event is necessary for cytoskeletal reorganization in these cells. Early studies of Gα12/13-mediated cytoskeletal reorganization suggested that tyrosine kinases may also play a role in regulating Rho activation. Specifically, tyrosine kinase inhibitors can prevent neurite retraction and cell rounding induced by Gα13 (Katoh et al. 1998), the assembly of stress fibers and focal adhesions in response to LPA stimulation (Gohla, Harhammer and Schultz 1998), and Rho activation itself (Kranenburg et al. 1999). Additionally, non-receptor tyrosine kinases which can synergistically enhance Gα13-mediated gene transcription have been reported (Mao et al. 1998a). Subsequent studies demonstrated that the non-receptor tyrosine kinase FAK (*f*ocal *a*dhesion *k*inase) can phosphorylate both LARG and PDZ-RhoGEF, but not p115RhoGEF in response to thrombin stimulation in HEK293 cells (Chikumi, Fukuhara and Gutkind 2002). LARG can also be phosphorylated by another non-receptor tyrosine kinase, Tec, while p115RhoGEF is not a substrate for this kinase (Suzuki et al. 2003). Importantly, this study demonstrated that direct phosphorylation of LARG by Tec in vitro does not affect its basal RhoGEF activity, but greatly enhances its response to activated Gα12. In comparison, activated Gα13 stimulates phosphorylated and non-phosphorylated LARG equally well under the same experimental conditions, suggesting Gα13-mediated regulation of LARG is independent of its phosphorylation status. However, the ability of both Gα13 and Gα12 to stimulate gene transcription from the SRE is greatly enhanced by the presence of Tec (Mao et al. 1998a; Suzuki et al. 2003). Thus tyrosine phosphorylation of PDZ-RhoGEF and LARG, but not p115RhoGEF, appears to be an important mechanism of regulation.

Subcellular localization. RGS-RhoGEF function may also be regulated by subcellular localization. When over-expressed with wild-type Gα13, p115RhoGEF is predominantly located in the cytoplasm; however, when expressed with a constitutively active mutant of Gα13, a significant fraction can be found at the plasma membrane (Bhattacharyya and Wedegaertner 2000). The same study demonstrated that a mutant of Gα13, which is unable to be palmitoylated, fails to localize to the plasma membrane. This mutant is unable to induce p115RhoGEF translocation or initiate Rho-dependent signaling. However, the mutant Gα13 and p115RhoGEF could still physically associate as analyzed by co-immunoprecipitation, suggesting that Gα13-induced membrane localization of p115RhoGEF may be an important regulatory mechanism. As mentioned earlier, deletion of the C-terminus of p115RhoGEF results in increased activity in cells as compared to the full-length exchange factor. Interestingly, this truncation mutant is located almost exclusively in the membrane fraction as judged by immunoblotting of cell lysates, suggesting a relationship between constitutive membrane localization and enhanced GEF activity (Wells et al. 2001). Further analysis revealed that translocation of p115RhoGEF could be induced by stimulating the Gα12/13-coupled thromboxane A_2 receptor, supporting the notion that activated Gα12/13 can stimulate the redistribution of p115RhoGEF to the plasma membrane (Bhattacharyya and Wedegaertner 2003a). Bhattacharyya and Wedegaertner (2003a) also analyzed the ability of truncated versions of p115RhoGEF to redistribute to the membrane fraction and identified the RGS and PH domains as particularly important for mediating plasma membrane association. Since elements N-terminal to the RGS domain of p115RhoGEF have been shown to be important for the interaction with Gα13 (Chen et al. 2003; Chen et al. 2005), the role of this region in Gα13-induced plasma membrane localization was also investigated. In particular, two point mutations, Glu27Ala and Glu29Ala, in the acidic patch N-terminal to the RGS domain, were analyzed for their effect on Gα13 binding and Gα13-induced translocation. Interestingly, these mutations impaired the ability of p115RhoGEF to bind to Gα13 but did not affect its ability to localize to the plasma membrane (Bhattacharyya and Wedegaertner 2003b).

These data thus highlight the separation of two functions of p115RhoGEF's N-terminus: direct binding to activated Gα13 and Gα13-mediated translocation to the plasma membrane. Compared with p115RhoGEF, mechanisms controlling the subcellular distribution of PDZ-RhoGEF and LARG are less well understood. In differentiated HL60 cells, PDZ-RhoGEF tagged with yellow fluorescent protein is found mainly at the cell periphery, with smaller amounts present in the cytosol (Wong, Van Keymeulen, and Bourne 2007). Upon stimulation with the agonist formyl-Met-Leu-Phe, PDZ-RhoGEF concentrates at the "back" end of these cells suggesting subcellular distribution may be important for regulation of activity. Both Gα12/13 and myosin II appear to play a role in controlling PDZ-RhoGEF localization in this case, but the precise mechanisms controlling this process in other cell types remains unknown. When overexpressed as a fusion protein with green fluorescent protein in COS7 cells, PDZ-RhoGEF is located in the cytoplasm but also exhibits peri-plasma membrane localization (Grabocka and Wedegaertner 2007). This is consistent with a previous report demonstrating that PDZ-RhoGEF interacts with cortical actin (Banerjee and Wedegaertner 2004). Among the RGS-RhoGEFs, association with the actin cytoskeleton appears to be a unique property of PDZ-RhoGEF. Fluorescently tagged LARG is predominantly found in the cytoplasm, as is endogenous LARG, but upon co-expression with a constitutively active mutant of Gα13, it redistributes to the plasma membrane (Grabocka and Wedegaertner 2007). Interestingly, the same study demonstrated that upon disruption of homo-oligomerization, LARG monomers undergo nucleo-cytoplasmic shuttling. Thus, regulation of LARG activity may be achieved in part by controlling its access to various subcellular compartments.

Structural Basis of RGS-RhoGEF Function

RGS-like domains. Among the members of the RGS-RhoGEF family, p115RhoGEF and LARG have been shown to act as specific *G*TPase *a*ctivating *p*roteins (GAPs) for Gα12 and Gα13 in vitro while the RGS domain of PDZ-RhoGEF lacks detectable GAP activity (Kozasa et al. 1998; Wells et al. 2002; Suzuki et al. 2003). High resolution crystal structures of the RGS domains from both PDZ-RhoGEF and p115RhoGEF have been reported (Chen et al. 2001; Longenecker et al. 2001). Although the sequence identity between the RGS domains from RGS-RhoGEFs and other RGS family members is between 10% and 15% (Kozasa et al. 1998), all of these domains share similar tertiary folds composed of an all-helical bundle. One interesting point of structural divergence between these RGS domains and RGS4 is the extended C-terminus found in the RGS-RhoGEFs. This extension of the RGS domain also forms α helices that are tightly associated with the core RGS domain via hydrophobic contacts and removal of these helices is predicted to expose a large hydrophobic surface. This probably explains the necessity of these C-terminal elements for the stable expression of the RGS domain of p115RhoGEF in *E. coli* (Wells et al. 2002). A crystal structure of the RGS domain of p115RhoGEF bound to a Gαi-13 chimera has also been published and is discussed below (Chen et al. 2005). As the most detailed biochemical analysis of GAP activity has been performed on p115RhoGEF, the following discussion of GAP activity is based mainly on analysis of this protein.

Biochemical analysis of p115RhoGEF GAP activity. Like all other heterotrimeric G protein α subunits, Gα13 cycles between GDP and GTP bound states and possesses an intrinsic ability to hydrolyze GTP to GDP. In vitro analysis of recombinant Gα13 protein has demonstrated that it has a relatively slow rate of GDP dissociation and GTPγS binding (Singer, Miller, and Sternweis 1994). Analysis of the agonist-stimulated binding of a photoreactive GTP analog to Gα13 in platelet membranes confirms this low basal nucleotide exchange rate (Offermanns et al. 1994). The slow nucleotide exchange rate of Gα13 has hindered a direct determination of the rate of GTP hydrolysis in vitro. However, it has been estimated that Gα13 hydrolyzes GTP at a rate of at least $0.2\ min^{-1}$ at 30°C in the presence of 1 μM free Mg^{2+} (Singer, Miller, and Sternweis 1994).

The crystal structures of Gαi1-Gα13 chimeras provide an explanation for the slow rate of GDP release (Chen et al. 2005; Kreutz et al. 2006). In the GDP bound state, residue E172 of Gα13, located in the αD-αE1 loop near the nucleotide binding pocket, forms a salt bridge with K292 and is packed against the nucleotide, which may serve to physically hinder the dissociation of GDP. A comparison of the conformation of E172 in the inactive GDP bound state (Gαi/13 structure, Kreutz et al. 2006) versus the AlF_4^--activated state (Gα13-i5 structure, Chen et al. 2005) reveals that activation causes E172 to adopt a more "puckered" conformation, which lacks contact with the nucleotide. Furthermore, in the deactivated Gαi/13 structure, GDP appears to be less solvent accessible than in the AlF_4^- activated form of Gα13-i5, which may also influence the kinetics of guanine nucleotide exchange.

Site-directed mutagenesis of both Gα13 and p115RhoGEF has provided important insight into the mechanism of GTPase acceleration by p115RhoGEF's RGS domain. In terms of Gα13, lysine 204, located in its Switch I region, has been shown to be important for interaction with the RGS domains of both p115RhoGEF and LARG (Nakamura et al. 2004; Grabocka and Wedegaertner 2005). Interestingly, mutation of glycine 205, which is a highly conserved RGS contact point among other Gα subunits, has little impact on the Gα13-p115RhoGEF interaction (Grabocka and Wedegaertner 2005). Detailed analysis of the RGS-like domain (residues 43–160) has demonstrated that residues N-terminal to this region are required for binding to Gα13 and for GAP activity. Residues 1–252 of p115RhoGEF stimulate Gα13's GTPase activity similarly to the full-length protein (Kozasa et al. 1998; Wells et al. 2002). However, removal of the first 24 residues reduces GAP activity approximately 99% while deletion of the first 41 residues produces a construct which still binds to Gα13 but lacks measurable GAP activity (Wells et al. 2002). In particular, residues 27–31 are crucial for p115's GTPase activity. Mutations in this acidic patch almost completely abolish both RGS-stimulated GTP hydrolysis and binding to Gα13 (Bhattacharyya and Wedegaertner 2003b; Chen et al. 2003). The high-resolution crystal structure of a Gαi1–13 chimera bound to the RGS-like domain of p115RhoGEF supports this biochemical analysis

(Chen et al. 2005). The region N-terminal to the RGS box, specifically the βN-αN hairpin (residues 17–39), forms close contacts with the switch regions of Gαi1–13, while the RGS domain itself contacts the switch II region and α3 helix of Gαi1–13. These binding surfaces are similar to other Gα-effector interfaces (Tesmer et al. 1997; Slep et al. 2001; Tesmer et al. 2005).

DH/PH Domains

PDZ-RhoGEF. A crystal structure of the DH/PH domains of PDZ-RhoGEF bound to nucleotide-free RhoA has also been solved (Derewenda et al. 2004). The DH domain assumes the classical "chaise lounge" configuration, with the long α helices forming the "seat" and the short α helices comprising the "seatback." As with other RhoGEFs, CR1 (*c*onserved *r*egion) and CR3 of the DH domain contact switch I in RhoA, while switch II is also engaged by CR3 and in addition, the α6 helix of the RhoGEF. The latter set of interactions is believed to dictate the specificity for the GTPase. Subsequent mutational analysis of RhoA has helped to clarify the precise residues dictating its interaction with PDZ-RhoGEF (Oleksy et al. 2006). The PH domain of PDZ-RhoGEF also assumes a fold common to this domain family in the form of an antiparallel β sandwich capped with an α helix. Unlike other Dbl family members, however, the β4 strand of PDZ-RhoGEF's PH domain contains an additional 18-residue insertion, which includes an uncharacteristic proline residue. This results in a β strand which protrudes into the putative phosphoinositide binding pocket. Additionally, this binding pocket is lined with fewer positively charged residues than are found in the PH domains of other RhoGEFs like Dbs. Together, these two pieces of data suggest that binding to phosphoinositides may not be a major function of this particular PH domain, though direct evidence in support of this notion is still lacking. Interestingly, the PH domain also makes direct contact with RhoA, though the interaction surface is small. Thus, it remains unclear to what extent the PH domain influences nucleotide exchange although, in vitro, the DH/PH domains are a more efficient GEF than the DH domain alone (Derewenda et al. 2004; Oleksy et al. 2006).

DH/PH domains of LARG. High-resolution crystal structures of the DH/PH domains of LARG alone, and in complex with RhoA, have been solved as well (Kristelly, Gao and Tesmer 2004). The conformation of the DH/PH domains of LARG is comparable to other solved structures of Dbl family members. There are, however, some structural divergences that appear to be unique to the RGS-RhoGEF family. LARG's DH domain contains a unique extension at its N-terminus, which binds directly to the Switch I region of RhoA. Deletion or mutation of this region decreases the GEF activity of the DH/PH domains in vitro compared to the wild-type domains. LARG also contains an insertion of 17 amino acids in the β4 strand of its PH domain. This strand is projected to be adjacent to the membrane, and thus may play a role in lipid binding or membrane localization. Upon binding to RhoA the PH domains also undergo a significant conformational change, rotating ~30°

relative to the DH domain, such that it now engages RhoA directly. As with PDZ-RhoGEF, this interaction buries a small amount of surface area, but the isolated DH domain catalyzes nucleotide exchange less efficiently in vitro than the tandem DH/PH domains (Reuther et al. 2001; Kristelly, Gao, and Tesmer 2004). For a comprehensive review of structure-function analysis of Dbl family members see Rossman, Der, and Sondek (2005).

Physiological Functions of Gα12/13-RhoA Signaling

Analysis of RGS-RhoGEF pathways in model organisms. Strong genetic evidence supporting a Gα12/13-RGSRhoGEF-Rho signaling pathway has also been obtained using the model organisms *C. elegans* and *D. melanogaster*. The conservation of this pathway through the course of evolution supports its physiological significance. Biochemical experiments have demonstrated that GPA-12, the ortholog of Gα12 in *C. elegans*, and a RGS-RhoGEF, CeRhoGEF, can interact in an activation-dependent manner (Yau et al. 2003). Furthermore, the same study revealed that in *C. elegans*, GPA-12 and CeRhoGEF are co-expressed in some ventral cord motor neurons. If expression of either GPA-12 or CeRhoGEF is silenced using RNA interference (RNAi), the result is a similar phenotype consisting of defects in egg laying and embryonic lethality, further suggesting these two proteins function in the same pathway (Yau et al. 2003). Additional evidence implicating GPA-12 and CeRhoGEF in Rho activation came from studies examining the role of these proteins in acetylcholine release from motor neurons. Genetic studies have demonstrated that a constitutively active mutant of Rho1, the RhoA ortholog in *C. elegans*, increases the release of acetylcholine (McMullan et al. 2006). Interestingly, expression of a constitutively active mutant of GPA-12 also causes an increase in acetylcholine release, an effect which is dependent on Rho1 and can be suppressed by mutating the *CeRhoGEF* gene (Hiley, McMullan and Nurrish 2006). In *D. melanogaster*, overexpression of Rho1 results in a “rough eye” phenotype, which can be partially suppressed by mutations in the *DRhoGEF2* gene, which was shown to encode a protein containing the tandem DH/PH domains characteristic of RhoGEFs (Barrett, Leptin and Settleman 1997). Subsequently, the N-terminus of DRhoGEF2 was shown to have sequence homology with other putative RhoGEFs containing RGS-like domains in their N-terminus (Kozasa et al. 1998). Embryos lacking functional DRhoGEF2 show similar defects in gastrulation to embryos without functional Concertina, the *Drosophila* Gα12/13 ortholog, suggesting that Concertina may propagate signals from an upstream ligand to Rho1 via DRhoGEF2 (Parks and Wieschaus 1991; Barrett, Leptin, and Settleman 1997).

Phenotypes of RGS-RhoGEF null mice. The first RGS-RhoGEF to be analyzed by simple knockout in the mouse was Lsc (murine p115RhoGEF) (Girkontaite et al. 2001; Rubtsov et al. 2005). Defects in Lsc null mice are primarily found in the immune system, which is reasonable as p115RhoGEF/Lsc is strongly expressed in hematopoietic tissue (Hart et al. 1996; Whitehead et al. 1996; Aasheim, Pedeutour, and Smeland 1997; Girkontaite et al. 2001). These mice have reduced T cell populations in their spleen and lymph nodes, as well as reduced numbers of marginal zone B cells in the spleen. Lsc null mice also display defects in lymphocyte migration, pseudopod formation, integrin-mediated adhesion, and immune responses, suggesting that it is required for normal B and T lymphocyte function (Girkontaite et al. 2001; Francis et al. 2006). Recently, a phenotype associated with LARG knockout has also been reported (Wirth et al. 2008). Aortic segments from LARG-deficient mice show impaired contractile responses to the vasoconstrictors angiotensin II and endothelin I. Furthermore, these mice are almost completely protected from salt-induced hypertension, while their basal blood pressure is unaffected. Similar phenotypes were observed in mice lacking Gα12 and Gα13 in smooth muscle cells. These findings suggest that the Gα13-LARG pathway is a key regulator of vascular smooth muscle tone in the context of hypertension. The phenotype(s) associated with PDZ-RhoGEF knockout have yet to be reported.

Phenotypes of Gα12 and Gα13 null mice. Differential physiological functions of Gα12 and Gα13 during embryonic development have been clearly demonstrated using gene knockout techniques in the mouse. A homozygous deficiency of Gα13 in mice impairs the development of the vascular system, and is embryonic lethal (Offermanns et al. 1997). In order to study the role of Gα13 in the adult mouse, Cre/loxP-mediated recombination has been utilized to conditionally inactivate the Gα13 gene in a tissue-specific manner. Studies using mice lacking Gα13 in platelets revealed that this α subunit is involved in both hemostasis and thrombosis. These mice show impaired responses to multiple platelet activators, fail to form stable thrombi ex vivo, and the mice exhibit a large increase in tail-bleeding times (Moers et al. 2003). When Gα13 is conditionally inactivated in endothelial cells, the resulting phenotype is similar to that of Gα13 null animals. However, reintroducing Gα13 expression in endothelial cells of Gα13 simple knockouts fails to completely rescue the phenotype, suggesting that Gα13 expression in other cell types is necessary during embryonic development (Ruppel et al. 2005). In contrast to the phenotype observed in Gα13 null mice, mice lacking Gα12 develop normally and do not exhibit any gross morphological or behavioral defects. A double knockout of Gα12 and Gα13 produces developmental defects in the headfold, somites, and neural tube and these embryos arrest earlier than Gα13 null embryos suggesting that the function of Gα12 is not completely redundant to that of Gα13 during embryonic development (Gu et al. 2002). In *D. melanogaster*, loss of the Gα13 ortholog Concertina inhibits gastrulation by impairing cell shape change and movement, and also results in embryonic lethality (Parks and Wieschaus 1991).

RGS-RhoGEFs and Human Disease

Despite interest in the possible functions of RGS-RhoGEFs in human disease, the extent to which these proteins contribute to cancer is still unknown. As mentioned previously, LARG was originally identified as a fusion partner with the *MLL* gene in a patient with acute myeloid leukemia (Kourlas et al. 2000). The MLL-LARG fusion protein retains LARG's RGS and DH/PH domains, thus it is tempting to speculate that aberrant signaling from this portion of the molecule may have contributed to leukemogenesis. However, further studies will be necessary to confirm this conjecture. Although LARG's role in leukemia remains unclear, a recent analysis of bone marrow samples from patients with Shwachman-Diamond syndrome, which often develops into acute myeloid leukemia, demonstrated that in these patients LARG expression is dramatically increased (Rujkijyanont et al. 2007). Given that LARG is abundant in hematopoietic stem cells, further investigation is needed into its precise function in this cell type (Zinovyeva et al. 2004). While relatively little is known about the biochemistry of PDZ-RhoGEF, an epidemiological study examining the risk of lung cancer associated with a nonsynonymous SNP (*s*ingle *n*ucleotide *p*olymorphism) in PDZ-RhoGEF revealed that among certain populations, a Ser1416Gly mutation significantly reduced the risk for lung cancer (Gu et al. 2006). How this mutation confers a protective effect and the mechanisms by which PDZ-RhoGEF may contribute to the onset or progression of this disease remain unknown. Interestingly, a recent genome-wide analysis of colon cancers revealed a somatic Met165Val mutation in the RGS domain of p115RhoGEF (Sjoblom et al. 2006). The functional significance of this mutation is currently unknown.

Summary

RGS-RhoGEFs are a unique subclass of guanine nucleotide exchange factors for the monomeric GTPase RhoA whose form and function have been conserved through the course of evolution. These proteins are effectors of the heterotrimeric G proteins $G\alpha12$ and $G\alpha13$. $G\alpha12/13$ stimulate the guanine nucleotide exchange activity of the DH/PH domains of RGS-RhoGEFs to facilitate activation of RhoA. As their name implies, RGS-RhoGEFs contain a RGS domain in their N-terminus which acts as a specific GTPase activating protein for $G\alpha12$ and $G\alpha13$. Thus, RGS-RhoGEFs serve as the direct link between $G\alpha12/13$ and cellular events controlled by RhoA, such as rearrangement of the actin cytoskeleton and gene transcription. A thorough understanding of the mechanisms regulating this signaling pathway will provide important insights into a variety of physiological processes.

Acknowledgments We would like to apologize to our colleagues whose contributions to this field were inadvertently omitted or could not be properly acknowledged due to space limitations.

References

Aasheim, H. C., Pedeutour, F. and Smeland, E. B. 1997. Characterization, expression and chromosomal localization of a human gene homologous to the mouse Lsc oncogene, with strongest expression in hematopoetic tissues. Oncogene 14: 1747–1752.

Aurandt, J. et al. 2002. The semaphorin receptor plexin-B1 signals through a direct interaction with the Rho-specific nucleotide exchange factor, LARG. Proc Natl Acad Sci U S A 99: 12085–12090.

Avraham, H. and Weinberg, R. A. 1989. Characterization and expression of the human rhoH12 gene product. Mol Cell Biol 9: 2058–2066.

Banerjee, J. and Wedegaertner, P. B. 2004. Identification of a novel sequence in PDZ-RhoGEF that mediates interaction with the actin cytoskeleton. Mol Biol Cell 15: 1760–1775.

Barrett, K., Leptin, M. and Settleman, J. 1997. The Rho GTPase and a putative RhoGEF mediate a signaling pathway for the cell shape changes in Drosophila gastrulation. Cell 91: 905–915.

Bhattacharyya, R. and Wedegaertner, P. B. 2000. Galpha 13 requires palmitoylation for plasma membrane localization, Rho-dependent signaling, and promotion of p115-RhoGEF membrane binding. J Biol Chem 275: 14992–14999.

Bhattacharyya, R. and Wedegaertner, P. B. 2003a. Characterization of G alpha 13-dependent plasma membrane recruitment of p115RhoGEF. Biochem J 371: 709–720.

Bhattacharyya, R. and Wedegaertner, P. B. 2003b. Mutation of an N-terminal acidic-rich region of p115-RhoGEF dissociates alpha13 binding and alpha13-promoted plasma membrane recruitment. FEBS Lett 540: 211–216.

Boettner, B. and Van Aelst, L. 2002. The role of Rho GTPases in disease development. Gene 286: 155–174.

Booden, M. A., Siderovski, D. P. and Der, C. J. 2002. Leukemia-associated Rho guanine nucleotide exchange factor promotes G alpha q-coupled activation of RhoA. Mol Cell Biol 22: 4053–4061.

Bos, J. L. 1988. The ras gene family and human carcinogenesis. Mutat Res 195: 255–271.

Bourguignon, L. Y. et al. 2006. Hyaluronan-CD44 interaction with leukemia-associated RhoGEF and epidermal growth factor receptor promotes Rho/Ras co-activation, phospholipase C epsilon-Ca2 + signaling, and cytoskeleton modification in head and neck squamous cell carcinoma cells. J Biol Chem 281: 14026–14040.

Chen, Z. et al. 2005. Structure of the p115RhoGEF rgRGS domain-Galpha13/i1 chimera complex suggests convergent evolution of a GTPase activator. Nat Struct Mol Biol 12: 191–197.

Chen, Z. et al. 2003. Mapping the Galpha13 binding interface of the rgRGS domain of p115RhoGEF. J Biol Chem 278: 9912–9919.

Chen, Z. et al. 2001. Structure of the rgRGS domain of p115RhoGEF. Nat Struct Biol 8: 805–809.

Chikumi, H. et al. 2004. Homo- and hetero-oligomerization of PDZ-RhoGEF, LARG and p115RhoGEF by their C-terminal region regulates their in vivo Rho GEF activity and transforming potential. Oncogene 23: 233–240.

Chikumi, H., Fukuhara, S. and Gutkind, J. S. 2002. Regulation of G protein-linked guanine nucleotide exchange factors for Rho, PDZ-RhoGEF, and LARG by tyrosine phosphorylation: evidence of a role for focal adhesion kinase. J Biol Chem 277: 12463–12473.

Clark, E. A. et al. 2000. Genomic analysis of metastasis reveals an essential role for RhoC. Nature 406: 532–535.

Derewenda, U. et al. 2004. The crystal structure of RhoA in complex with the DH/PH fragment of PDZRhoGEF, an activator of the Ca(2+) sensitization pathway in smooth muscle. Structure 12: 1955–1965.

Doupnik, C. A. et al. 1997. RGS proteins reconstitute the rapid gating kinetics of gbetagamma-activated inwardly rectifying K + channels. Proc Natl Acad Sci U S A 94: 10461–10466.

Eisenhaure, T. M. et al. 2003. The Rho guanine nucleotide exchange factor Lsc homo-oligomerizes and is negatively regulated through domains in its carboxyl terminus that are absent in novel splenic isoforms. J Biol Chem 278: 30975–30984.

Francis, S. A. et al. 2006. Rho GEF Lsc is required for normal polarization, migration, and adhesion of formyl-peptide-stimulated neutrophils. Blood 107: 1627–1635.

Fritz, G. et al. 2002. Rho GTPases in human breast tumours: expression and mutation analyses and correlation with clinical parameters. Br J Cancer 87: 635–644.

Fritz, G., Just, I. and Kaina, B. 1999. Rho GTPases are over-expressed in human tumors. Int J Cancer 81: 682–687.

Fukuhara, S., Chikumi, H. and Gutkind, J. S. 2000. Leukemia-associated Rho guanine nucleotide exchange factor (LARG) links heterotrimeric G proteins of the G(12) family to Rho. FEBS Lett 485: 183–188.

Fukuhara, S., Chikumi, H. and Gutkind, J. S. 2001. RGS-containing RhoGEFs: the missing link between transforming G proteins and Rho? Oncogene 20: 1661–1668.

Fukuhara, S. et al. 1999. A novel PDZ domain containing guanine nucleotide exchange factor links heterotrimeric G proteins to Rho. J Biol Chem 274: 5868–5879.

Girkontaite, I. et al. 2001. Lsc is required for marginal zone B cells, regulation of lymphocyte motility and immune responses. Nat Immunol 2: 855–862.

Gohla, A., Harhammer, R. and Schultz, G. 1998. The G-protein G13 but not G12 mediates signaling from lysophosphatidic acid receptor via epidermal growth factor receptor to Rho. J Biol Chem 273: 4653–4659.

Grabocka, E. and Wedegaertner, P. B. 2005. Functional consequences of G alpha 13 mutations that disrupt interaction with p115RhoGEF. Oncogene 24: 2155–2165.

Grabocka, E. and Wedegaertner, P. B. 2007. Disruption of oligomerization induces nucleocytoplasmic shuttling of leukemia-associated rho Guanine-nucleotide exchange factor. Mol Pharmacol 72: 993–1002.

Gu, J. et al. 2006. A nonsynonymous single-nucleotide polymorphism in the PDZ-Rho guanine nucleotide exchange factor (Ser1416Gly) modulates the risk of lung cancer in Mexican Americans. Cancer 106: 2716–2724.

Gu, J. L. et al. 2002. Interaction of G alpha(12) with G alpha(13) and G alpha(q) signaling pathways. Proc Natl Acad Sci U S A 99: 9352–9357.

Hall, A. 1998. Rho GTPases and the actin cytoskeleton. Science 279: 509–514.

Hart, M. J. et al. 1998. Direct stimulation of the guanine nucleotide exchange activity of p115 RhoGEF by Galpha13. Science 280: 2112–2114.

Hart, M. J. et al. 1996. Identification of a novel guanine nucleotide exchange factor for the Rho GTPase. J Biol Chem 271: 25452–25458.

Hiley, E., McMullan, R. and Nurrish, S. J. 2006. The Galpha12-RGS RhoGEF-RhoA signalling pathway regulates neurotransmitter release in C. elegans. Embo J 25: 5884–5895.

Hirotani, M. et al. 2002. Interaction of plexin-B1 with PDZ domain-containing Rho guanine nucleotide exchange factors. Biochem Biophys Res Commun 297: 32–37.

Holinstat, M. et al. 2003. Protein kinase Calpha-induced p115RhoGEF phosphorylation signals endothelial cytoskeletal rearrangement. J Biol Chem 278: 28793–28798.

Jackson, M. et al. 2001. Modulation of the neuronal glutamate transporter EAAT4 by two interacting proteins. Nature 410: 89–93.

Katoh, H. et al. 1998. Constitutively active Galpha12, Galpha13, and Galphaq induce Rho-dependent neurite retraction through different signaling pathways. J Biol Chem 273: 28700–28707.

Khosravi-Far, R. et al. 1995. Activation of Rac1, RhoA, and mitogen-activated protein kinases is required for Ras transformation. Mol Cell Biol 15: 6443–6453.

Kitzing, T. M. et al. 2007. Positive feedback between Dia1, LARG, and RhoA regulates cell morphology and invasion. Genes Dev 21: 1478–1483.

Kourlas, P. J. et al. 2000. Identification of a gene at 11q23 encoding a guanine nucleotide exchange factor: evidence for its fusion with MLL in acute myeloid leukemia. Proc Natl Acad Sci U S A 97: 2145–2150.

Kozasa, T. et al. 1998. p115 RhoGEF, a GTPase activating protein for Galpha12 and Galpha13. Science 280: 2109–2111.

Kranenburg, O. et al. 1999. Activation of RhoA by lysophosphatidic acid and Galpha12/13 subunits in neuronal cells: induction of neurite retraction. Mol Biol Cell 10: 1851–1857.

Kreutz, B. et al. 2007. Distinct regions of Galpha13 participate in its regulatory interactions with RGS homology domain-containing RhoGEFs. Cell Signal 19: 1681–1689.

Kreutz, B. et al. 2006. A new approach to producing functional G alpha subunits yields the activated and deactivated structures of G alpha(12/13) proteins. Biochemistry 45: 167–174.

Kristelly, R., Gao, G. and Tesmer, J. J. 2004. Structural determinants of RhoA binding and nucleotide exchange in leukemia-associated Rho guanine-nucleotide exchange factor. J Biol Chem 279: 47352–47362.

Longenecker, K. L. et al. 2001. Structure of the RGS-like domain from PDZ-RhoGEF: linking heterotrimeric g protein-coupled signaling to Rho GTPases. Structure 9: 559–569.
Majumdar, M. et al. 1999. A rho exchange factor mediates thrombin and Galpha(12)-induced cytoskeletal responses. J Biol Chem 274: 26815–26821.
Mao, J. et al. 1998a. Tec/Bmx non-receptor tyrosine kinases are involved in regulation of Rho and serum response factor by Galpha12/13. Embo J 17: 5638–5646.
Mao, J. et al. 1998b. Guanine nucleotide exchange factor GEF115 specifically mediates activation of Rho and serum response factor by the G protein alpha subunit Galpha13. Proc Natl Acad Sci U S A 95: 12973–12976.
McMullan, R. et al. 2006. Rho is a presynaptic activator of neurotransmitter release at pre-existing synapses in C. elegans. Genes Dev 20: 65–76.
Moers, A. et al. 2003. G13 is an essential mediator of platelet activation in hemostasis and thrombosis. Nat Med 9: 1418–1422.
Nakamura, S. et al. 2004. Critical role of lysine 204 in switch I region of Galpha13 for regulation of p115RhoGEF and leukemia-associated RhoGEF. Mol Pharmacol 66: 1029–1034.
Offermanns, S. et al. 1994. G proteins of the G12 family are activated via thromboxane A2 and thrombin receptors in human platelets. Proc Natl Acad Sci U S A 91: 504–508.
Offermanns, S. et al. 1997. Vascular system defects and impaired cell chemokinesis as a result of Galpha13 deficiency. Science 275: 533–536.
Oleksy, A. et al. 2006. The molecular basis of RhoA specificity in the guanine nucleotide exchange factor PDZ-RhoGEF. J Biol Chem 281: 32891–32897.
Parks, S. and Wieschaus, E. 1991. The Drosophila gastrulation gene concertina encodes a G alpha-like protein. Cell 64: 447–458.
Perrot, V., Vazquez-Prado, J. and Gutkind, J. S. 2002. Plexin B regulates Rho through the guanine nucleotide exchange factors leukemia-associated Rho GEF (LARG) and PDZ-RhoGEF. J Biol Chem 277: 43115–43120.
Qiu, R. G. et al. 1995. A role for Rho in Ras transformation. Proc Natl Acad Sci U S A 92: 11781–11785.
Radhika, V. and Dhanasekaran, N. 2001. Transforming G proteins. Oncogene 20: 1607–1614.
Reuther, G. W. et al. 2001. Leukemia-associated Rho guanine nucleotide exchange factor, a Dbl family protein found mutated in leukemia, causes transformation by activation of RhoA. J Biol Chem 276: 27145–27151.
Rossman, K. L., Der, C. J. and Sondek, J. 2005. GEF means go: turning on RHO GTPases with guanine nucleotide-exchange factors. Nat Rev Mol Cell Biol 6: 167–180.
Rubtsov, A. et al. 2005. Lsc regulates marginal-zone B cell migration and adhesion and is required for the IgM T-dependent antibody response. Immunity 23: 527–538.
Rujkijyanont, P. et al. 2007. Leukaemia-related gene expression in bone marrow cells from patients with the preleukaemic disorder Shwachman-Diamond syndrome. Br J Haematol 137: 537–544.
Rumenapp, U. et al. 1999. Rho-specific binding and guanine nucleotide exchange catalysis by KIAA0380, a dbl family member. FEBS Lett 459: 313–318.
Ruppel, K. M. et al. 2005. Essential role for Galpha13 in endothelial cells during embryonic development. Proc Natl Acad Sci U S A 102: 8281–8286.
Sahai, E. and Marshall, C. J. 2002. RHO-GTPases and cancer. Nat Rev Cancer 2: 133–142.
Sahai, E., Olson, M. F. and Marshall, C. J. 2001. Cross-talk between Ras and Rho signalling pathways in transformation favours proliferation and increased motility. Embo J 20: 755–766.
Saitoh, O. et al. 1997. RGS8 accelerates G-protein-mediated modulation of K + currents. Nature 390: 525–529.
Singer, W. D., Miller, R. T. and Sternweis, P. C. 1994. Purification and characterization of the alpha subunit of G13. J Biol Chem 269: 19796–19802.
Sjoblom, T. et al. 2006. The consensus coding sequences of human breast and colorectal cancers. Science 314: 268–274.

Slep, K. C. et al. 2001. Structural determinants for regulation of phosphodiesterase by a G protein at 2.0 A. Nature 409: 1071–1077.
Suzuki, N. et al. 2003. Galpha 12 activates Rho GTPase through tyrosine-phosphorylated leukemia-associated RhoGEF. Proc Natl Acad Sci U S A 100: 733–738.
Suzuki, N. et al. 2009. Activation of leukemia-associated RhoGEF by Ga13 with significant rearrangements in the interface. J Biol Chem 284: 5000–5009.
Swiercz, J. M. et al. 2002. Plexin-B1 directly interacts with PDZ-RhoGEF/LARG to regulate RhoA and growth cone morphology. Neuron 35: 51–63.
Taya, S. et al. 2001. Direct interaction of insulin-like growth factor-1 receptor with leukemia-associated RhoGEF. J Cell Biol 155: 809–820.
Tesmer, J. J. et al. 1997. Crystal structure of the catalytic domains of adenylyl cyclase in a complex with Gsalpha.GTPgammaS. Science 278: 1907–1916.
Tesmer, V. M. et al. 2005. Snapshot of activated G proteins at the membrane: the Galphaq-GRK2-Gbetagamma complex. Science 310: 1686–1690.
Togashi, H. et al. 2000. Functions of a rho-specific guanine nucleotide exchange factor in neurite retraction. Possible role of a proline-rich motif of KIAA0380 in localization. J Biol Chem 275: 29570–29578.
Van Aelst, L. and D'Souza-Schorey, C. 1997. Rho GTPases and signaling networks. Genes Dev 11: 2295–2322.
Vazquez-Prado, J. et al. 2004. Chimeric G alpha i2/G alpha 13 proteins reveal the structural requirements for the binding and activation of the RGS-like (RGL)-containing Rho guanine nucleotide exchange factors (GEFs) by G alpha 13. J Biol Chem 279: 54283–54290.
Wells, C. D. et al. 2001. Identification of potential mechanisms for regulation of p115 RhoGEF through analysis of endogenous and mutant forms of the exchange factor. J Biol Chem 276: 28897–28905.
Wells, C. D. et al. 2002. Mechanisms for reversible regulation between G13 and Rho exchange factors. J Biol Chem 277: 1174–1181.
Whitehead, I. P. et al. 1996. Expression cloning of lsc, a novel oncogene with structural similarities to the Dbl family of guanine nucleotide exchange factors. J Biol Chem 271: 18643–18650.
Wirth, A. et al. 2008. G12-G13-LARG-mediated signaling in vascular smooth muscle is required for salt-induced hypertension. Nat Med 14: 64–68.
Wong, K., Van Keymeulen, A. and Bourne, H. R. 2007. PDZRhoGEF and myosin II localize RhoA activity to the back of polarizing neutrophil-like cells. J Cell Biol 179: 1141–1148.
Xu, N. et al. 1993. A mutant alpha subunit of G12 potentiates the eicosanoid pathway and is highly oncogenic in NIH 3T3 cells. Proc Natl Acad Sci U S A 90: 6741–6745.
Xu, N., Voyno-Yasenetskaya, T. and Gutkind, J. S. 1994. Potent transforming activity of the G13 alpha subunit defines a novel family of oncogenes. Biochem Biophys Res Commun 201: 603–609.
Yamada, T. et al. 2005. Physical and functional interactions of the lysophosphatidic acid receptors with PDZ domain-containing Rho guanine nucleotide exchange factors (RhoGEFs). J Biol Chem 280: 19358–19363.
Yau, D. M. et al. 2003. Identification and molecular characterization of the G alpha12-Rho guanine nucleotide exchange factor pathway in Caenorhabditis elegans. Proc Natl Acad Sci U S A 100: 14748–14753.
Zhang, H. et al. 1999. Functional interaction between the cytoplasmic leucine-zipper domain of HIV-1 gp41 and p115-RhoGEF. Curr Biol 9: 1271–1274.
Zhong, H. et al. 2003. A spatial focusing model for G protein signals. Regulator of G protein signaling (RGS) protein-mediated kinetic scaffolding. J Biol Chem 278: 7278–7284.
Zinovyeva, M. et al. 2004. Molecular cloning, sequence and expression pattern analysis of the mouse orthologue of the leukemia-associated guanine nucleotide exchange factor. Gene 337: 181–188.

Chapter 5
Vav Proteins in Cancer

Daniel D. Billadeau

Introduction

Rho family GTP-binding proteins participate in multiple cellular processes including cell cycle progression, cell survival, gene transcription, cytoskeletal reorganization, adhesion, cell migration, cell polarization, cytokinesis, phagocytosis, and neurite extension and retraction (Bustelo et al. 2007). In addition, there is increasing experimental evidence to the fact that many members of this family also play a significant role in the development of human cancer by promoting cell motility and invasion, cell survival, proliferation, and angiogenesis. Although mutational activation of Ras is associated with more than 30% of human malignancies, similar activating mutations have not been identified for Rho family genes in human cancers. However, several Rho family members are overexpressed in human malignancies (Rossman et al. 2005). The significance of this overexpression is not clear, but the fact that Ras signaling pathways converge in the activation of Rho family GTPases, coupled with the observation that Rho proteins are required for Ras-mediated cellular transformation, highlights the importance of Rho proteins in cancer and identifies them and their effector pathways as potential therapeutic targets.

Mechanistically, increased GTP loading of Rho family GTPases in cancer can occur through the activation of signaling pathways that stimulate GEF activity, or through the inactivation of GTPase activating proteins (GAPs). Indeed, it is becoming increasingly apparent that GEFs are major contributors to the hyperactivation of Rho proteins in cancer. The first RhoGEF was identified over 20 years ago as a transforming gene in *D*iffuse *B* cell *L*ymphoma and was termed Dbl (Rossman et al. 2005). Dbl was found to harbor a polypeptide domain containing ~200 amino acids that were similar to the yeast Cdc42 activating protein Cdc24, now known as the Dbl homology (DH) domain, which is responsible for the activation of Rho family GTPases (Hart et al. 1994). Distal to the DH of Dbl, is ~100 amino acid

D.D. Billadeau (✉)
Department of Immunology and Division of Oncology Research,
College of Medicine, Mayo Clinic, Rochester, MN, 55905, USA
e-mail: Billadeau.Daniel@mayo.edu

K. van Golen (ed.), *The Rho GTPases in Cancer*,
DOI 10.1007/978-1-4419-1111-7_5, © Springer Science+Business Media, LLC 2010

domain showing structural similarity to pleckstrin homology (PH) domains. Subsequently, 69 DH-PH-containing Dbl family RhoGEFs have been identified in humans and the vast majority of those that have been analyzed are transforming when overexpressed as full-length or truncated constitutively active proteins, suggesting an oncogenic role for this family of proteins (Rossman et al. 2005). Significantly, point mutations within the DH domain that abolish GEF activity, abrogate the transforming potential of these proteins, highlighting the role for Rho GTPases in Dbl family-mediated oncogenic transformation.

The participation of Dbl family RhoGEFs in cancer has become an active area of investigation, as they represent the conduit for conveying perturbed intracellular signaling pathways into deregulated Rho GTPase activation during tumorigenesis. Although most Dbl family RhoGEFs have been isolated as transforming oncogenes through expression cloning of RNA and DNA from human tumors, most of these were due to cloning artifacts resulting in truncated proteins displaying constitutive GEF activity. For the most part, mutational activation of RhoGEFs has not been observed in human cancers, yet translocations of the Dbl GEF BCR with the Abl tyrosine kinase in chronic myelogenous leukemia, LARG with the Mixed Lineage Leukemia (MLL) gene in acute myelogenous leukemia, and in a case of renal cell cancer, an activating point mutation in the PH domain of Tiam1, resulting in enhanced transforming potential, have been identified (Rossman et al. 2005). Finally, in addition to the overexpression of GEFs identified in some human cancers, misexpression of Dbl GEFs could also play a role in cancer development or progression. Indeed, Vav1 is ectopically expressed in pancreatic adenocarcinoma (Fernandez-Zapico et al. 2005), neuroblastoma (Hornstein et al. 2003), and melanoma (Bartolome et al. 2006) and overexpressed in B-cell chronic lymphocytic leukemia (Prieto-Sanchez et al. 2006). Additionally, Vav3 becomes overexpressed in prostate cancer during the progression to androgen independence (Dong et al. 2006; Lyons et al. 2006; Lyons et al. 2007) and Vav2 has been found to be a critical regulator of growth factor-stimulated motility in human cancers (Bourguignon et al. 2001; Lai et al. 2007; Patel et al. 2007). Based on our understanding of the molecular mechanisms regulating Vav GEF activity, which in part requires the relief of an autoinhibitory intramolecular interaction by tyrosine phosphorylation (Fig. 5.1), overexpression of RTKs and their ligands in cancer, as well as the activation of Src kinases may facilitate Vav-mediated activation of Rho/Rac GTP-binding proteins and their effector pathways in these human cancers.

Structure, Function, and Regulation of Vav Family GEFs

Vav (now referred to as Vav1), the prototypical Vav family member was serendipitously isolated in a screen for oncogenes from esophageal tumor DNA, due to a cloning artifact that replaced the first 67 amino acids of its N-terminus with sequences from the selectable marker plasmid (Katzav et al. 1991; Katzav et al. 1989). Interestingly, full-length Vav1 is not nearly as transforming as the deleted

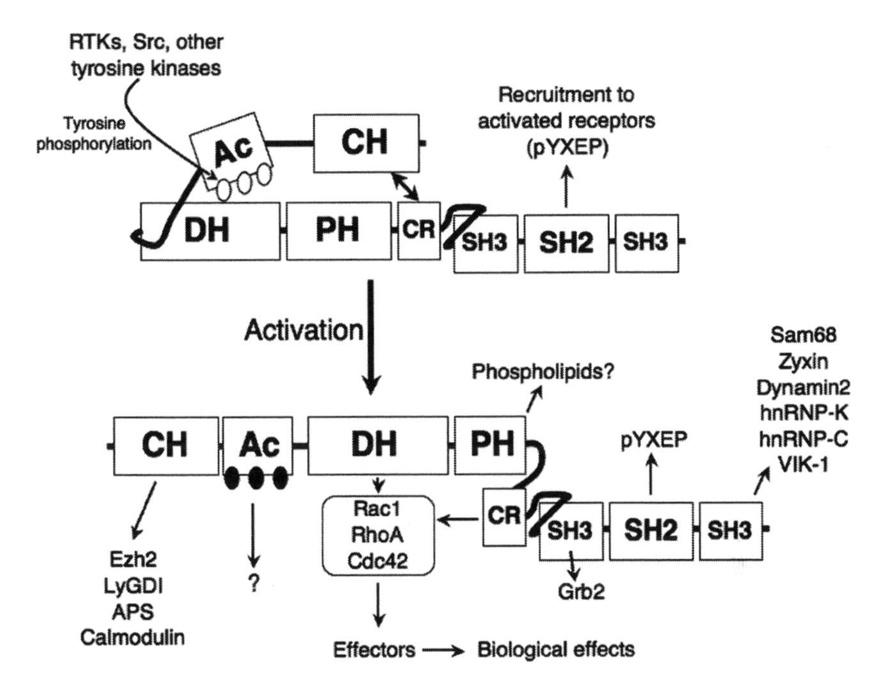

Fig. 5.1 Structure and activation of the Vav family of proto-oncogenes. Vav proteins are thought to exist in an autoinhibited conformation until activated by tyrosine phosphorylation. In the autoinhibited state, the calponin homology (CH) domain of Vav makes contact with the cysteine-rich (CR) domain. Additionally, a highly conserved tyrosine residue within the acidic (Ac) region, Y174 in Vav1 (white oval in Ac domain), interacts in the Dbl homology (DH) domain, thus preventing binding to Rho family GTPases. Upon recruitment to activated receptors via its src homology (SH) 2 domain, which interacts with phosphorylated tyrosines in the context pYXEP, Vav becomes phosphorylated at the three-tyrosine residues within its Ac domain (black ovals) by activated receptor tyrosine kinases (RTKs), Src family tyrosine kinases, or other tyrosine kinases. This results in its activation, thus releasing the Ac and CH domains from their autoinhibited conformation, allowing the DH domain, along with the CR and PH to coordinate activation of Rho GTPase family members such as Rac1, RhoA, and Cdc42. A proline-rich sequence following the CR interacts with the N-terminal SH3 domain providing a docking site for the adaptor protein Grb2. Other known protein interactions are shown and discussed in more detail in the text

oncogenic version, providing the first clue into the mechanism by which the Vav GEFs are regulated. Subsequent to the identification of wild-type Vav1, two additional Vav family members were isolated, Vav2 and Vav3, that were similar in domain architecture and oncogenic activation to Vav1 (Movilla et al. 1999; Schuebel et al. 1996). In contrast to Vav1, which is primarily expressed in bone marrow-derived cell lineages, Vav2 and Vav3 show a wider pattern of expression. Because of its initial characterization as a hematopoietic cell-specific signaling protein, Vav1 regulation and signaling has been extensively studied in immune cells and mice lacking Vav1 show dramatic defects in immune system development and function (Turner et al. 2002). Despite being more broadly expressed, targeted disruption of either Vav2 or Vav3 results in normal embryonic development, and only mild effects on immune cell functions (Doody et al. 2001; Tedford et al. 2001). The triple Vav knockout (Vav1–3) is also viable, displaying exaggerated defects in immune cells compared to the Vav1 knockout (Fujikawa et al. 2003), suggesting that Vav2 and Vav3 can compensate for the loss of Vav1 to some extent. Interestingly, more recent studies have identified that Vav2 or Vav3 deficiency result in tachycardia, hypertension, and in the case of Vav2-deficiency, defects in the heart arterial walls and kidney are observed, whereas Vav3-deficient mice show extensive cardiovascular remodeling and hyperactivity of the sympathetic nervous system (Sauzeau et al. 2007; Sauzeau et al. 2006). This indicates that Vav isoforms regulate signaling processes required for hematopoietic function, cardiovascular physiology, and the sympathetic nervous system and thus might be involved in associated human pathologies.

Vav proteins contain an N-terminal calponin homology (CH) domain followed by an acidic region (Ac), the DH-PH cassette characteristic of Dbl family proteins, a cysteine-rich domain (CR), a short proline-rich motif, and an src homology 3 (SH3)-SH2-SH3 module (Fig. 5.1). An overwhelming amount of evidence indicates that Vav proteins couple to multiple cell surface receptors and undergo receptor-induced tyrosine phosphorylation by RTKs, Src, and Syk family kinases. The ability of Vav proteins to become tyrosine phosphorylated requires an intact SH2 domain, which interacts with specific phospho-tyrosine residues in the context of the surrounding three to six amino acids located C-terminal to the phosphorylated tyrosine, thereby recruiting Vav to areas of tyrosine kinase activation. As a result, mutations in the SH2 domain of Vav proteins prevent interaction with their substrates and abrogate transformation of NIH3T3 cells by oncogenic Vav1 (Katzav 1993). The presence of multiple protein-protein interaction domains indicates that Vav proteins can link to multiple signaling cascades where they can not only participate in the activation of Rho/Rac family GTPases, but also coordinate other interactions that may be required for efficient transmission of signals from activated receptors.

The ~116 amino acid CH domain has been implicated in the auto-regulation of Vav proteins. Early studies of Vav3 found its CH domain interacted with its own CR domain, thereby stabilizing an inactive conformation (Movilla et al. 1999). Subsequent experimental evidence demonstrated that further deletion of the CH domain and an adjoining ~50 amino acid Ac region, which harbors 3 highly conserved tyrosine residues found in all three Vav isoforms, results in enhanced GEF activity and transforming potential (Movilla et al. 1999; Schuebel et al. 1996). In fact, Tyr^{174} in Vav1 was a known Src kinase substrate and work by Crespo and colleagues demonstrated that tyrosine phosphorylation of Vav1 by Lck could increase GEF activity on Rac1 (Crespo et al. 1997). Subsequently, Src-dependent phosphorylation of Vav2 and Vav3 was also found to increase their GEF activity and transforming potential (Movilla et al. 1999; Schuebel et al. 1998). Thus, all three Vav family members are regulated by tyrosine phosphorylation, implying a common mechanism of activation. However, removal of the CH domain itself is more transforming than mutation of the three conserved tyrosine residues (Zugaza et al. 2002), indicating that removal of the inhibitory tyrosines is not sufficient for the full activation of Vav GEF activity and that the CH domain also contributes to the inhibition of Vav.

Although still a matter of debate, Vav proteins have been shown to activate several Rho family GTPases in vitro including Rac1, Cdc42, and RhoA, as well as RhoB and RhoG. The transforming potential of Vav proteins is clearly linked to their ability to activate these small GTPases, through which they regulate cytoskeletal remodeling, cell migration, MAPK cascades, transcription factors, and ultimately cell survival and proliferation. Characteristic of other DH domains, nuclear magnetic resonance (NMR) structural studies of Vav1 indicated that the murine Vav1 DH domain is composed of 11 α-helices that form a flattened elongated bundle, with an exposed surface containing many residues required for GEF activity (Aghazadeh et al. 2000). Additionally, this study found that a peptide containing the Tyr^{174} phosphorylation motif formed an α-helix, which packed into the DH domain active site making interactions with the side-chains of residues known to be involved in the interaction of other DH domains and their small GTPase targets. Consistent with this notion, overexpression of the CH-Ac of Vav1 (1–185) inhibits activation of serum response factor by a truncated constitutively active Vav1 protein (Δ186), indicating that this inhibition can occur in trans. This highly conserved tyrosine residue likely functions similarly in Vav 2 and 3. Together, these data suggest that Vav binding and GEF activity are regulated by an autoinhibitory fold through the occlusion of important residues involved in GTPase binding and exchange activity.

Dbl family proteins may utilize other domains outside the DH to facilitate GTPase recognition and exchange. Based on structural and biochemical studies it has been suggested that the PH domains of Dbl GEFs may participate in the regulation of GEF activity by inhibiting the interaction of the DH domain with the GTPase (e.g., mSos), through a direct interaction with the GTPases (e.g., LARG and Dbs) or via membrane localization through an interaction with phosphoinositides within the inner leaflet of the plasma membrane (e.g., Dbl). Interestingly, the addition of PIP3 moderately enhanced Vav1 GEF activity in vitro, whereas, PIP2 inhibited exchange (Han et al. 1998), suggesting the possibility that binding to phospholipids may affect Vav GEF activity in vivo. Indeed, a mutation in the PH domain of oncogenic Vav1 or Vav2 that renders it unable to bind PIP3 inhibits transformation of 3T3 cells (Booden et al. 2002; Palmby et al. 2002). Moreover, oncogenic versions of Vav1 and Vav2 can synergize with constitutively active PI3-kinase in transformation (Booden et al. 2002; Palmby et al. 2002), and EGF-stimulated activation of Rac1 requires PI3-kinase and Vav2 (Marcoux et al. 2003). Interestingly, mutation of the PH domain of Vav3 does not affect its in vitro exchange activity or ability to induce cytoskeletal reorganization, suggesting that the Vav3 PH domain has little role in Vav3 regulation (Movilla et al. 1999). While these results might suggest that Vav proteins are recruited to the plasma membrane via their PH domain, a GFP-fusion of the PH domain of Vav1 does not localize to the cell membrane and artificially targeting of an oncogenic Vav1 containing a PH domain mutant to the plasma membrane does not restore its transforming potential (Palmby et al. 2002). Thus, at least in the case of Vav1, the PH domain may directly affect GEF activity. Indeed, the catalytic activity of DH-PH domains is nearly 100-fold greater than that of the isolated DH domain (Liu et al. 1998).

In addition to the DH-PH cassette, the CR domain of Vav proteins seems critical for the biological functions of Vav1–3. Mutations of residues within this region result in decreased transforming potential. As mentioned, the CH domain can interact with the CR domain and probably contributes to stabilization of the autoinhibitory α-helix within the DH domain (Fig. 5.1). However, the Vav family may also utilize the CR domain for target recognition since the CR of Vav1 and Vav3 were found to interact with RhoA and Rac1. Moreover, using fluorescence anisotropy and NMR-based chemical shift mapping, it was shown that the Vav1 CR domain preferentially interacts with Rac1-GDP making multiple contacts with Rac1 (Heo et al. 2005). Moreover, addition of the CR to the DH-PH increases not only the kinetics of nucleotide exchange, but also the number of interactions between the Vav2 DH domain and Rac1. Finally, mutagenesis of the CR region of Vav1 provided some insight into functional mechanism (Zugaza et al. 2002). Although some CR mutations inhibited transformation by oncogenic Vav1 and caused diminished GEF activity in vitro, others were not transforming even with retained GEF activity. Thus, this region of Vav harbors three distinct regulatory functions. First, it inhibits wild-type Vav function by stabilizing an autoinhibitory interaction with the CH domain. Second, it coordinates binding to the GTPase for efficient exchange through the DH domain. Finally, it is involved in a third function, unrelated to GEF activity, which is essential for transformation.

Linking Vav Proteins to Intracellular Signaling

In addition to activating Rho GTPases, Vav proteins have the capacity to couple to numerous signaling pathways via interactions with several intracellular signaling molecules. Aside from the domains directly involved in Rho GTPase activation (Ac-DH-PH-CR), other domains within Vav (CH domain, the SH2 domain, and the C-terminal SH3) are involved in Vav-mediated signaling and/or transformation of 3T3 fibroblasts.

As indicated above, the CH domain plays a key role in the regulation of Vav GEF activity. This domain, however, has also been shown to interact with several signaling molecules including LyGDI (Groysman et al. 2000), a Rho guanosine dissociation inhibitor, Ezh2 (Hobert et al. 1996a), a histone tri-methylase involved in the epigenetic regulation of genes, calmodulin, a calcium-binding protein, Socs1, a regulator of cytokine signaling, and APS, an adaptor molecule. In theory, LyGDI could bring unloaded Rho proteins to activated Vav1, leading to the local accumulation of Vav substrates while Socs1 could help recruit Vav proteins to cytokine receptors. Interestingly, despite the fact that Vav1 nuclear localization has been described (Houlard et al. 2002), cytosolic Vav1-Ezh2 complexes were shown to regulate growth factor-induced cytoskeletal reorganization in a methyltransferase-dependent manner (Su et al. 2005). Surprisingly, T cell receptor (TCR)-stimulated activation of Cdc42 was impaired in T cells lacking Ezh2. Thus, Vav1 may serve as a scaffold molecule to organize Ezh2 signaling complexes in the cytoplasm and nucleus. Calmodulin was shown to couple Vav1 to calcium signaling and NFAT activation in T cells, a process regulated by wild-type Vav1, but not a mutant lacking the CH domain (Zhou et al. 2007). Finally, the APS adaptor protein results in increased tyrosine phosphorylation of Vav3 and increases its transforming potential (Yabana et al. 2002). However, since Vav proteins lacking the CH domain are transforming, the significance of its interaction partners is unclear in the context of 3T3 transformation, but may be relevant to normal physiological signaling pathways regulated by Vav. However, the fact that only overexpression and ectopic expression of Vav proteins, but not truncated Vav proteins, have been found in human malignancies may suggest that interactions in the CH domain could indeed be important for the regulation of Vav GEF activity and Vav regulated signaling pathways in human cancers harboring wild-type Vav proteins.

The SH2 domain of Vav proteins interacts with specific phospho-tyrosine containing motifs. In this regard, Vav proteins favor pYXEP, a motif that is found in numerous signaling proteins including RTKs, cytosolic kinases, and phosphorylated adaptor proteins. It is interesting, that when compared to wild-type Vav1, neither onco-Vav1 nor the 3YF mutant increases cell proliferation in pancreatic cancer cells, yet it does require an intact SH2 domain. Indeed, Vav1 from pancreatic cancer cell lines is tyrosine phosphorylated basally, and this phosphorylation can be further induced by stimulation with soluble growth factors including EGF, insulin, and IGF-1. Thus, one should not overlook the possibility that besides regulating the GEF activity of Vav proteins, phosphorylated tyrosines within the Ac region may serve as docking sites for other SH2-containing proteins, creating a signaling node, leading to signal cascade propagation. Interestingly, several proteins, including the Src kinase Lck and Syk tyrosine kinase can interact with phosphorylated Tyr^{174}. It is of interest that, at least in T cells, some of the effects of Vav1 are GEF independent (Miletic et al. 2006), but do require the SH2 domain. Therefore, the SH2 domain of Vav1 may be required for its inducible tyrosine phosphorylation, and subsequent scaffolding/adaptor effects.

Two SH3 domains flank the Vav1 SH2 domain and are involved in Vav1 signaling. The N-terminal SH3 domain of Vav1 has been found to interact with a proline-rich motif localized proximal to this SH3 domain (Nishida et al. 2001). This interaction exposes a hydrophobic motif within the SH3 domain that is recognized and bound by the C-terminal SH3 domain of the adaptor protein Grb2 (Nishida et al. 2001; Ogura et al. 2002). The functional significance of this interaction in Vav1 signaling was recently shown to be important for the recruitment of Vav1 to DAP10, a component of the natural killer cell receptor NKG2D (Upshaw et al. 2006). Interestingly, in the context of onco-Vav1, neither mutation of the proline-rich motif nor mutation of the N-terminal SH3 domain (inhibiting Grb2 binding) alters its transforming potential (Zugaza et al. 2002), suggesting that the Vav1-Grb2 interaction is dispensable during transformation.

In contrast to the N-terminal SH3 domain, the C-terminal SH3 domain is required for onco-Vav1-mediated transformation (Groysman et al. 1998) and Vav1-mediated signaling in immune cells. In fact, inactivation of this SH3 domain, uncouples the ability of Vav1 to induce TCR-stimulated NFAT-mediated gene transcription. While this SH3 domain has been observed to interact with numerous proteins, it is unclear which target(s) is required for the normal versus transforming potential of Vav1. We have recently identified the large GTPase Dynamin 2 as Vav1-SH3 interacting protein involved in Vav1-dependent actin remodeling in T cells (Gomez et al. 2005). Dynamin 2 is involved in receptor-mediated endocytosis, vesicle trafficking, mitosis, and cell migration through its effects on the actin cytoskeleton. Interestingly, this interaction with Dynamin 2 is specific to Vav1, as neither Vav2 nor Vav3 can interact with Dynamin 2.

In addition to Dynamin 2, Vav1 also interacts with the actin regulatory scaffold protein Zyxin (Hobert et al. 1996b). Zyxin is involved in cell adhesion, cell migration, and integrin function. Interestingly, T cells lacking Vav1 demonstrate defects in cell adhesion and spreading on fibronectin, events that require functional integrins. Thus, the interaction of Vav1 with Zyxin could impact cancer cell motility through integrins. In addition to actin regulation, Zyxin has also been found to interact with LATS1, an evolutionarily conserved serine/threonine kinase involved in mitotic regulation (Hirota et al. 2000; Iida et al. 2004). In fact, Zyxin and LATS1 can be found on the mitotic apparatus and ectopic expression of a truncated LATS protein, which lacks the Zyxin binding site, inhibits localization of Zyxin to the mitotic apparatus, and increases the length of mitosis (Hirota et al. 2000). Thus Zyxin and LATS1 play a crucial role in controlling mitosis. Whether Vav proteins localize with Zyxin to the mitotic apparatus and regulate the length of mitosis is unknown.

Interestingly, in addition to the two proteins highlighted above, many of the proposed SH3-binding proteins are nuclear proteins, which include Ku-70, a subunit of the DNA-dependent protein kinase complex, the ribonucleoproteins hnRNP-K and hnRNP-C, and Kruppel-like zinc-finger protein VIK-1(Katzav 2007). Intriguingly, the C-terminal SH3 domain of Vav1 may be involved in the localization of Vav1 to the nucleus (Houlard et al. 2002). Although the majority of Vav1 is cytosolic, Vav1 contains two cryptic nuclear localization signal sequences within its PH domain, which are thought to be masked by the C-terminal SH3 domain, based on the observation that Vav1 mutants lacking the C-terminal SH3 domain exhibit nuclear accumulation. Indeed, Vav1 was found to accumulate in the nucleus of mast cells and regulate NFκB and NFAT-mediated gene transcription following stimulation through the IgE receptor (Houlard et al. 2002). More recently, Sam68, an RNA-binding protein, was found to interact with the Vav1 SH3 domain and be involved in the nuclear accumulation of Vav1 (Lazer et al. 2007). Interestingly, co-expression of Sam68 enhanced Vav1-mediated transformation suggesting that Sam68 may contribute to the tumorigenic phenotype induced by Vav1. Thus, while Vav1 is activated at the cell membrane in response to stimulation, this activation can lead to nuclear accumulation of Vav1 and affects on gene transcription, cell cycle progression, or other nuclear-regulated pathways.

The Role of Vav Proteins in Human Cancer

All three Vav family members can induce transformation of 3T3 cells when overexpressed as truncated constitutively active proteins, and both Vav2 and Vav3 also result in multi-nucleation, suggesting a role for these family members in cytokinesis

(Fujikawa et al. 2002; Movilla et al. 1999; Schuebel et al. 1996; Zeng et al. 2000). Oncogenic Vav1 can also cooperate with Ras in transformation in a myc-dependent manner (Katzav et al. 1995). Moreover, all three Vav proteins can be regulated by RTKs and Src kinases. In fact, Vav2 is involved in the activation of Rac1 in Src-transformed cells (Servitja et al. 2003), the activation of Cdc42 by EGF (Tu et al. 2003), and the upregulation of c-myc expression induced by PDGF stimulation (Chiariello et al. 2001). Finally, the Bcr-Abl fusion protein identified in chronic myelogenous leukemia activates Vav1 leading to Rac1-mediated cell motility (Daubon et al. 2007) and the nucleophosmin-anaplastic lymphoma kinase found in anaplastic large cell lymphomas regulates Rac1 activation through Vav3 (Colomba et al. 2007). Although truncated Vav mutants have not been observed in human cancer, the influence of deregulated RTKs and Src kinases as well as Ras gene mutations in human malignancy provide Vav proteins with the potential to participate in various aspects of tumor progression when simply expressed as wild-type proteins. In fact, as described below, Vav proteins have now been associated with cell migration, survival, and proliferation of human cancer cells, as well as regulation of angiogenesis.

The Vav1 gene maps to chromosome 19p12-p13.2, a region of karyotypic abnormalities in hematological malignancies and neuroblastomas. Because of this association, two groups investigated the expression of Vav1 in human neuroblastomas, despite its exclusive expression in bone marrow-derived cell lineages. Upon examination of neuroblastoma cell lines and primary patient samples, wild-type Vav1 expression was observed with a frequency of 75% (Betz et al. 2003; Hornstein et al. 2003). Ectopic expression of Vav1 was also found in pancreatic cancer cell lines and primary pancreatic adenocarcinomas (PDA) with a frequency of 53% (Fernandez-Zapico et al. 2005). It is noteworthy that Vav1-positive pancreatic tumors had a lower survival rate than Vav1-negative tumors. Despite the fact that greater than 90% of the PDAs harbor activating mutations in K-Ras, suppression of Vav1 by RNA interference inhibits cellular proliferation and survival in vitro and in vivo (Fernandez-Zapico et al. 2005), indicating that Vav1-positive tumors may become reliant on Vav1-dependent intracellular signaling pathways for cell survival and proliferation. Vav1 has also been shown to couple to numerous RTKs in PDA cell lines, regulating the activation of Rac1, NFκB, and the cyclin D1 promoter (Fernandez-Zapico et al. 2005). It is of significant interest that expression of wild-type Vav1 in a Vav1-negative pancreatic tumor cell line was as effective at inducing cell proliferation as were onco- or 3YF-Vav1 (Fernandez-Zapico et al. 2005), suggesting that the acquisition of wild-type Vav1 was sufficient to drive cellular proliferation. Vav1 is also expressed in a large number of melanoma cell lines and primary patient melanoma specimens. Interestingly, chemokine stimulation leads to Jak-mediated tyrosine phosphorylation of Vav1 and Vav2, activation of both Rac1 and RhoA, as well as membrane-type-1 matrix metalloproteinase (MMP)/MMP-2-dependent melanoma cell invasion, which seems to be Vav1/2 dependent (Bartolome et al. 2006). Finally, Vav1 is overexpressed in a subset of B cell chronic lymphocytic leukemia's harboring loss on 13q (Prieto-Sanchez et al. 2006).

As indicated earlier, Vav1 can activate several transcription factors (e.g., Elk-1, NFAT, NFκB), which in turn may regulate the expression of genes involved in cellular transformation or tumor progression. One such gene that was found to be upregulated in onco-Vav1 transformed 3T3 cells is osteopontin (Schapira et al. 2006), a marker of tumor invasion, progression, and metastasis. It is of interest that osteopontin expression was required for the invasion and anchorage-independent growth of the transformed 3T3 cells (Schapira et al. 2006). Osteopontin expression can be regulated by growth factors and therefore is overexpressed in multiple human tumors (Katzav 2007). Whether osteopontin expression is regulated by Vav proteins in response to extracellular stimuli in human cancers, remains to be determined.

To date, there have been no reports regarding the overexpression of Vav2 in human malignancies. However, recent studies in several human cancers have implicated Vav2 signaling in cellular invasion and angiogenesis. The prolactin receptor (PRL) contributes to the progression and motility of human breast cancer cells and was recently shown to activate Rac1 through Vav2 (Miller et al. 2005). In fact PRL stimulation results in the formation of a complex involving Vav2 and the serine kinase Nek3. Nek3 was found to be required for Vav2-mediated activation of Rac1, as well as survival and motility of breast cancer cells (Miller et al. 2007; Miller et al. 2005). In ovarian cancer cell lines, it was found that Vav2 associates with the hyaluronan receptor CD44v3 and stimulates the activation of Ras and Rac1 signaling through complex formation with the EGFR following CD44v3 binding to hyaluronan (Bourguignon et al. 2001). In addition, squamous cell carcinomas of the head and neck are highly invasive and show increased levels of Rac1-GTP (Lai et al. 2007; Patel et al. 2007). This Rac1 activation is correlated with constitutive tyrosine phosphorylation of Vav2, which is likely due to an EGFR-stimulated autocrine loop (Patel et al. 2007). Importantly, overexpression of Vav2 in the immortalized keratinocyte cell line HaCat increased the levels of active Rac1, Cdc42 and cellular invasion (Lai et al. 2007). Consistent with this, RNAi toward Vav2 results in diminished Rac1-GTP levels and invasion (Patel et al. 2007). Taken together, these data provide evidence that wild-type Vav2 can regulate signaling pathways in cancer leading to cell survival and invasion.

In addition to regulating cancer cell motility and survival, Vav2 has recently emerged as an important regulator of vascular endothelial growth factor (VEGF) receptor-induced cell motility and permeability. Angiogenesis occurs during tumor development in order to provide oxygen and nutrients to the tumor. In general, the formation of this tumor vasculature arises from signaling through VEGF receptor (VEGFR) on endothelium resulting in the migration and proliferation of the endothelial cells within and around the tumor. Recently, both the VEGFR and EphA receptor were found to activate Rac1 in a Vav2-dependent manner (Garrett et al. 2007; Hunter et al. 2006). Importantly, Vav2 was tyrosine phosphorylated in a Src-dependent manner in response to VEGFR stimulation and RNAi toward Vav2 impaired VEGFR-stimulated activation of Rac1 and endothelial cell migration (Garrett et al. 2007). Interestingly, EphA receptor stimulation results in the phosphorylation of both Vav2 and Vav3 and either RNAi depletion of these GEFs, or fibroblasts lacking Vav2/3 fail to activate Rac1, spread in response to ephrin-A1 stimulation, form lamellipodia, or induce an angiogenic response in vivo or in vitro

(Hunter et al. 2006). In addition to stimulating proliferation and migration, VEGF also leads to vascular permeability resulting in leaky blood vessels, a common phenomenon observed in tumor microvasculature. Recent work has now uncovered a novel-signaling pathway stimulated by VEGF leading to the internalization of the endothelial cadherin VE-cadherin. This pathway involves VEGF-stimulated activation of a Src-Vav2-Rac1-Pak1 axis, resulting in the phosphorylation of serine 665 on the intracellular tail of VE-cadherin by Pak1, recruitment of β-arrestin2 and ultimately internalization (Gavard et al. 2006). It is of interest that S665 is highly conserved throughout evolution, but a similar residue/motif is not present in either E- or N-cadherin. Taken together, these data highlight a novel role for Vav2 in angiogenesis and underscore the possibility for small molecule inhibitors toward Vav2 or Rac1 in the treatment of cancer cell proliferation, survival, and invasion, as well as in anti-angiogenic therapies.

Vav3 has recently emerged as an important Vav GEF in prostate cancer tumorigenesis. It was initially observed that Vav3 overexpression occurs in many prostate cancer cell lines, 32% of primary prostate cancer specimens, and in a genetically engineered mouse model of human prostate cancer (Banach-Petrosky et al. 2007; Dong et al. 2006; Lyons et al. 2006). In the mouse model, it was found that administration of lower physiological concentrations of androgen promoted prostate cancer progression and the tumors that developed showed dramatically increased Vav3 expression, suggesting the possibility that Vav3 may be involved in hormone refractory prostate cancer (Banach-Petrosky et al. 2007). Indeed, Vav3 overexpression was found to increase androgen receptor (AR) nuclear localization and transcriptional activity even in the absence of exogenous androgen and to promote the growth of the androgen-dependent cell line LNCaP (Dong et al. 2006; Lyons et al. 2006; Lyons et al. 2007). Additionally, RNAi toward Vav3 or Rac1 inhibited the growth of both androgen-dependent and androgen-independent prostate cancer cell lines in vitro and in vivo. This Vav3-mediated activation of AR was found to be Rac1, PI3K/Akt, and MAPK/ERK dependent (Dong et al. 2006; Lyons et al. 2006; Lyons et al. 2007). Although it is unclear how Vav3 GEF activity is regulated in prostate cancer cells, it might occur through increased growth factor receptor signaling. Also, the mechanism by which Vav3 mRNA levels increase during the progression to androgen-independence is unclear. In all, these data identify Vav3 as an important modulator of ligand-independent AR transcriptional activity in prostate cancer progression.

Conclusion

Vav proteins have emerged as important signaling mediators in multiple human malignancies and during the process of angiogenesis. Despite the fact that mutated or truncated oncogenic versions of Vav family members have not been isolated from human cancers to date, their regulation by tyrosine phosphorylation may circumscribe the necessity for this type of alteration to hyperactivate the protein, since many human malignancies harbor constitutive RTK and/or Src tyrosine kinase activity. In addition, due to the complex structure of Vav proteins, it would not be surprising that, at least in the case of Vav1, ectopic expression would result in the activation of unique signaling pathways that drive cancer cell proliferation, survival, and migration. Clearly, inhibition of Rac activity may represent an appropriate means of Vav signaling intervention in both cancer and angiogenesis. In fact, a small molecule that displaces nucleotide from Rac1 has recently been identified (Shutes et al. 2007), but whether it can inhibit Vav-mediated signaling in human cancer cells, or prevent the migration and VE-cadherin internalization by VEGF in endothelial cells, remains to be determined. Another possible route for therapeutic intervention would be to prevent binding of Vav proteins to their target GTPases. Although a compound has been identified that impairs the activation of Rac1 by certain Rac GEFs (e.g., Tiam and Trio-N) (Gao et al. 2004), and consequently, transformation mediated by oncogenic Tiam1, this compound does not impair the activation of Rac1 by Vav GEFs, due to a different mechanism of recognition of Rac1 by the Vav DH domain. Regardless, a more thorough understanding of the signaling pathways that Vav proteins regulate (GEF dependent and GEF independent) during cancer progression will provide the necessary insight to develop small molecule inhibitors to antagonize Vav signaling in cancer.

References

Aghazadeh, B., Lowry, W. E., Huang, X. Y., and Rosen, M. K. (2000). Structural basis for relief of autoinhibition of the Dbl homology domain of proto-oncogene Vav by tyrosine phosphorylation. *Cell* **102**, 625–633.

Banach-Petrosky, W., Jessen, W. J., Ouyang, X., Gao, H., Rao, J., Quinn, J., Aronow, B. J., and Abate-Shen, C. (2007). Prolonged exposure to reduced levels of androgen accelerates prostate cancer progression in Nkx3.1; Pten mutant mice. *Cancer research* **67**, 9089–9096.

Bartolome, R. A., Molina-Ortiz, I., Samaniego, R., Sanchez-Mateos, P., Bustelo, X. R., and Teixido, J. (2006). Activation of Vav/Rho GTPase signaling by CXCL12 controls membrane-type matrix metalloproteinase-dependent melanoma cell invasion. *Cancer research* **66**, 248–258.

Betz, R., Sandhoff, K., Fischer, K. D., and van Echten-Deckert, G. (2003). Detection and identification of Vav1 protein in primary cultured murine cerebellar neurons and in neuroblastoma cells (SH-SY5Y and Neuro-2a). *Neuroscience letters* **339**, 37–40.

Booden, M. A., Campbell, S. L., and Der, C. J. (2002). Critical but distinct roles for the pleckstrin homology and cysteine-rich domains as positive modulators of Vav2 signaling and transformation. *Molecular and cellular biology* **22**, 2487–2497.

Bourguignon, L. Y., Zhu, H., Zhou, B., Diedrich, F., Singleton, P. A., and Hung, M. C. (2001). Hyaluronan promotes CD44v3-Vav2 interaction with Grb2-p185(HER2) and induces Rac1 and Ras signaling during ovarian tumor cell migration and growth. *The Journal of biological chemistry* **276**, 48679–48692.

Bustelo, X. R., Sauzeau, V., and Berenjeno, I. M. (2007). GTP-binding proteins of the Rho/Rac family: regulation, effectors and functions in vivo. *Bioessays* **29**, 356–370.

Chiariello, M., Marinissen, M. J., and Gutkind, J. S. (2001). Regulation of c-myc expression by PDGF through Rho GTPases. *Nature cell biology* **3**, 580–586.

Colomba, A., Courilleau, D., Ramel, D., Billadeau, D. D., Espinos, E., Delsol, G., Payrastre, B., and Gaits-Iacovoni, F. (2007). Activation of Rac1 and the exchange factor Vav3 are involved in NPM-ALK signaling in anaplastic large cell lymphomas. *Oncogene*.

Crespo, P., Schuebel, K. E., Ostrom, A. A., Gutkind, J. S., and Bustelo, X. R. (1997). Phosphotyrosine-dependent activation of Rac-1 GDP/GTP exchange by the vav proto-oncogene product. *Nature* **385**, 169–172.

Daubon, T., Chasseriau, J., Ali, A. E., Rivet, J., Kitzis, A., Constantin, B., and Bourmeyster, N. (2007). Differential motility of p190(bcr-abl)- and p210(bcr-abl)-expressing cells: respective roles of Vav and Bcr-Abl GEFs. *Oncogene*.

Dong, Z., Liu, Y., Lu, S., Wang, A., Lee, K., Wang, L. H., Revelo, M., and Lu, S. (2006). Vav3 oncogene is overexpressed and regulates cell growth and androgen receptor activity in human prostate cancer. *Molecular endocrinology (Baltimore, Md* **20**, 2315–2325.

Doody, G. M., Bell, S. E., Vigorito, E., Clayton, E., McAdam, S., Tooze, R., Fernandez, C., Lee, I. J., and Turner, M. (2001). Signal transduction through Vav-2 participates in humoral immune responses and B cell maturation. *Nature immunology* **2**, 542–547.

Fernandez-Zapico, M. E., Gonzalez-Paz, N. C., Weiss, E., Savoy, D. N., Molina, J. R., Fonseca, R., Smyrk, T. C., Chari, S. T., Urrutia, R., and Billadeau, D. D. (2005). Ectopic expression of VAV1 reveals an unexpected role in pancreatic cancer tumorigenesis. *Cancer cell* **7**, 39–49.

Fujikawa, K., Inoue, Y., Sakai, M., Koyama, Y., Nishi, S., Funada, R., Alt, F. W., and Swat, W. (2002). Vav3 is regulated during the cell cycle and effects cell division. *Proceedings of the National Academy of Sciences of the United States of America* **99**, 4313–4318.

Fujikawa, K., Miletic, A. V., Alt, F. W., Faccio, R., Brown, T., Hoog, J., Fredericks, J., Nishi, S., Mildiner, S., Moores, S. L., Brugge, J., Rosen, F. S., and Swat, W. (2003). Vav1/2/3-null mice define an essential role for Vav family proteins in lymphocyte development and activation but a differential requirement in MAPK signaling in T and B cells. *The Journal of experimental medicine* **198**, 1595–1608.

Gao, Y., Dickerson, J. B., Guo, F., Zheng, J., and Zheng, Y. (2004). Rational design and characterization of a Rac GTPase-specific small molecule inhibitor. *Proceedings of the National Academy of Sciences of the United States of America* **101**, 7618–7623.

Garrett, T. A., Van Buul, J. D., and Burridge, K. (2007). VEGF-induced Rac1 activation in endothelial cells is regulated by the guanine nucleotide exchange factor Vav2. *Experimental cell research* **313**, 3285–3297.

Gavard, J., and Gutkind, J. S. (2006). VEGF controls endothelial-cell permeability by promoting the beta-arrestin-dependent endocytosis of VE-cadherin. *Nature cell biology* **8**, 1223–1234.

Gomez, T. S., Hamann, M. J., McCarney, S., Savoy, D. N., Lubking, C. M., Heldebrant, M. P., Labno, C. M., McKean, D. J., McNiven, M. A., Burkhardt, J. K., and Billadeau, D. D. (2005). Dynamin 2 regulates T cell activation by controlling actin polymerization at the immunological synapse. *Nature immunology* **6**, 261–270.

Groysman, M., Nagano, M., Shaanan, B., and Katzav, S. (1998). Mutagenic analysis of Vav reveals that an intact SH3 domain is required for transformation. *Oncogene* **17**, 1597–1606.

Groysman, M., Russek, C. S., and Katzav, S. (2000). Vav, a GDP/GTP nucleotide exchange factor, interacts with GDIs, proteins that inhibit GDP/GTP dissociation. *FEBS letters* **467**, 75–80.

Han, J., Luby-Phelps, K., Das, B., Shu, X., Xia, Y., Mosteller, R. D., Krishna, U. M., Falck, J. R., White, M. A., and Broek, D. (1998). Role of substrates and products of PI 3-kinase in regulating activation of Rac-related guanosine triphosphatases by Vav. *Science (New York, N.Y* **279**, 558–560.

Hart, M. J., Eva, A., Zangrilli, D., Aaronson, S. A., Evans, T., Cerione, R. A., and Zheng, Y. (1994). Cellular transformation and guanine nucleotide exchange activity are catalyzed by a common domain on the dbl oncogene product. *The Journal of biological chemistry* **269**, 62–65.

Heo, J., Thapar, R., and Campbell, S. L. (2005). Recognition and activation of Rho GTPases by Vav1 and Vav2 guanine nucleotide exchange factors. *Biochemistry* **44**, 6573–6585.

Hirota, T., Morisaki, T., Nishiyama, Y., Marumoto, T., Tada, K., Hara, T., Masuko, N., Inagaki, M., Hatakeyama, K., and Saya, H. (2000). Zyxin, a regulator of actin filament assembly, targets the mitotic apparatus by interacting with h-warts/LATS1 tumor suppressor. *The Journal of cell biology* **149**, 1073–1086.

Hobert, O., Jallal, B., and Ullrich, A. (1996a). Interaction of Vav with ENX-1, a putative transcriptional regulator of homeobox gene expression. *Molecular and cellular biology* **16**, 3066–3073.

Hobert, O., Schilling, J. W., Beckerle, M. C., Ullrich, A., and Jallal, B. (1996b). SH3 domain-dependent interaction of the proto-oncogene product Vav with the focal contact protein zyxin. *Oncogene* **12**, 1577–1581.

Hornstein, I., Pikarsky, E., Groysman, M., Amir, G., Peylan-Ramu, N., and Katzav, S. (2003). The haematopoietic specific signal transducer Vav1 is expressed in a subset of human neuroblastomas. *The Journal of pathology* **199**, 526–533.

Houlard, M., Arudchandran, R., Regnier-Ricard, F., Germani, A., Gisselbrecht, S., Blank, U., Rivera, J., and Varin-Blank, N. (2002). Vav1 is a component of transcriptionally active complexes. *The Journal of experimental medicine* **195**, 1115–1127.

Hunter, S. G., Zhuang, G., Brantley-Sieders, D., Swat, W., Cowan, C. W., and Chen, J. (2006). Essential role of Vav family guanine nucleotide exchange factors in EphA receptor-mediated angiogenesis. *Molecular and cellular biology* **26**, 4830–4842.

Iida, S., Hirota, T., Morisaki, T., Marumoto, T., Hara, T., Kuninaka, S., Honda, S., Kosai, K., Kawasuji, M., Pallas, D. C., and Saya, H. (2004). Tumor suppressor WARTS ensures genomic integrity by regulating both mitotic progression and G1 tetraploidy checkpoint function. *Oncogene* **23**, 5266–5274.

Katzav, S. (1993). Single point mutations in the SH2 domain impair the transforming potential of vav and fail to activate proto-vav. *Oncogene* **8**, 1757–1763.

Katzav, S. (2007). Flesh and blood: the story of Vav1, a gene that signals in hematopoietic cells but can be transforming in human malignancies. *Cancer letters* **255**, 241–254.

Katzav, S., Cleveland, J. L., Heslop, H. E., and Pulido, D. (1991). Loss of the amino-terminal helix-loop-helix domain of the vav proto-oncogene activates its transforming potential. *Molecular and cellular biology* **11**, 1912–1920.

Katzav, S., Martin-Zanca, D., and Barbacid, M. (1989). vav, a novel human oncogene derived from a locus ubiquitously expressed in hematopoietic cells. *The EMBO journal* **8**, 2283–2290.

Katzav, S., Packham, G., Sutherland, M., Aroca, P., Santos, E., and Cleveland, J. L. (1995). Vav and Ras induce fibroblast transformation by overlapping signaling pathways which require c-Myc function. *Oncogene* **11**, 1079–1088.

Lai, S. Y., Ziober, A. F., Lee, M. N., Cohen, N. A., Falls, E. M., and Ziober, B. L. (2007). Activated Vav2 modulates cellular invasion through Rac1 and Cdc42 in oral squamous cell carcinoma. *Oral Oncol.*

Lazer, G., Pe'er, L., Schapira, V., Richard, S., and Katzav, S. (2007). The association of Sam68 with Vav1 contributes to tumorigenesis. *Cellular signalling* **19**, 2479–2486.

Liu, X., Wang, H., Eberstadt, M., Schnuchel, A., Olejniczak, E. T., Meadows, R. P., Schkeryantz, J. M., Janowick, D. A., Harlan, J. E., Harris, E. A., Staunton, D. E., and Fesik, S. W. (1998). NMR structure and mutagenesis of the N-terminal Dbl homology domain of the nucleotide exchange factor Trio. *Cell* **95**, 269–277.

Lyons, L. S., and Burnstein, K. L. (2006). Vav3, a Rho GTPase guanine nucleotide exchange factor, increases during progression to androgen independence in prostate cancer cells and potentiates androgen receptor transcriptional activity. *Molecular endocrinology (Baltimore, Md* **20**, 1061–1072.

Lyons, L. S., Rao, S., Balkan, W., Faysal, J., Maiorino, C. A., and Burnstein, K. L. (2007). Ligand-independent Activation of Androgen Receptors by Rho GTPase Signaling in Prostate Cancer. *Molecular endocrinology (Baltimore, Md.*

Marcoux, N., and Vuori, K. (2003). EGF receptor mediates adhesion-dependent activation of the Rac GTPase: a role for phosphatidylinositol 3-kinase and Vav2. *Oncogene* **22**, 6100–6106.

Miletic, A. V., Sakata-Sogawa, K., Hiroshima, M., Hamann, M. J., Gomez, T. S., Ota, N., Kloeppel, T., Kanagawa, O., Tokunaga, M., Billadeau, D. D., and Swat, W. (2006). Vav1 acidic region tyrosine 174 is required for the formation of T cell receptor-induced microclusters and is essential in T cell development and activation. *The Journal of biological chemistry* **281**, 38257–38265.

Miller, S. L., Antico, G., Raghunath, P. N., Tomaszewski, J. E., and Clevenger, C. V. (2007). Nek3 kinase regulates prolactin-mediated cytoskeletal reorganization and motility of breast cancer cells. *Oncogene* **26**, 4668–4678.

Miller, S. L., DeMaria, J. E., Freier, D. O., Riegel, A. M., and Clevenger, C. V. (2005). Novel association of Vav2 and Nek3 modulates signaling through the human prolactin receptor. *Molecular endocrinology (Baltimore, Md* **19**, 939–949.

Movilla, N., and Bustelo, X. R. (1999). Biological and regulatory properties of Vav-3, a new member of the Vav family of oncoproteins. *Molecular and cellular biology* **19**, 7870–7885.

Nishida, M., Nagata, K., Hachimori, Y., Horiuchi, M., Ogura, K., Mandiyan, V., Schlessinger, J., and Inagaki, F. (2001). Novel recognition mode between Vav and Grb2 SH3 domains. *The EMBO journal* **20**, 2995–3007.

Ogura, K., Nagata, K., Horiuchi, M., Ebisui, E., Hasuda, T., Yuzawa, S., Nishida, M., Hatanaka, H., and Inagaki, F. (2002). Solution structure of N-terminal SH3 domain of Vav and the recognition site for Grb2 C-terminal SH3 domain. *Journal of biomolecular NMR* **22**, 37–46.

Palmby, T. R., Abe, K., and Der, C. J. (2002). Critical role of the pleckstrin homology and cysteine-rich domains in Vav signaling and transforming activity. *The Journal of biological chemistry* **277**, 39350–39359.

Patel, V., Rosenfeldt, H. M., Lyons, R., Servitja, J. M., Bustelo, X. R., Siroff, M., and Gutkind, J. S. (2007). Persistent activation of Rac1 in squamous carcinomas of the head and neck: evidence for an EGFR/Vav2 signaling axis involved in cell invasion. *Carcinogenesis* **28**, 1145–1152.

Prieto-Sanchez, R. M., Hernandez, J. A., Garcia, J. L., Gutierrez, N. C., San Miguel, J., Bustelo, X. R., and Hernandez, J. M. (2006). Overexpression of the VAV proto-oncogene product is associated with B-cell chronic lymphocytic leukaemia displaying loss on 13q. *British journal of haematology* **133**, 642–645.

Rossman, K. L., Der, C. J., and Sondek, J. (2005). GEF means go: turning on RHO GTPases with guanine nucleotide-exchange factors. *Nature reviews* **6**, 167–180.

Sauzeau, V., Jerkic, M., Lopez-Novoa, J. M., and Bustelo, X. R. (2007). Loss of Vav2 proto-oncogene causes tachycardia and cardiovascular disease in mice. *Molecular biology of the cell* **18**, 943–952.

Sauzeau, V., Sevilla, M. A., Rivas-Elena, J. V., de Alava, E., Montero, M. J., Lopez-Novoa, J. M., and Bustelo, X. R. (2006). Vav3 proto-oncogene deficiency leads to sympathetic hyperactivity and cardiovascular dysfunction. *Nature medicine* **12**, 841–845.

Schapira, V., Lazer, G., and Katzav, S. (2006). Osteopontin is an oncogenic Vav1- but not wild-type Vav1-responsive gene: implications for fibroblast transformation. *Cancer research* **66**, 6183–6191.

Schuebel, K. E., Bustelo, X. R., Nielsen, D. A., Song, B. J., Barbacid, M., Goldman, D., and Lee, I. J. (1996). Isolation and characterization of murine vav2, a member of the vav family of proto-oncogenes. *Oncogene* **13**, 363–371.

Schuebel, K. E., Movilla, N., Rosa, J. L., and Bustelo, X. R. (1998). Phosphorylation-dependent and constitutive activation of Rho proteins by wild-type and oncogenic Vav-2. *The EMBO journal* **17**, 6608–6621.

Servitja, J. M., Marinissen, M. J., Sodhi, A., Bustelo, X. R., and Gutkind, J. S. (2003). Rac1 function is required for Src-induced transformation. Evidence of a role for Tiam1 and Vav2 in Rac activation by Src. *The Journal of biological chemistry* **278**, 34339–34346.

Shutes, A., Onesto, C., Picard, V., Leblond, B., Schweighoffer, F., and Der, C. J. (2007). Specificity and mechanism of action of EHT 1864, a novel small molecule inhibitor of Rac family small GTPases. *The Journal of biological chemistry* **282**, 35666–35678.

Su, I. H., Dobenecker, M. W., Dickinson, E., Oser, M., Basavaraj, A., Marqueron, R., Viale, A., Reinberg, D., Wulfing, C., and Tarakhovsky, A. (2005). Polycomb group protein ezh2 controls actin polymerization and cell signaling. *Cell* **121**, 425–436.

Tedford, K., Nitschke, L., Girkontaite, I., Charlesworth, A., Chan, G., Sakk, V., Barbacid, M., and Fischer, K. D. (2001). Compensation between Vav-1 and Vav-2 in B cell development and antigen receptor signaling. *Nature immunology* **2**, 548–555.

Tu, S., Wu, W. J., Wang, J., and Cerione, R. A. (2003). Epidermal growth factor-dependent regulation of Cdc42 is mediated by the Src tyrosine kinase. *The Journal of biological chemistry* **278**, 49293–49300.

Turner, M., and Billadeau, D. D. (2002). VAV proteins as signal integrators for multi-subunit immune-recognition receptors. *Nat Rev Immunol* **2**, 476–486.

Upshaw, J. L., Arneson, L. N., Schoon, R. A., Dick, C. J., Billadeau, D. D., and Leibson, P. J. (2006). NKG2D-mediated signaling requires a DAP10-bound Grb2-Vav1 intermediate and phosphatidylinositol-3-kinase in human natural killer cells. *Nature immunology* **7**, 524–532.

Yabana, N., and Shibuya, M. (2002). Adaptor protein APS binds the NH2-terminal autoinhibitory domain of guanine nucleotide exchange factor Vav3 and augments its activity. *Oncogene* **21**, 7720–7729.

Zeng, L., Sachdev, P., Yan, L., Chan, J. L., Trenkle, T., McClelland, M., Welsh, J., and Wang, L. H. (2000). Vav3 mediates receptor protein tyrosine kinase signaling, regulates GTPase activity, modulates cell morphology, and induces cell transformation. *Molecular and cellular biology* **20**, 9212–9224.

Zhou, Z., Yin, J., Dou, Z., Tang, J., Zhang, C., and Cao, Y. (2007). The calponin homology domain of Vav1 associates with calmodulin and is prerequisite to T cell antigen receptor-induced calcium release in Jurkat T lymphocytes. *The Journal of biological chemistry* **282**, 23737–23744.

Zugaza, J. L., Lopez-Lago, M. A., Caloca, M. J., Dosil, M., Movilla, N., and Bustelo, X. R. (2002). Structural determinants for the biological activity of Vav proteins. *The Journal of biological chemistry* **277**, 45377–45392.

Chapter 6
Rho GTPase-Activating Proteins in Cancer

Matthew W. Grogg and Yi Zheng

Introduction

Rho GTPases are intracellular signaling proteins involved in the regulation of multiple cell functions including cell growth, cell-cycle progression, cytoskeletal remodeling, gene expression, apoptosis, and membrane trafficking (Van Aelst and D'Souza-Schorey 1997; Hall 1998; Bishop and Hall 2000). Cycling between the active GTP-bound and the inactive GDP-bound states, they operate as molecular switches functioning in diverse signal transduction pathways. Rho GTPases in the GTP-bound active state can interact with a number of effectors to mediate signal flow from adhesion receptors, cytokine receptors, growth factor receptors, and G-protein coupled receptors. When the bound GTP is hydrolyzed to GDP, the Rho GTPases return to the inactive basal state. The regulation of the GTP-binding/GTP-hydrolysis cycle is controlled by three families of proteins: GEFs (guanine nucleotide-exchange factors), GAPs (GTPase-activating proteins), and GDIs (guanine nucleotide-dissociation inhibitors). GEFs, which promote the release of bound GDP and facilitate GTP binding, lead to activation of Rho GTPases (Zheng 2001). GAPs recognize the GTP-bound conformation of Rho GTPases and cause an increased intrinsic GTPase activity, which results in an inactive GTPase (Lamarche and Hall 1994; Tcherkezian and Lamarche-Vane 2007). GDIs function to sequester the GDP-bound form of Rho GTPases and may also regulate their intracellular localization (Olofsson 1999). An emerging theme from recent studies of Rho GTPase regulation is that there is a balancing act between the activation and the deactivation signals, and the cycling of Rho proteins between the GTP- and GDP-bound states may be required for effective signal flow through Rho GTPases to elicit downstream biological functions which may involve the concerted action of all classes of the regulatory proteins (Symons and Settleman 2000).

M.W. Grogg and Y. Zheng (✉)
Division of Experimental Hematology and Cancer Biology,
Children's Research Foundation, Cincinnati Children's Hospital Medical Center,
University of Cincinnati, 3333 Brunet Avenue, Cincinnati, OH, 45229, USA
e-mail: yi.zheng@cchmc.org

K. van Golen (ed.), *The Rho GTPases in Cancer*,
DOI 10.1007/978-1-4419-1111-7_6, © Springer Science+Business Media, LLC 2010

It has been almost two decades since the discovery of the first RhoGAP activity from cell lysates (Garrett et al. 1989). More than 70 RhoGAPs have since been identified in eukaryotes, and over 70 proteins in the human genome are known to contain a RhoGAP domain (Kandpal 2006). It is interesting to note that RhoGAPs outnumber their Rho GTPase substrates by >threefold, inevitably leading to the question of why there is a need for such an abundance of RhoGAPs. Studies over the last decade have provided several meaningful, if not concrete, explanations for this question including tissue-specific expression, divergence in the availability of Rho substrates, and multiple modes of regulation of RhoGAPs. Here we focus on our current understanding of the role of RhoGAPs in cancer. For a thorough review and discussion of the regulation and function of the RhoGAP family, several excellent reviews are available (Moon and Zheng 2003; Tcherkezian and Lamarche-Vane 2007).

Biochemical Function

RhoGAP family members contain a conserved RhoGAP domain that shares ~200 amino acid sequences with >20% sequence identity with other family members (Ahmadian et al. 1996; Moon and Zheng 2003). The RhoGAP domain is distinct from the other GAP domains responsible for inactivating GTPases of other subclasses of the Ras superfamily (i.e., Ras, Ran, or ARF). The RhoGAP domain binds to the GTP-bound Rho proteins and stimulates their intrinsic GTPase activity. Examination of the crystal structures of several RhoGAPs, including a few RhoGAP domains in complex with their Rho GTPase substrates, revealed the topology and critical residues in the active site that participate in catalysis (Rittinger et al. 1997). While the sequences of RhoGAP domains differ from those of other classes of GAPs, such as Ras GTPase-activating proteins (RasGAPs), the tertiary folding pattern as well as the basic GTPase-activating mechanism of the RhoGAP domain appears to be similar to that of RasGAP (Bax 1998; Rittinger et al. 1998). The RhoGAP domain consists of nine helices and a highly conserved arginine residue located in a loop structure (Gamblin and Smerdon 1998). Conformational changes during the GAP reaction place the catalytic arginine residue of RhoGAP into the active site of Rho GTPase and stabilize charges developed during the formation of the transitional state. Confirmation of the significance of a catalytic arginine of RhoGAP in accelerating Rho GTP hydrolysis has been shown through mutational approaches (Li, Zhang and Zheng 1997; Nassar et al. 1998). An issue requiring continuing attention by structural studies and lending to the complexity of RhoGAP specificity is how substrate specificity of RhoGAPs is achieved in RhoGAP-Rho GTPase pairwise interactions, because mutagenesis studies have also shown that residues outside the switch regions and the P-loop of Rho GTPases are clearly involved in directing RhoGAP specificity (Li, et al. 1997; Longenecker et al. 2000).

In addition to the RhoGAP domain, RhoGAPs also contain functional domains including PDZ, SH2, SH3, and SEC14 that participate in protein-protein interactions, membrane targeting, and cellular localization (Fig. 1) (Kandpal 2006). In fact, RhoGAPs

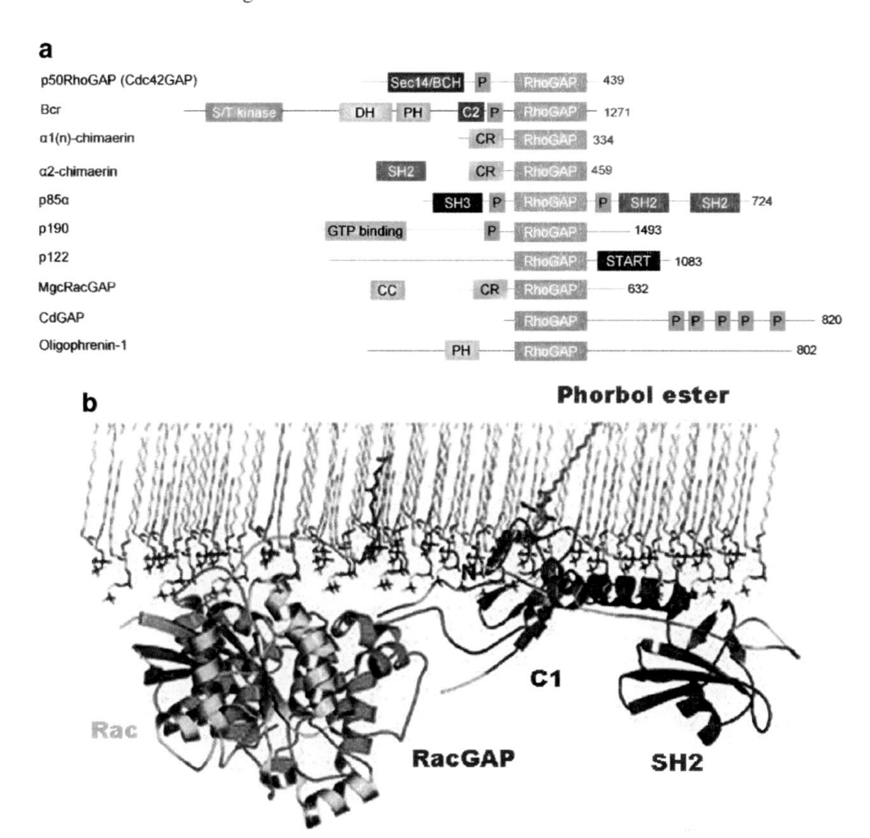

Fig. 6.1 (a) Multifunctional domain features of RhoGAP family members based on linear amino acid sequences. Abbreviations: *C2*, calcium-dependent lipid binding; *CC*, coiled coil; *CR*, cysteine rich, zinc butterfly motif, binds diacylglycerol and phorbol ester; *DH*, Dbl homology; *PH*, pleckstrin homology; *P*, praline-rich SH3 binding motif; *RhoGAP*, Rho GTPase-activating protein motif; *Sec14/BCH*, Sec14 homology/BNIP2 and Cdc42 GAP homology; *SH2*, Src homology 2; *SH3*, Src homology 3; *S/T kinase*, serine and threonine kinase; *START*, Star-related lipid transport domain. (b) Structural model for membrane-docked, active β2-Chimaerin, and its substrate, Rac1. The full-length molecule structure depicts complex interdomain and membrane interactions that are subjected to signaling modulation in regulating the RhoGAP activity. Domains colored red (SH2), blue (C1), green (RacGAP), and grey (linkers). The geranylgeranyl group is taken from the structure of Rac (Yellow) bound to its guanine nucleotide dissociation inhibitor. Adapted from ref Canagarajah et al. (2004)

may contain anywhere from one to as many as nine domains (Kandpal 2006). The biochemical activity of RhoGAPs relating to Rho GTPases varies widely in vitro, and the specificity of RhoGAPs differs in both expression and function. While many RhoGAPs are ubiquitous, several RhoGAPs show differential tissue expression and may have tissue-specific functions. This is the case with p73RhoGAP which is restricted to vascular endothelium cells and functions to regulate Rho GTPases during angiogenesis (Su et al. 2004). Grit (GTPase regulator interacting with TrkA – also known as p200RhoGAP) is another example showing a brain-specific expression pattern that is involved in neuritogenesis (Nakamura et al. 2002; Shang, Moon and Zheng 2007). Some RhoGAPs preferentially recognize a single Rho GTPase substrate, whereas others display a broader range of specificity capable of interacting with multiple Rho substrates including three commonly tested Rho substrates - RhoA, Rac1, and Cdc42. In cells, however, the specificity of the RhoGAP domain can be further enhanced by pathway specific regulation and compartmentalized distribution. Microinjection of the RhoGAP domain of p122RhoGAP or Graf, for example, blocked the lysophosphatidic acid-induced actin stress fiber formation in fibroblasts that is mediated by RhoA in spite of their indiscriminate activity towards Rac1 or Cdc42 *in vitro* (Sekimata et al. 1999; Taylor et al. 1999), implicating another layer of regulation of the RhoGAP specificity in cells that can affect substrate selection. Moreover, given the multidomain nature of many RhoGAPs (Fig. 1), various pathways might converge on them by interaction with the regulatory motifs thereby contributing to the tight regulation of the GAP activity.

Cellular Functions

While they act as negative regulators of Rho GTPases, RhoGAPs serve to regulate individual Rho GTPase activity at the appropriate time and place in cells. With a large number of GAPs available for specific regulatory purposes, the multifunctional domains add sophistication to accomplish this. Not only do some RhoGAPs exhibit differences in tissue expression and target selection, they also play roles in specific biological functions such as endocytosis, exocytosis, cytokinesis, and cell migration, that may underlie the mechanisms of diseases they are associated with (Tcherkezian and Lamarche-Vane 2007). An example is Oligophrenin-1, a RhoGAP highly expressed in the human fetal brain that acts to stimulate the GTPase activity of multiple Rho GTPases and was found to be associated with X-linked mental retardation (Billuart et al. 1998). Another example is the MgcRacGAP, a GTPase-activating protein for RhoA that regulates cytokinesis. Cells with MgcRacGAP RNA interference (RNAi) failed to go through cytokinesis and produced no cleavage furrow (Zhao and Fang 2005). MgcRacGAP interacts with ECT2, a GEF for RhoA, and regulates its localization for the proper RhoA activation to allow assembly of the contractile ring and cytokinesis. Recently, it was shown in mouse embryonic fibroblasts (MEFs) generated from Cdc42GAP–/– cells that Cdc42GAP, through its negative regulation of Cdc42, is critical in filopodia induction, directed migration, and proliferation (Yang et al. 2006).

While the negative role that GAPs play in Rho GTPase regulation has been well documented, the role of GAPs, by virtue of their ability to interact with GTP-bound, activated Rho GTPases to act as an effector or mediator of cross-talk has been an issue of continuing discussion. Several observations support such a possibility. For example, full-length α1-chimaerin lacking GAP activity but retaining the ability to bind GTPases cooperates with Rac1 and Cdc42 to promote formation of lamellipodia and filopodia suggesting a double function as a RhoGAP and a Rho GTPase effector (Kozma et al. 1996; Tcherkezian and Lamarche-Vane 2007). TCGAP has been shown to bind to Rho GTPases through its RhoGAP domain and is involved in insulin-mediated glucose-transport signaling, but the GAP activity does not seem to be required (Chiang et al. 2003). Another interesting example is p200 RhoGAP (also termed RICS or Grit). In this case, p200 RhoGAP can promote cell proliferation by mediating cross-talk between the Ras and Rho signaling pathways by interacting with the SH3 domain of p120 RasGAP through its proline-rich motif (Shang et al. 2007). In addition, the p85 subunits (p85α and p85β) of PI3K (phosphoinositide 3-kinase) contain RhoGAP domains that allow interaction with Rac1 and Cdc42 but show no detectable GAP activity (Zheng et al. 1994). It remains an open question whether these putative effector functions of RhoGAPs are physiologically or pathologically relevant.

Physiological Studies

In vitro biochemical and cell biological studies have provided much information on the mechanism, function, and regulation of RhoGAP activity; however, our understanding of the physiological role of most RhoGAPs still trails. Some of the convincing data of RhoGAP functions come from a thorough *Drosophila* study using systematic knockdown of each RhoGAP by siRNA in the fruit fly brain. The fruit fly genome encodes 6 Rho family genes while there are at least 21 Rho family GAPs. The study revealed that while the knockdown of most GAPs failed to show an obvious phenotype, several GAPs were required for normal development and survival (Billuart et al. 2001). One such GAP was p190 RhoGAP. Knockdown of p190 RhoGAP in *Drosophila* led to a retraction of axonal branches through the upregulation of RhoA activity and its effector Drok (Billuart et al. 2001). The results of this study in general support those from studies of p190 RhoGAP in mammals. Mouse gene targeting studies have found that p190 RhoGAP is required for axon outgrowth and guidance as well as neuronal morphogenesis (Brouns et al. 2000; Brouns, Matheson, and Settleman 2001). p190 deficient cells of the neural tube floor plate had large accumulations of polymerized actin, suggesting a negative role of p190 in the regulation of Rho-mediated actin assembly in the neuroepithelium (Brouns et al. 2000). Further support for the role of RhoGAPs in the nervous system is found in a recent study involving the deletion of RICS in mice. Hippocampal and cerebellar granule neurons from RICS−/− mice bore longer neurites than those from wild-type mice (Nasu-Nishimura et al. 2006). Interestingly, RICS was previously shown to contain GAP activity for Cdc42, Rac1, and/or RhoA and interact with β-catenin via its armadillo repeats thereby lending further support to the multiple functions for RhoGAPs theory (Okabe et al. 2003). Another RhoGAP, p190-B, demonstrates Rho-specific GAP activity, and mice lacking p190B display reduced cell size and organ size caused in part by reduced activity of the CREB transcription factor (Sordella et al. 2002). p190-B RhoGAP is also involved in the regulation of cell fate decision as a component of a switch in which cells undergo either adipogenesis or myogenesis (Sordella et al. 2003), and has recently been implicated in mammary bud formation through modulation of mesenchymal proliferation and differentiation (Heckman et al. 2007).

Cdc42GAP (also termed p50RhoGAP), one of the founding members of the RhoGAP family which preferentially down-regulates the activity of Cdc42, has been extensively studied in the last few years through mouse gene targeting methods. Gene targeting of Cdc42GAP resulted in constitutively elevated Cdc42-GTP levels in diverse tissues of adult mice. Cdc42GAP (–/–) mice displayed severe growth defects due to reduced cell number and increased c-Jun N-terminal kinase-mediated apoptosis (Wang et al. 2005). Cdc42GAP has a profound effect on erythropoiesis and multiple hematopoietic stem cell functions, as Cdc42GAP knockout mice were anemic and showed deficiencies in adhesion, migration, and engraftment of hematopoietic stem/progenitor cells (Wang et al. 2006). Interestingly, Cdc42GAP (–/–) mice had a significantly shortened life span and multiple premature aging-like phenotypes, including a reduction in body mass, a loss of subdermal adipose tissue, severe lordokyphosis, muscle atrophy, osteoporosis, and reduction of re-epithelialization ability in wound healing, suggesting a role of Cdc42 activity in regulating mammalian genomic stability and aging-related physiology (Wang et al. 2007). These findings of the Cdc42GAP studies provide an interesting case of a RhoGAP that is ubiquitously expressed, and underscore the complexity involved in Rho GTPase regulation and the global role of a RhoGAP.

Human Cancers

The importance of Rho GTPases in regulating many cell functions such as proliferation, adhesion, migration, and morphology make Rho GTPases and their regulators likely candidates in tumorigenesis and tumor progression. It has long been established that there is a deregulation of Rho GTPase activity in many types of human cancers and that, in fact, Rho GTPases contribute to the growth, invasiveness, and/or metastasis phenotypes of multiple tumor cells (Sahai and Marshall 2002; Gomez del Pulgar et al. 2005). While the physiologic function of most Rho GTPase regulators remains unclear, many GEFs have been identified as proto-oncogene products (i.e., VAV1, Trio, LARG, and Dbl), leading to the speculation that GAPs, working in opposition to the GEFs in the Rho GTPase regulatory cycle, may function as tumor suppressors. Indeed there is substantial evidence in support of such a hypothesis, most notably through the analysis of Rho GTPases and their regulators in tumors, tumor cell lines, and in cells following in vitro alterations of these proteins (Kandpal 2006). However, to date most RhoGAPs linked to cancers seem to show an alteration in expression rather than mutation. Mouse gene targeting studies have yet to reveal the loss of function of a RhoGAP, either by itself or in cooperation with other hits that could lead to tumorigenesis. Adding further to the complexity, there are multiple reports of RhoGAPs being either up- or down-regulated in tumors. A fair conclusion one can draw at this time is that the contribution of RhoGAPs to tumorigenicity involves the deregulation of Rho GTPases. Table 1 compiles a summary of RhoGAPs known to be associated with cancers which is further divided by the classification of RhoGAP mutation, deletion, up-regulation, down-regulation, or a combination of changes. It is clear that many RhoGAPs could play roles in different cancer types.

Deleted in Liver Cancer (DLC)

One of the closely implicated RhoGAPs in cancer is the deleted in liver cancer-1 (DLC-1). DLC-1 was identified as a genomic DNA segment under-represented in a human hepatocellular carcinoma (HCC) specimen (Durkin et al. 2007a). DLC-1 was mapped to the human chromosome 8p22 region that is frequently lost in HCC and other cancers and was termed deleted in liver cancer when found to be deleted in primary HCC and HCC cell lines (Durkin et al. 2007a). DLC-1 mRNA has since been found to be down-regulated or absent in many common human cancers, including HCC, breast, colon, ovarian, uterine, gastric, lung, pancreatic, prostate, renal, and nasopharyngeal tumors as well as several non-malignant conditions (Durkin et al. 2007a). At the protein level DLC-1 was recently shown to be decreased in prostate tumor samples when compared with normal prostate tissue (Guan et al. 2006). The screening of several cell lines derived from HCC, breast, colon, and prostate tumors found aberrant promoter methylation (Yuan, Durkin and Popescu 2003), which could be a major mechanism for inactivation of DLC-1 gene expression in many solid tumors in addition to gene deletion. Histone modifications may also serve to silence DLC-1 as it was shown that treatment with histone deacetylase inhibitor trichostatin A (TSA) increased DLC-1 expression in human gastric cancer, prostate cancer, multiple myeloma, and leukaemia cell lines (Durkin et al. 2007a).

While mutations of DLC-1 are rare in most types of cancer, several single nucleotide polymorphisms (SNPs) have been identified in the DLC-1 genomic sequence. However, these SNPs have yet to be associated with increased susceptibility to cancers (Durkin et al. 2007a). In a large-scale genomic screen, DLC1 was identified as a highly significant breast cancer susceptibility gene (Tang et al. 2004; van den Boom et al. 2004). In several experiments DLC-1 has been shown to inhibit tumor cell growth including in HCC, breast, lung, nasopharyngeal, and epithelial ovarian cancer cells (Durkin et al. 2007a). For example, over-expression of DLC1–1 was able to suppress the growth of tumors in athymic nude mouse xenograft models (Durkin et al, 2007a). Mechanistically, the inhibition of tumor cell growth by DLC-1 can be attributed to the ability of DLC-1 to down-regulate Rho GTPase activity, as RhoGAP-deficient mutants were less active in suppressing cell growth (Wong et al. 2005; Qian et al. 2007). However, because DLC-1$^{-/-}$ mice are embryonic lethal and the heterozygotes have a normal phenotype despite lower levels of

Table 6.1 Alterations of RhoGAPs found in primary human tumors

Gene name	Function	Gene association	GAP alteration
ARHGAP1	Tumor suppressor	Ovarian cancer cells	Downregulation
β2-Chaimaerin		Multiple cancers	
BCR		Philadelphia chromosome (CML; ALL)	Mutation; fusion
DLC1		Multiple cancers	Deletion; mutation; downregulation
DLC2		Multiple cancers	
DLC3		Multiple cancers	
GRAF(ARHGAP26)		Leukaemia (AML)	Mutation; fusion
ARHGAP21	Oncogene	Head neck squamous Cancer	Upregulation
p85-α(P13k)		Ovarian cancer	
p 85-β(P13k)		Ovarian cancer	
p 190B RhoGAP		Breast cancer	
Oligophrenin 1		Gastric cancer cells	
RacGAP1		Ovarian cancer	
SRGAP1(ARHGAP13)		Breast cancer cells	

DLC-1 mRNA (Durkin et al. 2005), direct effects of DLC-1 knockout on tumor development in an animal model have yet to be studied by conditional gene targeting approaches.

Closely related to DLC-1 are two RhoGAP family members DLC-2 and DLC-3. There is evidence indicating that they may also play a role in human cancers. The *STARD13/DLC2* locus (chromosome 13q13) is a frequent target of genomic deletions in multiple human cancers (Durkin et al. 2007a). As was the case with DLC-1, DLC-2 mRNA is found to be down-regulated in a number of primary lung, breast, kidney, and uterine tumors (Ullmannova and Popescu 2006). The GAP activity of DLC-2 was confirmed in several in vitro studies to inhibit the transformation of NIH3T3 cells when co-transfected with oncogenic Ras and to inhibit the growth of several cancer cell lines (Durkin et al. 2007a). Interestingly, when DLC-2γ (one of four isoforms of DLC-2) was transfected into HepG2 hepatoma cells, cell proliferation, motility, and transformation were markedly suppressed corresponding to a significant reduction of RhoA activity (Leung et al. 2005), suggesting RhoA could be a target of this RhoGAP.

Unlike DLC-1 and DLC-2, the *STARD8/DLC3* locus has not been associated with cancers. However, reduced DLC-3 mRNA levels have been observed in a high percentage of human prostate, kidney, lung, breast, and ovarian cancer tissues (Durkin et al. 2007a). In addition, two missense mutations in *STARD8/DLC3* were found in breast tumors in a large-scale analysis of mutations in human breast and colon cancers (Durkin et al. 2007a). More recently, it was shown that transfection of DLC-3α into human breast and prostate cancer cells inhibited cell proliferation, colony formation, and growth in soft agar, indicating that the deregulation of DLC-3 may contribute to breast and prostate tumorigenesis (Durkin et al. 2007b).

p190 RhoGAP

Long suspected as a possible regulator of oncogenic Ras signaling because of its tight binding activity to p120 RasGAP, p190 RhoGAP was shown to regulate Ras-induced transformation of NIH3T3 fibroblasts in overexpression studies a decade ago. Expression of both an antisense p190 RNA or a dominant negative mutant induced transformation of the NIH3T3 cells, while expression of the p190 N-terminal GTPase domain or C-terminal GAP domain suppressed transformation (Wang et al. 1997). p190 RhoGAP has also been implicated in inflammatory breast cancer (Zrihan-Licht et al. 2000). The increased EGFR activity seen in some breast cancers leads to phosphorylation and association of p190 RhoGAP and p120RasGAP resulting in cellular invasion. In fact, increased expression of both p190 RhoGAP and p120 RasGAP has been seen in human mammary epithelial (HME) and SUM149 inflammatory breast cancer cells, a finding consistent with that seen in aggressive mouse mammary tumors (Lin and van Golen 2004). More recently it was shown that p190 RhoGAP-mediated inactivation of Rho GTPases, specifically RhoA, reduced the invasion and metastasis of human pancreatic cancer cells (Kusama et al. 2006). A chimera of the RhoGAP domain of p190 and the C-terminus of RhoA was transfected into human pancreatic cancer cells, AsPC-1, resulting in a marked reduction of EGF-induced RhoA activation and EGF-induced invasion. Furthermore, intrasplenic injection of these chimera-transfected AsPC-1 cells resulted in a significant reduction in the number and size of metastatic nodules in the liver, suggesting that the inhibitory action of p190 RhoGAP toward RhoA may be a novel approach to treating invasion and metastasis in cancer cells (Kusama et al. 2006).

Chimaerins

Another RhoGAP member that may serve as a tumor suppressor is β2-chimaerin. Reduced mRNA levels of β2-chimaerin are seen in several cancers including breast cancer, duodenal adenocarcinomas, and malignant gliomas (Bruinsma and Baranski 2007). Overexpression of β2-chimaerin in a breast cancer cell line inhibits proliferation (Yang et al. 2005). Interestingly, when the GAP domain of β2-chimaerin is overexpressed in a mouse mammary cancer cell line there is a significantly reduced growth rate and metastatic potential of arising tumors (Menna et al. 2003). These observations support the conventional thinking that down-regulation of a RhoGAP can lead to metastatic transformation and that restoring normal levels of the RhoGAP could prevent tumorigenesis (Bruinsma and Baranski 2007). Again, data from a β2-chimaerin knockout mouse model are not available, precluding us from drawing a more direct conclusion whether reduced or absent β2-chimaerin levels would promote tumorigenesis. However, recent data from a study in *Drosophila melanogaster* might point to such a possibility. The fly genome contains a single chimaerin gene, *RhoGAP5A*, which is expressed in multiple tissues including the eye. Reduction of RhoGAP5A levels in the eye results in an increase in cell number and aberrant cell-cell adhesion, a phenotype that is consistent with tumorigenesis (Bruinsma, Cagan and Baranski 2007).

Other RhoGAPs Associated with Cancer

A number of other RhoGAPs have been linked to cancers. One of the most well-known cases of a RhoGAP associated with cancer is the fusion of Bcr and Abl genes in leukaemia-associated chromosomal translocations (Shtivelman et al. 1985). The RhoGAP domain of Bcr is deleted in the Bcr–Abl fusion proteins p190 and p210 while the mechanistic consequence of this loss remains unclear (Moon and Zheng 2003). ABR, a gene homologous to BCR, was found to be deleted in seven of eight cases of medulloblastoma (McDonald et al. 1994). ARHGAP8, possibly a gene duplication of ARHGAP1/CDC42GAP/p50RHOGAP, is located within a critical region of loss-of-heterozygosity on chromosome 22q13.31 in breast and colorectal carcinomas (Johnstone et al. 2004). Further analyses of individuals with colorectal or breast cancer revealed six germline missense variants of ARHGAP8, although no somatic mutations were identified. These observations are consistent with the hypothesis that at least some RhoGAPs may serve a tumor suppressor-like role. However, one surprise is that ARHGAP8 expression was up-regulated in the majority of primary colorectal tumors analyzed (Johnstone et al. 2004). A possible explanation could be that because the Rho GTPase substrates are known to be overexpressed in the tumors, the increase in ARHGAP8 may function in a feedback loop to control the activity of these GTPases. Another example along this line of observation is that of GRAF, the focal adhesion kinase associated RhoGAP, which was identified as a fusion partner of the mixed-lineage leukaemia gene by a unique chromosome translocation in juvenile myelomonocytic leukaemia (Moon and Zheng 2003).

Conclusions

It was not long ago that RhoGAPs were thought of as merely deactivators of Rho GTPases. It is clear now that the regulation of Rho GTPases by RhoGAPs may have multiple roles. The sheer number and complexity of RhoGAPs leaves much to be understood, but accumulating evidence has implicated them as potential targets in cancer therapy. The concept of targeting small GTPases such as Ras in cancer therapy has been conceived and practiced for over a decade. Farnesyltransferase inhibitors have been used to inhibit Ras function by blocking post-translational modifications by farnesyl isoprenoid (Midgley and Kerr 2002). In fact, simvastatin, by virtue of its prenylation inhibition properties, has been shown to exhibit anti-tumor activity in a mouse xenograft model of human Hodgkin's lymphoma (HL) (von Tresckow et al. 2007). The use of farnesyltransferase inhibitors was also shown to radiosensitize K-*ras* mutant pancreatic tumors, although not through the inhibition of K-ras prenylation (Brunner et al. 2005). Another example was the use of intracellular antibody capture technology to generate intrabodies that bind to the oncogenic Ras protein (Tanaka and Rabbitts 2003). In addition, antisense technology was used to specifically inhibit the expression of pathogenic Ras to reverse Ras-induced tumor proliferation (Midgley and Kerr 2002). While these strategies have shown promise, difficulties associated with specificity and efficacy of each of the approaches have presented a clear challenge to develop novel means that involve more innovative and sophisticated ways in targeted therapy.

The idea that individual members of the Rho family have specific roles in tumor development is becoming entrenched in the field. This makes Rho GTPases, their effectors, and their regulators attractive targets for cancer therapy. Unlike Ras, there are no constitutively active mutants of the Rho family found in human tumors. With the knowledge that multiple RhoGAPs may play an important role in some cancers, screens can be performed to elucidate the multiple partners that the RhoGAPs interact with, thereby providing more knowledge on their role in signaling. The generic formula that loss of RhoGAPs would lead to increased Rho GTPase signaling associated with cancer phenotypes is a good starting point for considering potential targeting strategies for therapy; however, the possibility that a gain of certain RhoGAP activity and the associated decrease in specific Rho GTPase activity may also have a tumor promoting effect have to be considered as well (Fig. 2). This will obviously depend on better understanding the tumor promoting or tumor

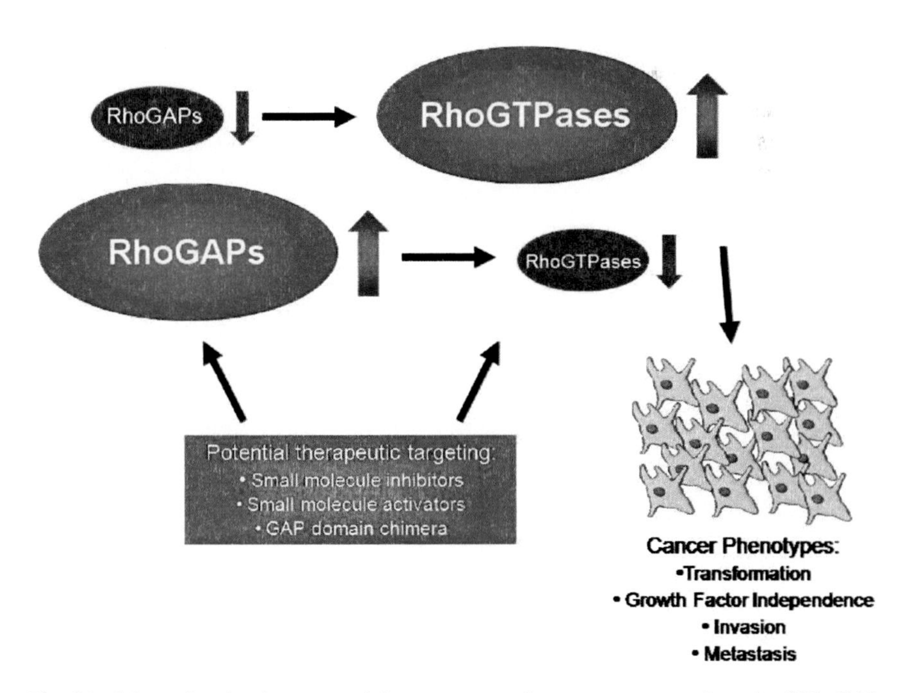

Fig. 6.2 Schematics showing a potential tumor suppressing or tumor promoting role of RhoGAPs and possible therapeutic targeting strategies

suppressing function of each RhoGAP in primary cancer settings. The use of small molecular weight chemicals to increase or decrease RhoGAP specificity and/or activity for Rho GTPase modulation is a tantalizing idea that has not yet been fully explored. In parallel, the use of the naturally occurring RhoGAP domains as negative regulatory modules of specific Rho GTPases by fusing chimeric molecules of a RhoGAP domain with the C-terminus hypervariable sequences of a Rho GTPase could allow for the regulated, intracellular location-specific targeting of the Rho GTPase (Wang et al. 2003). Previously such chimeric molecules have been shown to be useful in reversing the growth or invasion phenotypes in breast cancer cells (Wang et al. 2003). As our understanding of RhoGAP structure-function and regulation becomes more in-depth, and the implication of specific RhoGAP family members in cancer becomes concrete, certain RhoGAPs could emerge as diagnostic markers or therapeutic targets.

References

Ahmadian M.R., Wiesmüller, L., Lautwein, A., Bischoff, F.R., and Wittinghofer, A. 1996. Structural differences in the minimal catalytic domains of the GTPase-activating proteins p120GAP and neurofibromin. J. Biol. Chem. 271:16409–15.

Bax, B. 1998. Domains of rasGAP and rhoGAP are related. Nature. 392:447–448.

Billuart, P., Winter, C.G., Maresh, A., Zhao, X., and Luo, L. 2001. Regulating axon branch stability: The role of p190 RhoGAP in repressing a retraction signaling pathway. Cell. 107:195–207.

Billuart, P., Bienvu, T., Ronce, N., des Portes, V., Vinet, M.C., Zemni, R., Roest Crollius, H., Carrie, A., Fauchereau, F., Cherry, M., Briault, S., Hamel, B., Fryns, J.P., Beldjord, C., Kahn, A., Moraine, C., and Chelly, J. 1998. Oligophrenin-1 encodes a rhoGAP protein involved in X-linked mental retardation. Nature. 392:923–926.

Bishop, A.L., and Hall, A. 2000. Rho GTPases and their effector proteins. Biochem. J. 348:241–255.

Brouns, M.R., Matheson, S.F., Hu, K.Q., Delalle, I., Caviness, V.S., Silver, J., Bronson, R.T., and Settleman, J. 2000. The adhesion signaling molecule p190 RhoGAP is required for morphogenetic processes in neural development. Development. 127:4891–4903.

Brouns, M.R., Matheson, S.F., and Settleman, J. 2001. p190 RhoGAP is the principal Src substrate in brain and regulates axon outgrowth, guidance, and fasciculation. Nat. Cell Biol. 3:361–367.

Bruinsma, S.P., Cagan, R.L., and Baranski, T.J. 2007. Chimaerin and Rac regulate cell number, adherens junctions, and ERK MAP kinase signaling in the *Drosophila* eye. Proc. Natl. Acad. Sci. U.S.A. 104:7098–7103.

Bruinsma, S.P., and Baranski, T.J. 2007. β2-Chimaerin in cancer signaling: connecting cell adhesion and MAP kinase activation. Cell Cycle. 6:2440–2444.

Brunner, T.B., Cengel, K.A., Hahn, S.M., Wu, J., Fraker, D.L., McKenna, W.G., and Bernhard, E.J. 2005. Pancreatic cancer cell radiation survival and prenyltransferase inhibition: The role of K-Ras. Cancer Res. 65:8433–8441.

Canagarajah, B., Leskow, F.C., Ho, J.Y.S., Mischak, H., Saidi, L.F., Kazanietz, M.G., and Hurley, J.H. 2004. Sturctural Mechanism for Lipid Activation of the Rac-Specific GAP, β2-Chimaerin. Cell. 119:407–418.

Chiang, S.H., Hwang, J., Legendre, M., Zhang, M., Kimura, A., and Saltiel, A.R. 2003. TCGAP, a multidomain Rho GTPase-activating protein involved in insulin-stimulated glucose transport. EMBO J. 22:2679–2691.

Crooke, S.T. 1996. Proof of mechanism of antisense drugs. Antisense Nucleic Acid Drug Dev. 6:145–147.

Durkin, M.E., Avner, M.R., Huh, C.G., Yuan, B.Z., Thorgeirsson, S.S., and Popescu, N.C. 2005. DLC-1, a Rho GTPase-activating protein with tumor suppressor function, is essential for embryonic development. FEBS Lett. 579:1191–1196.

Durkin, M.E., Yuan, B.Z., Zhou, X., Zimonjic, D.B., Lowy, D.R., Thorgeirsson, S.S., and Popescu, N.C. 2007. DLC-1: a Rho GTPase-activating protein and tumour suppressor. J.Cell. Mol. Med. 11:1185–1207.

Durkin, M.E., Ullmannova, V., Guan, M., and Popescu, N.C. 2007. Deleted in liver cancer 3 (DLC-3), a novel Rho GTPase-activating protein, is downregulated in cancer and inhibits tumor cell growth. Oncogene. 26:4580–4589.

Gamblin, S.J., and Smerdon, S.J. 1998. GTPase-activating proteins and their complexes. Curr. Opin. Struct. Biol. 8:195–201.

Garrett, M.D., Self, A.J., van Oers, C., and Hall, A. 1989. Identification of distinct cytoplasmic targets for ras/R-ras and rho regulatory proteins. J. Biol. Chem. 264:10–13.

Gomez del Pulgar, T., Benitah, S.A., Valeron, P.F., Espina, C., and Lacal, J.C. 2005. Rho GTPase expression in tumourigenesis: evidence for a significant link. Bioessays. 27:602–613.

Guan, M., Zhou, X., Soulitzis, N., Spandidos, D.A., and Popescu, N.C. 2006. Aberrant methylation and deacetylation of deleted in liver cancer-1 gene in prostate cancer: potential clinical applications. Clin. Cancer Res. 12:1412–1419.

Hall, A. 1998. Rho GTPases and the actin cytoskeleton. Science. 279:509–514.

Heckman, B.M., Chakravarty, G., Vargo-Gogola, T., Gonzales-Rimbau, M., Hadsell, D.L., Lee, A.V., Settleman, J., and Rosen J.M. 2007. Crosstalk between the p190-B RhoGAP and IGF signaling pathways is required for embryonic mammary bud development. Dev. Biol. 309:137–149.

Johnstone, C.N., Castellvi-Bel, S., Chang, L.M., Bessa, X., Nakagawa, H., Harada, H., Sung, R.K., Pique, J.M., Castells, A., and Rustgi, A.K. 2004. ARHGAP8 is a novel member of the RHOGAP family related to ARHGAP1/CDC42GAP/p50RHOGAP: mutation and expression analyses in colorectal and breast cancers. Gene. 336:59–71.

Kandpal, R.P. 2006. Rho GTPase activating proteins in cancer phenotypes. Curr. Protein Pept. Sci. 7:355–365.

Kozma, R., Ahmed, S., Best, A., and Lim, L. 1996. The GTPase-activating protein n-chimaerin cooperates with Rac1 and Cdc42Hs to induce the formation of lamellipodia and filopodia. Mol. Cell. Biol. 16:5069–5080.

Kusama, T., Mukai, M., Endo, H., Ishikawa, O., Tatsuta, M., Nakamura, H., and Inoue, M. 2006. Inactivation of Rho GTPases by p190 RhoGAP reduces human pancreatic cancer cell invasion and metastasis. Cancer Sci. 97:848–853.

Lamarche, N., and Hall, A. 1994. GAPs for rho-related GTPases. Trends Genet. 10:436–440.

Leung, T.H., Ching, Y.P., Yam, J.W., Wong, C.M., Yau, T.O., Jin, D.Y., and Ng, I.O. 2005. Deleted in liver cancer 2 (DLC2) suppresses cell transformation by means of inhibition of RhoA activity. Proc Natl Acad Sci USA. 102:15207–15212.

Li, R., Zhang, B., and Zheng, Y. 1997. Structural determinants required for the interaction between RhoA and the GTPase-activating domain of p190. J. Biol. Chem. 272:32830–32835.

Lin, M., and van Golen, K.L. 2004. Rho-regulatory proteins in breast cancer cell motility and invasion. Breast Cancer Res. Treat. 84:49–60.

Longenecker, K.L., Zhang, B., Derewenda, U., Sheffield, P.J., Dauter, Z., Parsons, J.T., Zheng, Y., and Derewenda, Z.S. 2000. Structure of the BH domain from graf and its implications for Rho GTPase recognition. J. Biol. Chem. 275:38605–38610.

McDonald, J.D., Daneshvar, L., Willert, J.R., Matsumura, K., Waldman, F., and Cogen, P.H. 1994. Physical mapping of chromosome 17p13.3 in the region of a putative tumor suppressor gene important in medulloblastoma. Genomics. 23:229–232.

Menna, P.L., Skilton, G., Leskow, F.C., Alonso, D.F., Gomez, D.E., and Kazanietz, M.G. 2003. Inhibition of aggressiveness of metastatic mouse mammary carcinoma cells by the beta2-chimaerin GAP domain. Cancer Res. 63:2284–2291.

Midgley, R.S., and Kerr, D.J. 2002. Ras as a target in cancer therapy. Crit. Rev. Oncol. Hematol. 44:109–120.

Moon, S.Y. and Zheng, Y. 2003. Rho GTPase-activating proteins in cell regulation. Trends Cell Biol. 13:13–22.

Nakamura, T., Komiya, M., Sone, K., Hirose, E., Gotoh, N., Morii, H., Ohta, Y., and Mori, N. 2002. Grit, a GTPase-activating protein for the Rho family, regulates neurite extension through association with the TrkA receptor and N-Shc and CrkL/Crk adapter molecules. Mol. Cell. Biol. 22:8721–8734.

Nassar, N., Hoffman, G.R., Manor, D., Clardy, J.C., and Cerione, R.A. 1998. Structures of Cdc42 bound to the active and catalytically compromised forms of Cdc42GAP. Nat. Struct. Biol. 5:1047–1052.

Nasu-Nishimura Y, Hayashi T, Ohishi T, Okabe T, Ohwada S, Hasegawa Y, Senda T, Toyoshima C, Nakamura T, Akiyama T. 2006. Role of the Rho GTPase-activating protein RICS in neurite outgrowth. Genes Cells. 11:607–614.

Olofsson, B. 1999. Rho guanine dissociation inhibitors: pivotal molecules in cellular signaling. Cell. Signal. 11:545–554.

Okabe, T., Nakamura, T., Nishimura, Y.N., Kohu, K., Ohwada, S., Morishita, Y., and Akiyama, T. 2003. RICS, a novel GTPase-activating protein for Cdc42 and Rac1, is involved in the beta-catenin-N-cadherin and N-methyl-D-aspartate receptor signaling. *J. Biol. Chem.* 278:9920 – 9927.

Qian, X., Li, G., Asmussen, H.K., Asnaghi, L., Vass, W.C., Braverman, R., Yamada, K.M., Popescu, N.C., Papageorge, A.G., and Lowy, D.R. 2007. Oncogenic inhibition by a deleted in liver cancer gene requires cooperation between tensin binding and Rho-specific GTPase-activating protein activities. Proc. Natl. Acad. Sci. U.S.A. 104:9012–9017.

Rittinger, K., Walker, P.A., Eccleston, J.F., Smerdon, S.J., and Gamblin, S.J. 1997. Structure at 1.65 A of RhoA and its GTPase-activating protein in complex with a transition-state analogue. Nature. 389:758–762.

Rittinger, K., Taylor, W.R., Smerdon, S.J., and Gamblin, S.J. 1998. Support for shared ancestry of GAPs Nature. 392:448–449.

Sahai, E., and Marshall, C.J. 2002. Rho-GTPases and cancer. Nat. Rev. Cancer. 2:133–142.

Sekimata, M., Kabuyama, Y., Emori, Y., and Homma, Y. 1999. Morphological changes and detachment of adherent cells induced by p122, a GTPase-activating protein for Rho. J. Biol. Chem. 274:17757–17762.

Shang, X., Moon, S.Y., and Zheng, Y. 2007. p200 RhoGAP promotes cell proliferation by mediating cross-talk between Ras and Rho signaling pathways. J. Biol. Chem. 282:8801–8811.

Shtivelman, E., Lifshitz, B., Gale, R.P., and Canaani, E. 1985. Fused transcript of abl and bcr genes in chronic myelogenous leukaemia. Nature. 315:550–554.

Sordella, R., Classon, M., Hu, K.Q., Matheson, S.F., Brouns, M.R., Fine B., Zhang, L., Takami, H., Yamada, Y., and Settleman, J. 2002. Modulation of CREB activity by the Rho GTPase regulates cell and organism size during mouse embryonic development. Dev. Cell 2:553–565.

Sordella, R., Jiang, W., Chen G.C., Curto, M., and Settleman, J. 2003. Modulation of Rho GTPase signaling regulates a switch between adipogenesis and myogenesis. Cell. 113:147–158.

Su, Z.J., Hahn, C.N., Goodall, G.J., Reck, N.M., Leske, A.F., Davy, A., Kremmidiotis, G., Vadas, M.A., and Gamble, J.R. 2004. A vascular cell-restricted RhoGAP, p73RhoGAP, is a key regulator of angiogenesis. Proc. Natl. Acad. Sci. U.S.A. 101:12212–12217.

Symons, M., and Settleman, J. 2000. Rho family GTPases: more than simple switches. Trends Cell Biol. 10:415–419.

Tanaka, T., and Rabbitts, T.H. 2003. Intrabodies based on intracellular capture frameworks that bind the RAS protein with high affinity and impair oncogenic transformation. EMBO J. 22: 1025–1035.

Tang, K., Oeth, P., Kammerer, S., Denissenko, M.F., Ekblom, J., Jurinke, C., van den Boom, D., Braun, A., and Cantor, C.R. 2004. Mining disease susceptibility genes through SNP analyses and expression profiling using MALDI-TOF mass spectrometry. J. Proteome Res. 3:218–227.

Taylor, J.M., Macklem, M.M., and Parsons, J.T. 1999. Cytoskeletal changes induced by GRAF, the GTPase regulator associates with focal adhesion kinase, are mediated by Rho. J. Cell Sci. 112:231–242.

Tcherkezian, J., and Lamarche-Vane, N. 2007. Current knowledge of the large RhoGAP family of proteins. Biol. Cell 99:67–86.

von Tresckow, B., von Strandmann, E.P., Sasse, S., Tawadros, S., Engert, A., and Hansen, H.P. 2007. Simvastatin-dependent apoptosis in Hodgkin's lymphoma cells and growth impairment of human Hodgkin's tumors *in vivo*. Haematologica. 92:682–685.

Ullmannova, V., Popescu, N.C. 2006. Expression profile of the tumor suppressor genes DLC-1 and DLC-2 in solid tumors. Int J Oncol. 29:1127–1132.

Van Aelst, L., and D'Souza-Schorey, C. 1997. Rho GTPases and signaling networks. Genes Dev. 11:2295–2322.

van den Boom, D., Beaulieu, M., Oeth, P., Roth, R., Honisch, C., Nelson, M.R., Jurinke, C., and Cantor, C. 2004. MALDI-TOF MS: a platform technology for genetic discovery. Int. J. Mass Spectrom. 238:173–188.

Wang, D.Z., Nur-E-Kamal, M.S., Tikoo, A., Montague, W., and Maruta, H. 1997. The GTPase and Rho GAP domains of p190, a tumor suppressor protein that binds the M_r 120,000 Ras GAP, independently function as anti-Ras tumor suppressors. Cancer Res. 57:2478–2484.

Wang, L., Yang, L., Luo, Y., and Zheng, Y. 2003. A novel strategy for specifically down-regulating individual Rho GTPase activity in tumor cells. J. Biol. Chem. 278:44617–44625.

Wang, L., Yang, L., Burns, K., Kuan, C.Y., and Zheng, Y. 2005. Cdc42GAP regulates c-Jun N-terminal kinase (JNK)-mediated apoptosis and cell number during mammalian perinatal growth. Proc. Natl. Acad. Sci. U.S.A. 102:13484–13489.

Wang, L., Yang, L., Filippi, M.D., Williams, D.A., and Zheng, Y. 2006. Genetic deletion of Cdc42GAP reveals a role of Cdc42 in erythropoiesis and hematopoietic stem/progenitor cell survival, adhesion, and engraftment. Blood. 107:98–105.

Wang, L., Yang, L., Debidda, M., Witte, D., and Zheng, Y. 2007. Cdc42 GTPase-activating protein deficiency promotes genomic instability and premature aging-like phenotypes. Proc. Natl. Acad. Sci. U.S.A. 104:1248–1253.

Wong, C.M., Yam, J.W., Ching, Y.P., Yau, T.O., Leung, T.H., Jin, D.Y., Ng, I.O. 2005. Rho GTPase-activating protein deleted in liver cancer suppresses cell proliferation and invasion in hepatocellular carcinoma. Cancer Res. 65:8861–8868.

Yang, C., Liu, Y., Leskow, F.C., Weaver, V.M., and Kazanietz, M.G. 2005. Rac-GAP-dependent inhibition of breast cancer cell proliferation by beta2-chimerin. J. Biol. Chem. 280:24363–24370.

Yang, L., Wang, L., and Zheng, Y. 2006. Gene targeting of Cdc42 and Cdc42GAP affirms the critical involvement of Cdc42 in filopodia induction, directed migration, and proliferation in primary mouse embryonic fibroblasts. Mol. Biol. Cell. 17:4675–4685.

Yuan, B.Z., Durkin, M.E., and Popescu, N.C. 2003. Promoter hyper-methylation of DLC-1, a candidate tumor suppressor gene, in several common human cancers. Cancer Genet. Cytogenet. 140:113–117.

Zhao, W.M., and Fang, G. 2005. MgcRacGAP controls the assembly of the contractile ring and the initiation of cytokinesis. Proc. Natl. Acad. Sci. U.S.A. 102:13158–13163.

Zheng, Y., Bagrodia, S., and Cerione, R.A. 1994. Activation of phosphoinositide-3-kinase activity by Cdc42Hs binding to p85. J. Biol. Chem. 269:18727–18730.

Zheng, Y. 2001. Dbl family guanine nucleotide exchange factors. Trends Biochem. Sci. 26:724–732.

Zohn, I.M., Campbell, S.L., Khosravi-Far, R., Rossman, K.L., and Der, C.J. 1998. Rho family proteins and Ras transformation: the RHOad less traveled gets congested. Oncogene. 17:1415–1438.

Zrihan-Licht, S., Fu, Y., Settleman, J., Schinkman, K., Shaw, L., Keydar, I., Avraham, S., and Avraham, H. 2000. RAFTK/Pyk2 tyrosine kinase mediates the association of p190 RhoGAP with RasGAP and is involved in breast cancer cell invasion. Oncogene. 19:1318–1328.

Part III
The Rho GTPase Proteins and Cancer

Chapter 7
RhoBTB Proteins in Cancer

Caroline McKinnon and Harry Mellor

Introduction

The human family of Rho GTPases comprises 20 proteins (Boureux et al. 2007; Pellegrin and Mellor 2006). The majority of the family conforms to the canonical small GTPase structure – a single globular domain of approximately 21 kDa that shows strong homology to the prototypical Ras GTPase fold. The exception comes in the form of the RhoBTB proteins – three much larger multidomain proteins with an N-terminal Rho GTPase domain that shows reduced homology to the rest of the family (Fig. 7.1).

The RhoBTB proteins were first identified by Rivero et al. (2001) as part of a screen for Rho GTPase genes in the social amoeba *Dictyostelium discoideum*. *Dictyostelium* has a large number of Rho GTPases, each related to the human Rac protein (Pellegrin and Mellor 2006; Rivero et al. 2001). Rivero showed that *Dictyostelium* RacA is a multidomain Rho GTPase comprising an N-terminal Rho domain followed by a proline-rich region, two BTB domains and a unique C-terminal region (Fig. 7.2). Importantly, this study identified homologues of RacA in *Drosophila* and in mammals. The term "RhoBTB" protein was coined to denote the BTB domains that define this subfamily of Rho GTPases. Subsequent analysis showed that RhoBTB proteins are absent in yeast, plants and in the nematode worm *Caenorhabditis elegans* (Ramos et al. 2002); but present in sea anemones, sea urchins (HM, unpublished observation), in the primitive chordate *Cionna intesinalis* (Philips et al. 2003), fish, and in all mammalian species examined (Ramos et al. 2002). In humans, RhoBTB1 and RhoBTB2 are closely related (77% identity) and their genes share the same exon/intron boundaries. RhoBTB3 shares the same domain structure, but is highly divergent and has a different genomic organisation (Ramos et al. 2002). The three proteins exhibit overlapping patterns of expression – RhoBTB1 and RhoBTB3 are widely expressed, whereas RhoBTB2 shows much lower levels of expression and is highest in neural tissues (Ramos et al. 2002).

C. McKinnon and H. Mellor (✉)
Department of Biochemistry, School of Medical Sciences,
University of Bristol, University Walk, Bristol BS8 1TD, UK,
e-mail: h.mellor@bristol.ac.uk

K. van Golen (ed.), *The Rho GTPases in Cancer*,

DOI 10.1007/978-1-4419-1111-7_7,

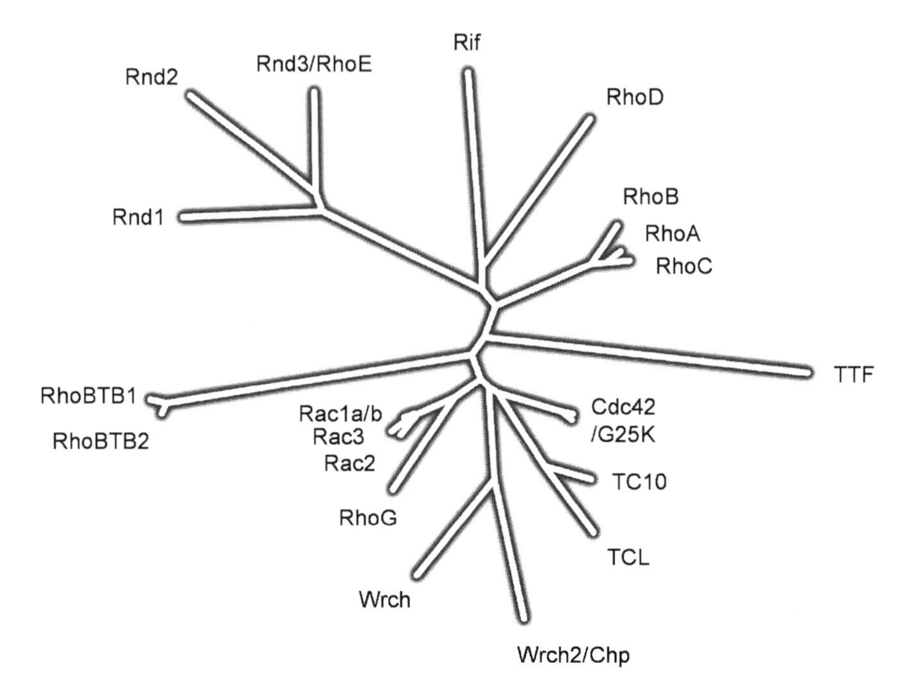

Fig. 7.1 The human Rho subfamily. The figure shows a rooted dendrogram of 19 of the 20 human Rho GTPase isoforms. Sequence alignments were performed using the Clustal W algorithm. The comparison is of the isolated Rho domains in each case. RhoBTB1 and RhoBTB2 form a separate branch. The Rho domain of RhoBTB3 is so highly divergent that it is not included here, although the protein clearly shares a common ancestor with other Rho GTPases. Cdc42 and Rac1 each have two alternative splice variants – G25K and Rac1b, respectively

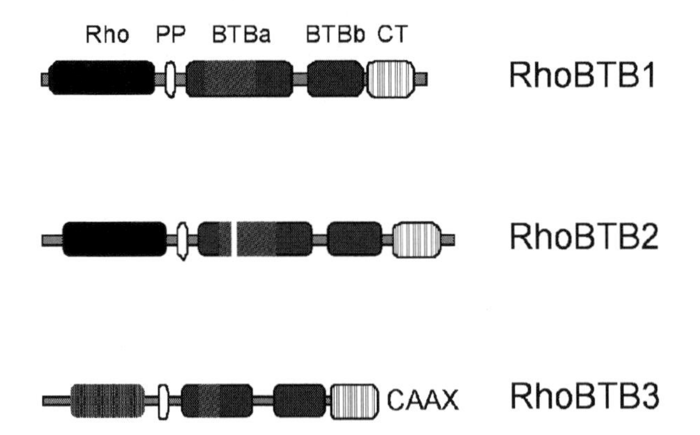

Fig. 7.2 The domain structure of the human RhoBTB proteins. The diagram shows the modular structure of the three human RhoBTB proteins. Each isoform has an N-terminal Rho domain followed by a conserved proline-rich region (*PP*), two BTB domains, and a conserved C-terminal domain (*CT*). The BTBa domain contains an extended loop in each case (shaded). In RhoBTB2, this includes a polyhistidine motif (white box). RhoBTB3 has a consensus sequence for C-terminal prenylation that is absent in the other isoforms (CAAX)

RhoBTB structure

BTB Domains

The immediately intriguing aspect of the RhoBTBs is their multimodular structure. Each contains two copies of the BTB ((Broad-Complex, Tramtrack, and Bric-a-brac) domain, sometimes also called the POZ (poxvirus and zinc finger) domain (Fig. 7.2). We refer to these two RhoBTB domains as BTBa and BTBb. BTB domains are units of protein–protein interaction that evolved in early eukaryotes. Humans have over 350 BTB domain containing proteins, making this one of the most frequently used protein modules (Perez-Torrado et al. 2006). BTB domains form dimers and have been shown to allow both homodimerisation and heterodimerisation of proteins. High-resolution X-ray crystallographic studies of BTB domain dimers have provided detailed insight into the way in which these modules work. The crystal structure of dimerised BTB domains from the promyelocytic leukemia zinc finger protein (PLZF) was the first such structure, revealing an intimate interaction where the dimerised BTB domains exchange strands to create an interface that buries over 25% of the surface area of the monomer (Ahmad et al. 1998). Dimerisation also creates a binding pocket between the two domains, and this has been shown to allow interaction with a wide range of binding partners (Stogios et al. 2005). A schematic representation of the BTB dimer structure is shown in Fig. 7.3.

The majority of BTB-containing proteins have a single domain. RhoBTBs are unusual in having two domains – potentially allowing both inter- and intra-molecular interactions. BTBa has a highly extended structure with insertion of an approximately 115 amino acid loop between elements $\alpha 2$ and $\beta 3$. The location of this loop suggests that it would contribute to the binding site formed on dimerisation (Fig. 7.3). In RhoBTB2, this loop contains a run of nine histidines (Fig. 7.2). The function of this polybasic region is unclear, but it is intriguing to speculate that it may allow non-specific interaction with DNA. The second BTB domain, BTBb, is more standard and lacks this insertion. It would seem likely that these domains allow dimerisation and/or heterodimerisation of RhoBTBs, although this has not been formally demonstrated at this time.

Rho Domain

The N-terminal Rho domain of the RhoBTBs is highly divergent, but most closely related to the Rac GTPase members of the Rho subfamily. Sequence alignment with Rac1 highlights the regions of significant change (Fig. 7.4). The Rho domain fold is comprised of five α-helices and six β-sheets that combine to form a globular domain. Between these secondary structural components are five highly conserved loops termed G1–G5 that interact with nucleotide and control the confirmation of the protein. G1, also called the P-loop, has the conserved sequence GxxxxGKS/T and binds the α- and β-phosphates of the guanine nucleotide. G4 has the consensus (N/T)(K/Q)xD and directly interacts with the purine group. G5 has the consensus

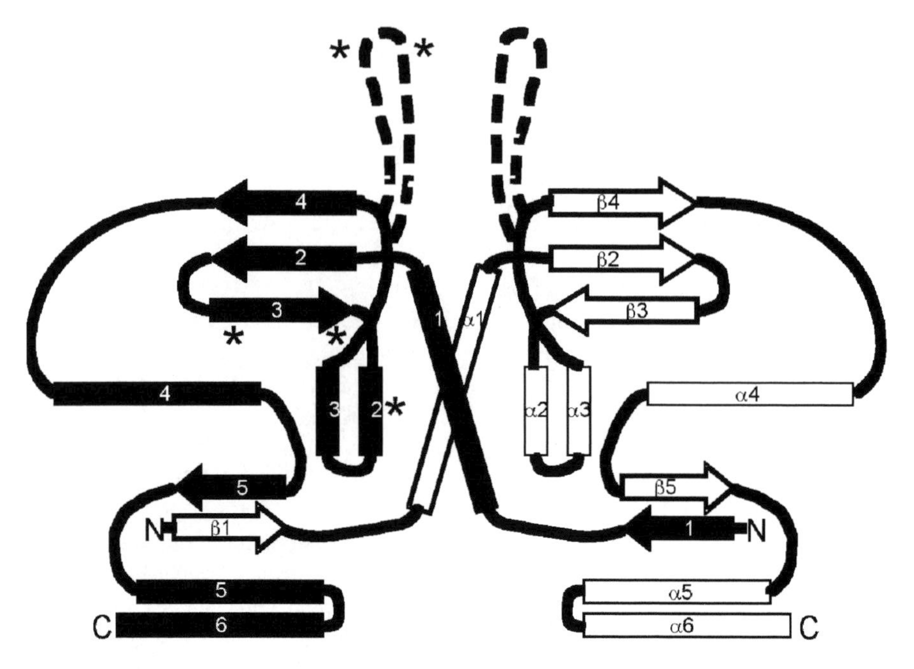

Fig. 7.3 The BTB domain dimer module. The figure shows a block diagram of the structural elements of a BTB dimer. This is based on the crystal structure of the homodimer of BTB domains from PLZF (Ahmad et al. 1998). The subunits are shown in black and white for clarity. The position of the RhoBTB extended loop region is indicated (dashed line), as are somatic missense mutations identified in RhoBTB2 in various cancer samples (asterisk). BTB domains form a tightly knit dimer that involved strand exchange. A partner binding pocket is formed in the cleft between the two dimers. The RhoBTB extended loop and many of the RhoBTB2 cancer mutations lie near this interface, suggesting that they may play a role in dimerisation and/or binding

(T/G/C)(C/S)A and is involved in recognising the guanine base. As with other small GTPases, most Rho family members switch between two confirmations – an inactive, closed confirmation where they are bound to GDP and an active, open conformation where they are bound to GTP. Confirmational change is restricted to two loops – G2 and G3, which are more commonly referred to as Switch I and Switch II. Each loop has a highly conserved residue that directly contacts the GTP γ-phosphate. In Switch I, this is a threonine; in Switch II it is an aspartate. When GTP is hydrolysed by the small GTPase, these interactions are lost and the protein flips to its inactive confirmation (Paduch et al. 2001).

Examination of the RhoBTB3 Rho domain quickly reveals that its sequence has undergone drastic drift from the conserved Rho GTPase domain (Fig. 7.4). Many critical residues are missing and it seems unlikely that this protein is able to bind nucleotide, let alone function as a nucleotide-regulated switch. RhoBTB1 and RhoBTB2 are much closer to the conserved sequence. There are insertions extending the P-loop, Switch I, and the α4 helix; however, most of the critical residues are conserved. Both RhoBTB1 and RhoBTB2 lack a conserved glutamine in Switch II. This residue (Q61 in Rac1) is

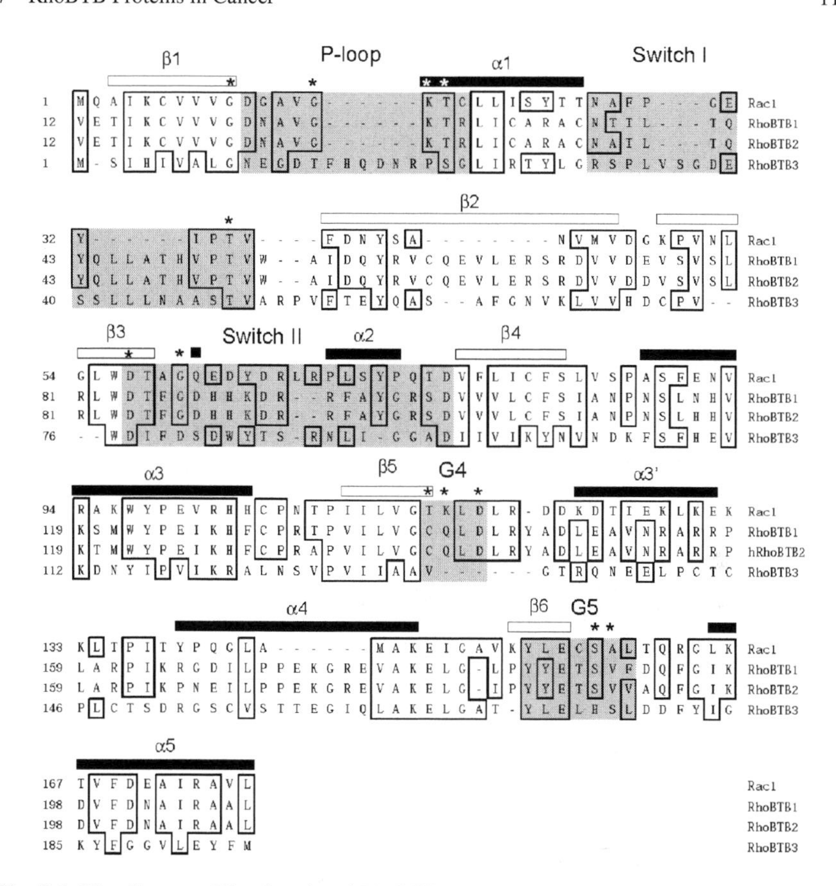

Fig. 7.4 The divergent Rho domain of RhoBTB proteins. (**a**) The diagram shows a sequence alignment of the Rho domain of the three human RhoBTBs against their nearest protein neighbour - Rac1. The positions of the five α-helices and six β-sheets that comprise the fold are indicated above the alignment. The five highly conserved loop regions (G1–G5) that connect these elements are shaded. G1 is more commonly referred to as the P-loop; G2 and G3 form the two Switch regions that contact the γ-phosphate and form the effector-binding region. Residues that are highly conserved in small GTPases are marked in each element (asterisk). The highly conserved Q61 residues (Rac1 numbering) is required for GTP hydrolysis and is absent in RhoBTBs. Its position is also marked (square). (**b**) Molecular modelling of the RhoBTB2 Rho domain

critical for GTP hydrolysis, suggesting that if these proteins can bind GTP, they are not able to hydrolyse it. This would be similar to the Rnd isoforms of Rho GTPases, which are unable to hydrolyse GTP and are instead regulated at the level of transcription (Chardin 2006). Work by Hamaguchi and colleagues has suggested that RhoBTB2 is unable to bind GTP in vitro (Chang et al. 2006). It is possible that these proteins have ceased to be functional nucleotide-binding domains and, instead, perhaps function solely as a protein–protein interaction module; however, homology modelling of the RhoBTB2 Rho domain suggests that the possibility of nucleotide binding is not excluded by the altered structure (Fig. 7.4).

C-Terminal Domain

RhoBTBs contain a unique C-terminal domain that is not found in other proteins. This region of approximately 80 amino acids is predicted to have high alpha-helical content and ends with a high concentration of charged amino acids. The majority of Rho family members undergo prenylation at the C-terminus immediately after the Rho domain, which provides a hydrophobic membrane anchor. RhoBTB1 and RhoBTB2 lack the consensus CAAX box motif required for post-translational modification by prenyl transferases; however, this motif is present at the C-terminus of RhoBTB3. Whether this is a functional prenylation site in vivo has not been tested.

Overall, it would seem that the RhoBTB proteins arose from a gene fusion between a Rac-like GTPase module and a BTB-containing module early in eukaryotic history. In some species, this gene was subsequently lost; however, this fusion was retained and subsequently amplified by gene duplication to create the three RhoBTB proteins present in humans and other mammals.

Genetic Alteration of RhoBTBs in Cancer

The first link between RhoBTB proteins and cancer came from the isolation of RhoBTB2 in a representational difference analysis (RDA) screen for genes lost in breast cancer. This study also gave RhoBTB2 the alternative nomenclature DBC2 for deleted in breast cancer 2 (Hamaguchi et al. 2002). RhoBTB2/DBC2 maps to chromosomal region 8p21, a region that shows loss of heterozygosity (LOH) in a number of cancers, especially head and neck tumours. RhoBTB2 was shown to exhibit a relatively low rate of homozygous deletion in breast cancer (3.5%); however, the expression of the gene was downregulated in approximately 50% of breast and lung cancer samples (Hamaguchi et al. 2002). Similar findings have been obtained in other cancers – RhoBTB2 expression is reduced in 75% of bladder cancer cell lines (Knowles et al. 2005) and we see reduced RhoBTB2 expression in more than 80% of cell lines derived from head and neck small cell carcinomas (HNSCC; unpublished observations).

Wigler and colleagues identified three somatic missense mutations of RhoBTB2 in breast and lung cancer specimens (Hamaguchi et al. 2002). Further mutations have been identified in bladder (Knowles et al. 2005) and gastric cancers (Cho et al. 2007). Mutation of RhoBTB2 appears to occur at a relatively low frequency; however, it is provocative that these mutations seem to cluster in the BTBa domain (Table 7.1). Analysis of the positions of these mutations in the BTBa structure shows that they are in the regions involved in forming the dimerisation interface and the binding pocket (Fig. 7.3). The Y284D mutation has been predicted to disrupt the proper folding of the BTBa domain and hence block dimerisation (Stogios et al. 2005). In support of this, the Y284D mutant RhoBTB2 fails to interact with its binding partner Cul3 (Wilkins et al. 2004). The functional consequences of the remaining RhoBTB2 mutations await full biochemical characterisation.

Table 7.1 Somatic RhoBTB2 mutations identified in human cancer samples

Mutation	Region	Source	Study
-121 C > T	promoter	Breast tumour	Ohadi et al. 2007
+48 G > A	5' UTR	Breast tumour	Ohadi et al. 2007
R275W	BTBa	Gastric tumour	Cho et al. 2007
Y284D	BTBa	Sk-Mes-1 lung cancer line	Hamaguchi et al. 2002
D299N	BTBa	Breast tumour	Hamaguchi et al. 2002
E349D	BTBa	Bladder tumour	Knowles et al. 2005
D368A	BTBa	BT483 breast cancer line	Hamaguchi et al. 2002
F647T	C-terminus	Breast tumour	Hamaguchi et al. 2002

Although there are currently no published studies on the cellular functions of RhoBTB1 or RhoBTB3, some evidence exists for a role of these proteins in cancer. RhoBTB1 has been identified as a candidate tumour suppressor in HNSCC. RhoBTB1 maps to chromosomal region 10q21, which is known to show LOH in 40% of HNSCC cases (Beder et al. 2003). Analysis of the *RhoBTB1* gene showed a matching (44%) LOH, and the expression of RhoBTB1 mRNA was downregulated in 37% of HNSCC samples (Beder et al. 2006). RhoBTB3 is an essentially uncharacterised protein; however, the COSMIC cancer sequencing project (Forbes et al. 2006) has identified a D72H mutation of RhoBTB3 in the NCI-H2009 lung cancer cell line. This mutation lies between the β2 and β3 elements in the Rho domain (Fig. 7.4).

Clearly these are early days in the analysis of the genetic alterations of RhoBTB proteins in cancer; however, there is compelling evidence for frequent changes to RhoBTB2 in a range of carcinomas. The most common event appears to be the downregulation of RhoBTB2 expression – either through allelic loss or gene silencing. Somatic missense mutation appears to be a relatively rare event, but considerably strengthens the case for loss of RhoBTB2 function being an important event and one that is under selective pressure. In addition, the clustering of sporadic mutations on the BTBa domain highlights this signalling module as of special interest. It would seem likely that signalling events mediated by the BTBa domain mediate the potential tumour suppressor functions of RhoBTB2.

RhoBTB Signalling

Unlike the majority of Rho family GTPases, RhoBTB proteins have no apparent effect on the organisation of the actin cytoskeleton (Aspenstrom et al. 2004). Instead, signalling appears to have more in common with the BTB family identity of these proteins. The BTB domain functions as a signalling module through recruitment of downstream signalling partners. Several BTB domain containing proteins have been shown to be cancer genes. These include BCL6, whose function is lost in a range of B-cell lymphomas (Jardin and Sahota 2005), and PLZF, which undergoes chromosomal fusion with the retinoic acid receptor in a subset of acute promyelocytic leukemias (McConnell and Licht 2007). In both cases, these proteins use their BTB domains to recruit transcriptional co-repressor complexes, and their deregulation in cancer leads to an altered transcriptional program.

There are many examples of BTB domain proteins that recruit transcriptional repressors in this way and whose cellular function is to regulate transcription. Most of these proteins, including BCL6 and PLZF, also contain a DNA-binding element that allows them to target co-repressor complex to specific gene targets (Perez-Torrado et al. 2006). The RhoBTB domains lack an obvious DNA–binding motif and instead fall into a separate category of BTB proteins – those that bind the E3 ubiquitin ligase Cul3. BTB proteins play critical roles in assembling protein degradation complexes – the Skp1 BTB protein binds Cul1 in the Skp1-Cul1-F-box complex and the BTB protein Elongin C binds Cul2 in the ElonginC-Cul2-SOCS-box complex. In each case, these proteins consist solely of a BTB domain and they bridge the E3 ubiquitin ligase to the protein that selects targets for ubiquitylation and degradation (Pintard et al. 2004). A recently discovered class of multimodular BTB proteins has been shown to bind to the Cul3 ubiquitin ligase. The archetypal member is the MEL-26 protein in *C. elegans*. This binds to Cul3 through its BTB domain and to the mitotic regulator MEI-1 through an N-terminal MATH domain (Pintard et al. 2003). MEI-1 is the ubiquitylation target of this complex and MEL-26 combines both cullin recruitment and target recruitment into one protein. Several other examples of multimodular Cul3-recruiting BTB proteins have now been identified, suggesting that this is a commonly used mechanism (Perez-Torrado et al. 2006). Carpenter and co-workers have shown that RhoBTB2 binds Cul3 through its BTBa domain. This interaction is disrupted in the Y284D lung cancer mutation (Wilkins et al. 2004). Given the domain structure of RhoBTB2, it is tempting to speculate that the function of the Rho domain is to recruit targets of ubiquitylation by Cul3 – directly analogous to MEI-1. The RhoBTB2-Cul3 complex is certainly functional as RhoBTB2 itself has been shown to be a target both in vitro and in vivo (Wilkins et al. 2004).

Recruitment of Cul3 by RhoBTB2 gives us a clear hint as to the possible general functions of the RhoBTB proteins. It is now important to determine whether RhoBTB1 and RhoBTB3 also are binding partners for Cul3. As yet we do not know the identities of the targets of the RhoBTB2-Cul3 complex. While RhoBTB2 itself is a target, this does not seem a satisfactory sole purpose for this partnership – it seems likely that other targets exist. Clearly, identification of the binding partners of the RhoBTB Rho domains and conserved C-terminal domains represents a likely route to solving this problem – these are obvious candidates for selective recruitment of proteins targets to the RhoBTB-Cul3 complex.

Transcriptional Regulation by RhoBTBs

The widespread role of BTB proteins in transcriptional regulation has prompted two groups to examine the consequences of downregulation of RhoBTB2 in cancer on gene expression. Hamaguchi and colleagues used siRNA silencing of RhoBTB2 in HeLa cells to mimic the loss of RhoBTB2 expression seen in cancer. They then used whole-genome microarray screening to identify genes whose expression was altered. The study found 247 genes whose expression was increased on RhoBTB2 silencing and 433 genes whose expression was decreased. These genes covered cellular pathways including cell-cycle control, cytoskeletal regulation, apoptosis and intracellular trafficking. Unfortunately neither the full list of genes nor their validation data was included in this paper (Siripurapu et al. 2005).

We have taken a similar approach, but with some modification. We targeted RhoBTB2 using siRNA in normal primary human bronchial epithelial cells. Our rational was that, as these cells had not accumulated the genetic alterations seen in cancer cell lines, they might serve as a blank canvas against which to see the specific changes occurring on RhoBTB2 loss. Interestingly, we observed only two genes whose expression changed significantly on silencing of RhoBTB2. One is the chemokine CXCL14/BRAK, which was not reported as a RhoBTB2 target in the Hamaguchi screen. Silencing of RhoBTB2 leads to a marked loss of CXCL14 expression in primary lung, breast and prostate epithelial cells (CM and HM, unpublished observations). CXCL14 is an interesting target. It is normally secreted by epithelial cells; however, expression is lost or downregulated in a wide range of carcinomas (Frederick et al. 2000). In breast tumours, CXCL14 is upregulated by stromal myoepithelial cells, suggesting that what is normally an autocrine signal becomes a paracrine one (Allinen et al. 2004). CXCL14 has been shown to be a chemo-attractant for dendritic cells (Shellenberger et al. 2004), monocytes (Kurth et al. 2001), and natural killer cells (Starnes et al. 2006). This has led to the suggestion that loss of CXCL14 expression in tumours may lead to a reduced level of immune surveillance and hence a survival advantage for the tumour cells. In keeping with this, head and neck tumours that lack CXCL14 expression grow faster and have fewer infiltrating dendritic cells (Shurin et al. 2005). We are currently exploring whether loss of RhoBTB2 function is the major cause of loss of CXCL14 expression in squamous cell carcinomas.

The basis for transcriptional regulation by RhoBTB2 is unclear. Both microarray studies were prompted no doubt by the links between other BTB-containing cancer genes and transcriptional regulation; however, RhoBTB2 has since been shown to bind Cul3, and no evidence as yet exists for the recruitment of co-repressor complexes by these proteins. Some of the degradation targets of other BTB-Cul3 complexes are transcriptional factors (Perez-Torrado et al. 2006). It seems possible that the transcriptional regulation mediated by RhoBTB2 involves the degradation of a protein factor involved in transcriptional regulation of those gene targets.

RhoBTBs and Cell Cycle

Initial studies showed that overexpression of RhoBTB2 could inhibit cell proliferation in a breast cancer cell line, whereas overexpression of RhoBTB cancer mutants did not (Hamaguchi et al. 2002). Recent work has identified RhoBTB2 as a direct target of the E2F1 transcription factor, an important regulator of cell cycle progression (Freeman et al. 2008). E2F activity is upregulated at the G1/S boundary and this study showed that RhoBTB2 expression increases in S-phase, remaining elevated until the end of mitosis. Analysis of the effects of overexpression of RhoBTB2 on cell cycle produced complicated results. RhoBTB2 overexpression initially caused an increase in the proportion of cells in S-phase and an increased level of DNA synthesis, as judged by BrdU incorporation. This did not translate into an increased number of cells, however, and the authors concluded that the long-term effects of RhoBTB2 overexpression were inhibitory to proliferation (Freeman et al. 2008). What happens to the cells that initially experience an increase in DNA synthesis is unclear. No significant level of apoptosis was detected; however, the distribution of cells in cell cycle was not measured under conditions of long-term expression in RhoBTB2 (Freeman et al. 2008). How RhoBTB2 might affect cell cycle is also unclear. Hamaguchi and colleagues have shown that overexpression of RhoBTB2 leads to a decrease in the levels of Cyclin D1 through post-translational regulation (Yoshihara et al. 2007). Whether this is a direct effect of RhoBTB2 or simply a reflection of the altered cell cycle profile of cells expressing RhoBTB2 is not known.

Are RhoBTBs Bona Fide Tumour Suppressors?

RhoBTB2/DBC2 has been described as a candidate tumour suppressor. It certainly bears many of the hallmarks of a tumour suppressor – it is downregulated by several distinct mechanisms in a wide range of carcinomas, it is regulated in a cell cycle dependent manner and it seems to be able to impact on cell cycle progression. Clearly it is premature to describe RhoBTB2 as a bona fide tumour suppressor as we still do not properly understand its cellular role and have no information about its effects on tumour progression. A second question is whether RhoBTBs are bona fide Rho GTPases. This is a semantic point – RhoBTBs are Rho GTPases by heritage alone, with a pedigree that can be traced back to an ancient fusion event between a Rac GTPase and a BTB-containing module; however, the RhoBTBs do not seem to *behave* like other Rho GTPases. In this respect, it is not helpful to think about the roles of RhoBTBs in cancer in the same way as those of other Rho GTPases where regulation of the actin cytoskeleton is a common theme. What is clear even at this early stage in the study of these intriguing signalling proteins is that all the evidence points towards the RhoBTBs as being bona fide cancer genes. Clarification of the role of the RhoBTB Rho GTPase domain in these events awaits further investigation.

References

Ahmad KF, Engel CK, Prive GG (1998) Crystal structure of the BTB domain from PLZF. Proc Natl Acad Sci U S A 95:12123–12128.

Allinen M, Beroukhim R, Cai L, et al (2004) Molecular characterization of the tumor microenvironment in breast cancer. Cancer Cell 6: 17–32.

Aspenstrom P, Fransson A, Saras J (2004) Rho GTPases have diverse effects on the organization of the actin filament system. Biochem J 377:327–337.
Beder LB, Gunduz M, Ouchida M, et al (2003) Genome-wide analyses on loss of heterozygosity in head and neck squamous cell carcinomas. Lab Invest 83:99–105.
Beder LB, Gunduz M, Ouchida M, et al (2006) Identification of a candidate tumor suppressor gene RHOBTB1 located at a novel allelic loss region 10q21 in head and neck cancer. J Cancer Res Clin Oncol 132:19–27.
Boureux A, Vignal E, Faure S, Fort P (2007) Evolution of the Rho family of ras-like GTPases in eukaryotes. Mol Biol Evol 24:203–216.
Chang FK, Sato N, Kobayashi-Simorowski N, Yoshihara T, Meth JL, Hamaguchi M (2006) DBC2 is essential for transporting vesicular stomatitis virus glycoprotein. J Mol Biol 364:302–308.
Chardin P (2006) Function and regulation of Rnd proteins. Nat Rev Mol Cell Biol 7:54–62.
Cho YG, Choi BJ, Kim CJ et al (2007) Genetic analysis of the DBC2 gene in gastric cancer. Acta Oncol 28:1–6.
Forbes S, Clements J, Dawson E et al (2006) Cosmic 2005. Br J Cancer 94:318–322.
Frederick MJ, Henderson Y, Xu X et al (2000) In vivo expression of the novel CXC chemokine BRAK in normal and cancerous human tissue. Am J Pathol 156:1937–1950.
Freeman SN, Ma Y, Cress WD (2008) RhoBTB2 (DBC2) Is a Mitotic E2F1 Target Gene with a Novel Role in Apoptosis. J Biol Chem 283:2353–2362.
Hamaguchi M, Meth JL, von Klitzing C et al (2002) DBC2, a candidate for a tumor suppressor gene involved in breast cancer. Proc Natl Acad Sci U S A 99:13647–13652.
Jardin F, Sahota SS (2005) Targeted somatic mutation of the BCL6 proto-oncogene and its impact on lymphomagenesis. Hematology 10:115–129.
Knowles MA, Aveyard JS, Taylor CF, Harnden P, Bass S (2005) Mutation analysis of the 8p candidate tumour suppressor genes DBC2 (RHOBTB2) and LZTS1 in bladder cancer. Cancer Lett 225:121–130.
Kurth I, Willimann K, Schaerli P, Hunziker T, Clark-Lewis I, Moser B (2001) Monocyte selectivity and tissue localization suggests a role for breast and kidney-expressed chemokine (BRAK) in macrophage development. J Exp Med 194:855–861.
McConnell MJ, Licht JD (2007) The PLZF gene of t (11;17)-associated APL. Curr Top Microbiol Immunol 313:31–48.
Ohadi M, Totonchi M, Maguire P et al (2007). Mutation analysis of the DBC2 gene in sporadic and familial breast cancer. Acta Oncol 46:770–772.
Paduch M, Jelen F, Otlewski J (2001) Structure of small G proteins and their regulators. Acta Biochim Pol 48:829–850.
Pellegrin S, Mellor H (2006) Evolution of the human Rho GTPase family – Conservation and diversity. In: Manser E (ed) RHO family GTPases, vol. 3. Springer, Netherlands, pp. 19–29.
Perez-Torrado R, Yamada D, Defossez PA (2006) Born to bind: the BTB protein-protein interaction domain. Bioessays 28:1194–1202.
Philips A, Blein M, Robert A et al (2003) Ascidians as a vertebrate-like model organism for physiological studies of Rho GTPase signaling. Biol Cell 95:295–302.
Pintard L, Willems A, Peter M (2004) Cullin-based ubiquitin ligases: Cul3-BTB complexes join the family. Embo J 23:1681–1687.
Pintard L, Willis JH, Willems A et al (2003) The BTB protein MEL-26 is a substrate-specific adaptor of the CUL-3 ubiquitin-ligase. Nature 425:311–316.
Ramos S, Khademi F, Somesh BP, Rivero F (2002) Genomic organization and expression profile of the small GTPases of the RhoBTB family in human and mouse. Gene 298:147–157.
Rivero F, Dislich H, Glockner G, Noegel AA (2001) The Dictyostelium discoideum family of Rho-related proteins. Nucleic Acids Res 29:1068–1079.
Shellenberger TD, Wang M, Gujrati M et al (2004) BRAK/CXCL14 is a potent inhibitor of angiogenesis and a chemotactic factor for immature dendritic cells. Cancer Res 64:8262–8270.
Shurin GV, Ferris RL, Tourkova IL et al (2005) Loss of new chemokine CXCL14 in tumor tissue is associated with low infiltration by dendritic cells (DC), while restoration of human CXCL14

expression in tumor cells causes attraction of DC both in vitro and in vivo. J Immunol 174:5490–5498.

Siripurapu V, Meth J, Kobayashi N, Hamaguchi M (2005) DBC2 significantly influences cell-cycle, apoptosis, cytoskeleton and membrane-trafficking pathways. J Mol Biol 346:83–89.

Starnes T, Rasila KK, Robertson MJ et al (2006) The chemokine CXCL14 (BRAK) stimulates activated NK cell migration: implications for the downregulation of CXCL14 in malignancy. Exp Hematol 34:1101–1105.

Stogios PJ, Downs GS, Jauhal JJ, Nandra SK, Prive GG (2005) Sequence and structural analysis of BTB domain proteins. Genome Biol 6:R82.

Wilkins A, Ping Q, Carpenter CL (2004) RhoBTB2 is a substrate of the mammalian Cul3 ubiquitin ligase complex. Genes Dev 18:856–861.

Yoshihara T, Collado D, Hamaguchi M (2007) Cyclin D1 down-regulation is essential for DBC2's tumor suppressor function. Biochem Biophys Res Commun 358:1076–1079.

Chapter 8
RhoC GTPase in Cancer Progression and Metastasis

Kenneth van Golen

Introduction

A variety of phenotypic characteristics are required for a cancer cell to successfully complete the metastatic cascade. The acquisitions of motile and invasive properties are required by a cell to become metastatically competent. As discussed in detail throughout this book, the Rho GTPases are a subfamily of small GTP-binding proteins, which are related to the Ras oncogene and thus comprise a subfamily of the Ras superfamily of small GTP-binding proteins. Nearly all cellular activities, particularly all aspects of cellular motility and invasion, are ultimately controlled by the Rho GTPases and are closely linked to signals from the extracellular environment, particularly in response to growth factors. Therefore, as reviewed in Chap. 2, the importance of Rho GTPases in the progression of cancer and the metastatic phenotype are becoming increasingly evident.

A great deal of excitement ensued over the discovery of Rho GTPases, as many investigators assumed that due to their homology to Ras, the Rho proteins would also be oncogenes in human cancer, harboring activating mutations similar to Ras. Three isoforms of Ras – Ha-Ras, Ki-Ras, and N-Ras – were identified early in the 1980s as oncogenes mutated in a variety of human cancers (Hall et al. 1983; Brown et al. 1984; McGrath et al. 1983; Capon et al. 1983; Feramisco et al. 1984; Stacey and Kung 1984). Approximately, 30% of human tumors carry an identifiable Ras mutation, which render the GTPase incapable of hydrolyzing bound GTP, thus remaining constitutively active (Rodenhuis 1992). Breast cancer is the exception, where approximately 5% tumors harbor an activating Ras mutation (Rochlitz et al. 1989). However, in contrast to Ras, no mutation in any of the Rho proteins has been identified in human tumors. Rather, overexpression of Rho proteins, particularly RhoA, RhoC, Rac1, and Cdc42, appears to be the rule in human cancers (Fritz et al. 1999;

K. van Golen (✉)
Center for Translational Cancer Research, Department of Biological Sciences, Laboratory of Cytoskeletal Physiology, University of Delaware
e-mail: klvg@udel.edu

K. van Golen (ed.), *The Rho GTPases in Cancer*,
DOI 10.1007/978-1-4419-1111-7_8,

Moscow et al. 1994; van Golen et al. 2000b, 1999; Imamura et al. 1999; Clark et al. 2000; del Peso et al. 1997; Avraham 1990).

RhoA GTPase Versus RhoC GTPase (What Is the Difference?)

RhoA, RhoB, and RhoC GTPases are highly homologous to each other and reside on the same branch of the Rho subfamily phylogenetic tree (reviewed in (Lin and van Golen 2003); also see Chap. 1). RhoA and RhoC share closest homology, being 84% identical on the mRNA level and 91% identical on the protein level (Takai et al. 2001). As shown in Fig. 8.1, the main differences in amino acid sequence homology between RhoA, -B, and -C occur in the C-terminal region known as the hypervariable region (Takai et al. 2001). Given the close homology between RhoA and RhoC, many investigators assumed that these two GTPases would have redundant activities and their activation state would be modulated by the same GAP and GEF proteins. This premise is partly based on the experiments that demonstrate that most of the same downstream effectors, Rho Kinase (aka Rock or Rok), mDia2, and rhotekin, are efficiently activated by both RhoA and RhoC (Sahai and Marshall 2002; Sander et al. 1999; Zondag et al. 2000; Bishop and Hall 2000). Whether this extends to the GAPs and GEFs is currently unknown and represents a gap in our understanding of how RhoA and RhoC are differentially regulated. Much of the information regarding which GAPs, GEFs, and effector proteins interact with RhoC were gleaned from in vitro pull-down assays and yeast two-hybrid experiments.

Despite this, it has become increasingly apparent that these two GTPases differ in their respective activities and exert distinctly different phenotypes when expressed. Expression of constitutively active RhoA into NIH3T3, Swiss3T3, and other similar cell types have resulted in the rapid formation of stress fibers and

Rho GTPase Sequence Homology

```
P08134_RHOC_HUM MAAIRKKLVIVGDGACGKTCLLIVFSKDQFPEVYVPTVFENYIADIEVDGKQVELALWDT
P61586_RHOA_HUM MAAIRKKLVIVGDGACGKTCLLIVFSKDQFPEVYVPTVFENYVADIEVDGKQVELALWDT
P62745_RHOB_HUM MAAIRKKLVVVGDGACGKTCLLIVFSKDEFPEVYVPTVFENYVADIEVDGKQVELALWDT
consensus       MAAIRKKLViVGDGACGKTCLLIVFSKDqFPEVYVPTVFENYvADIEVDGKQVELALWDT

P08134_RHOC_HUM AGQEDYDRLRPLSYPDTDVILMCFSIDSPDSLENIPEKWTPEVKHFCPNVPIILVGNKKD
P61586_RHOA_HUM AGQEDYDRLRPLSYPDTDVILMCFSIDSPDSLENIPEKWTPEVKHFCPNVPIILVGNKKD
P62745_RHOB_HUM AGQEDYDRLRPLSYPDTDVILMCFSVDSPDSLENIPEKWVPEVKHFCPNVPIILVANKKD
consensus       AGQEDYDRLRPLSYPDTDVILMCFSiDSPDSLENIPEKWtPEVKHFCPNVPIILVgNKKD

P08134_RHOC_HUM LRQDEHTRRELAKMKQEPVRSEEGRDMANRISAFGYLECSAKTKEGVREVFEMATRAGLQ
P61586_RHOA_HUM LRNDEHTRRELAKMKQEPVKPEEGRDMANRIGAFGYMECSAKTKDGVREVFEMATRAALQ
P62745_RHOB_HUM LRSDEHVRTELARMKQEPVRTDDGRAMAVRIQAYDYLECSAKTKEGVREVFETATRAALQ
consensus       LRqDEHtRrELAkMKQEPVrseeGRdMAnRI-AfgYlECSAKTKeGVREVFEmATRAaLQ

P08134_RHOC_HUM VRKNKRR---RGCPIL
P61586_RHOA_HUM ARRGKKK---SGCLVL
P62745_RHOB_HUM KRYGSQNGCINCCKVL
consensus       -Rkgkrr----gC-vL
```

Fig. 8.1 Comparison of the RhoA, RhoB, and RhoC protein sequences. All three proteins have a high degree of homology to one another with minor differences seen in the majority of the proteins. Significant differences are seen in the C-terminal hypervariable region

impediment of cellular motility and invasion (Hall 1990, 1998; Ridley and Hall 1992a, b; Ridley et al. 1992). Similarly, inactivation of p190RhoGAP (which catalyzes the hydrolysis of GTP to GDP on RhoA; please see Chap. 6) leads to increased accumulation of active GTP-bound RhoA and the formation of stress fibers (Arthur and Burridge 2001; Billuart et al. 2001; Haskell et al. 2001). This also leads to the inhibition of cellular motility and invasion. Conversely, expression of active, wild-type RhoC GTPase in immortalized human mammary epithelial cells (HMECs) results in increased cellular motility and invasion (van Golen et al. 2000b).

Given this data, it appears that RhoA GTPase may hinder, while RhoC GTPase promotes cellular motility and invasion. These data are supported circumstantially by studies that have demonstrated expression of RhoC in highly aggressive and metastatic tumors. While RhoA expression, as determined on the mRNA and protein level, has been found in nearly all tumors analyzed, RhoC expression correlates almost exclusively with locally advanced and metastatic tumors. Experimental evidence supports a potential reciprocal relationship between RhoA and RhoC in SUM159 breast cancer cells (Simpson et al. 2004). Using an siRNA approach, Simpson et al. selectively inhibited either RhoA or RhoC and measured the activity of the reciprocal GTPase and the ability of the cells to invade through a Matrigel™ coated filter. Interestingly, when RhoA levels are reduced, a concurrent increase in RhoC activity and a subsequent increase in invasiveness are observed. Likewise, downregulation of RhoC leads to an increase in RhoA activity and a significant decrease in invasiveness. Our laboratory has demonstrated similar results in the PC-3 prostate cancer cell line (Yao et al. 2006; Sequeira et al. 2008). Using a dominant negative version of RhoC we demonstrated an epithelial to mesenchymal transition (EMT) of PC-3 cells that was accompanied by a marked increase in linear motility and loss of directional invasion (Yao et al. 2006). Changes in all of the appropriate EMT markers and in the motile and invasive characteristics of the cells were due to an increased and sustained activation of Rac and Cdc42 GTPases. Similarly, using an shRNA approach to systematically downregulate RhoA, RhoC, and Rac GTPases, we demonstrated that RhoC is essential for prostate cancer cell invasion (Sequeira et al. 2008). Similarly, decreased expression of RhoA leads to a significant increase in the cell's invasive abilities.

Although this evidence suggests a nonredundant role for RhoA and RhoC, evidence from the MDA-MB-231 breast cancer cell line suggests that the opposite is true. Also using an siRNA approach, Pille et al. demonstrated that loss of expression of either RhoA or RhoC led to a significant decrease in the invasive and metastatic potential of MDA-MB-231 cells in vitro (Pille et al. 2005). Additional studies from the same group explored the efficacy of performing intravenous injections of siRNA-loaded nanoparticles into mice bearing MDA-MB-231 xenografts (Pille et al. 2008). The treatments proved not only to be nontoxic to the mice but also prevented growth of orthotopic MDA-MB-231 xenografts. In collaboration with Doug Yee's group, our laboratory group demonstrated that both RhoA and RhoC are required for IGF-I-stimulated invasion of the bone-adapted MDA-MB-231-BO cell line (Zhang et al. 2005). In a related study, we also demonstrated that RhoA and RhoC were required for IGF-I-induced production of proangiogenic factors by the MDA-MB-231 and MDA-MB-231-BO cell lines (Groh 2006).

The difference in how RhoA and RhoC mediate distinct cellular responses may lay within their guanine nucleotide binding domains. Differences in active-state conformational changes due to GTP binding may effect interactions with downstream effector proteins, modulate intrinsic and/or extrinsic GTP hydrolysis, or potentially effect the subcellular localization of the protein. The X-ray crystal structure of the active and inactive states of RhoA has been performed (Wei et al. 1997; Ihara et al. 1998; Longenecker et al. 2003). These structural analyses demonstrate two switch regions, analogous to Ras, lead to a single conformational change leading to a signaling-active state. An X-ray structure of RhoC, complexed with its effector protein mDia1, provided some, but limited insight into the mechanisms of its structural interaction and signaling state (Rose et al. 2005). Recently, a detailed study of RhoC structure in its active GDP-bound, and active state, bound to either nonhydrolyzable GppNHp or GTPγS gave significant insights of how it differs from RhoA (Dias and Cerione 2008). Unlike RhoA, RhoC undergoes two conformational changes during its conversion from an inactive to signaling-active state (Dias and Cerione 2008). These conformational changes could lead to interactions with downstream effectors unique from RhoA. This is briefly discussed in Chap.10 of this book.

Although many studies focus on a single Rho GTPase (e.g., RhoA or RhoC) that appears to be predominant, it is nearly certain that interaction and reciprocal activation of several GTPases dictate the motile, invasive, and metastatic phenotype of a cell (van Golen 2003; Moorman et al. 1999). To be fully effective in achieving the invasive phenotype, the Rho proteins likely need to complete a full GTPase cycle and continue to cycle, transitioning between an inactive and active state, thus allowing cells to effectively reorganize the cytoskeleton, and form lamellipodia and filipodia. The need for reciprocal activation during cell motility is perhaps the reason why no activating mutations, such as those identified for Ras, have been found for Rho in human cancers. RhoA may be a "foundation" GTPase – one that is required for cellular motility and invasion. Expression and activation of other GTPases, such as RhoC, augment and refine this motile phenotype, allowing the cell to respond and invade in the tumor microenvironment or at distant sites.

RhoC GTPase in Cancer and Metastasis

RhoC GTPase expression in human cancer was first described in 1998 where it was found in human pancreatic adenocarcinoma; however, no actual functional role was assigned to the protein (Suwa et al. 1994). In this study, RhoC expression correlated with increased metastasis, particularly perineural invasion and a worse prognosis. This same study also examined oncogenic K-Ras, RhoA, and RhoB expression but found no correlation between these GTPases and metastatic disease. In 2002, expression profiling of microdissected pancreatic tumors confirmed that RhoC was

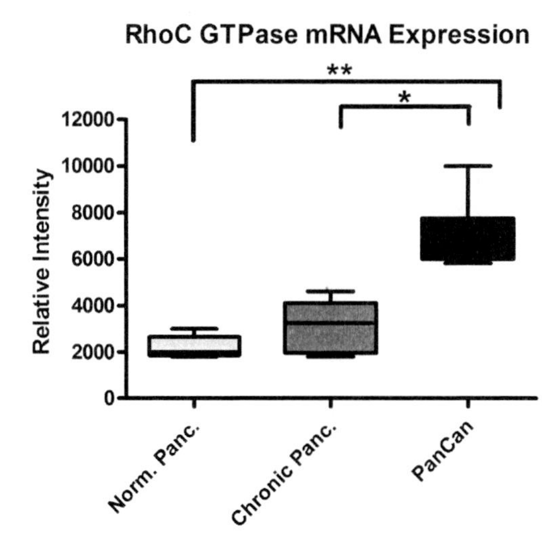

Fig. 8.2 Results of a cDNA microarray comparison of pancreatic tissue from ten normal, ten chronic pancreatitis, and ten pancreatic cancer patients. Laser capture microdissection was utilized to isolate pure epithelial cells and exclude stroma. Increased expression of RhoC GTPase is observed in both patients with chronic pancreatitis and pancreatic cancer

expressed in particularly invasive and metastatic pancreatic cancers (Crnogorac-Jurcevic et al. 2002). In another microarray analysis of ten microdissected pancreatic tumors, ten chronic pancreatitis, and ten normal pancreatic epithelium (Logsdon et al. 2003), an average of 3.1- and 2-fold increase in RhoC mRNA levels in each of the disease states, respectively, was observed (Fig. 8.2). Chronic pancreatitis is a major risk factor for the development of pancreatic cancer and these data suggest that increased RhoC expression in pancreatitis may predicate aggressive cancer. Immunohistochemical staining of preinvasive (PanIn) lesions (Klein et al. 2002), which often progress to invasive adenocarinoma, isolated from patients, indicate strong, heterogenous RhoC staining, which suggests a potential prognostic role of the GTPase in pancreatic cancer (Lucey and van Golen, unpublished data).

Evidence suggests a role for RhoC in the motile and invasive phenotype of pancreatic cancer cells (Lin et al. 2005). We have demonstrated that inhibition of RhoC activity through the use of a dominant negative RhoC abrogates the motile, invasive, and metastatic properties of the highly aggressive HPAF-II cell line. Paradoxically, the ability of these cells to grow under anchorage-independent conditions or to form ectopic, subcutaneous tumors in nude mice was actually enhanced. Although mice grew larger tumors at a significantly faster rate, no ascites fluid, micro-, or macrometastases were observed (van Golen, unpublished data).

In the HPAF-II pancreatic cancer cell line, interaction of RhoC with the scaffolding protein, caveolin-1 (cav-1), led to decreased RhoC activity and signaling through the Erk arm of the mitogen-activated protein kinase (MAPK) pathway along with attenuated migration and invasion (Lin et al. 2005). Loss of cav-1 expression in HPAF-II cells results in increased RhoC activity and p38 MAPK-dependent migration.

Together, these data suggest a distinct role for RhoC in conferring a metastatic phenotype, which may be modulated by other signaling and regulatory molecules.

Like pancreatic cancer, inflammatory breast cancer (IBC) is a particularly aggressive form of locally advanced cancer, which invades the dermal lymphatics of the skin overlaying the breast (reviewed in (Jaiyesimi et al. 1992; Kleer et al. 2000)). Because of this, IBC is highly metastatic and carries with it the worst prognosis of all breast cancers (Jaiyesimi et al. 1992; Kleer et al. 2000). In an attempt to identify a molecular profile of IBC,,we found that RhoC was overexpressed two- to eightfold compared with normal breast epithelium in higher than 90% of IBC tumor specimens compared with stage-matched non-IBC specimens (van Golen et al. 1999). These data were recently validated through the use of tissue microarray analysis comparing IBC with non-IBC, stage-matched patient samples (Van Laere et al. 2005).

Subsequently, RhoC has been identified as a prognostic marker for small breast tumors, less than 1 cm in diameter, which have a propensity to metastasize (Kleer et al. 2002a). Studies that introduced RhoC GTPase into immortalized HMECs (either the 6–15 HMECs or MCF10A cells) validated the role of RhoC as mammary oncogene (van Golen et al. 2000a, b, 2002b; Wu et al. 2004). Expression of wild-type RhoC in HMECs leads to anchorage-independent growth, increased motility and invasion, production of angiogenic factors, and induction of metastases in an experimental metastasis model (van Golen et al. 2000a, b, 2002b; Wu et al. 2004). Furthermore, cDNA microarray comparison of RhoC-expressing versus control HMECs demonstrates differential activation of genes involved in increased growth and invasion (Wu et al. 2004). Significant increases in mRNA levels of cyclin D1, vacular endotheial growth factor-C (VEGF-C), CXC cytokine ligand-1 (CXCL1), caveolin-2 (cav-2), and insulin-like growth factor binding protein-2 (IGFBP-2) were also observed (Wu et al. 2004).

The link between RhoC and IGF-signaling in IBC extends further than increased IGFBP-2 mRNA levels. Wisp3 (aka IGFBP-related protein-9; IGFBP-rp-9 and LIBC) was found to be consistently and concordantly lost in IBC tumors expressing RhoC (van Golen et al. 1999; Kleer et al. 2002b). Wisp3 is predominantly lost in IBC and is, therefore, a tumor suppressor that cooperates with RhoC to form IBC tumors and influences tumor growth by modulating insulin-like growth factor-I (IGF-I) signaling (Kleer et al. 2002a, b, 2004b). Thus, these data suggest intricate growth factor regulation of RhoC.

A similar link between the IGF-I signaling axis and RhoC exists in prostate cancer cells. PC-3 prostate cancer cells bound to type I collagen and treated with low-dose IGF-I exhibit increased RhoC activity and are able to invade through a type I collagen or Matrigel™ coated filter (Yao et al. 2006). Similarly, when LNCaP cells were selected by virtue of their ability to bind type I collagen, they displayed increased levels of active and total RhoC protein in comparison with parental and noncollagen-binding LNCaP cells (Hall et al. 2006). Interestingly, the C4–2B4 cell line, a bone-adapted subclone of the LNCaP progression model (Thalmann et al. 2000), also displays increased RhoC expression and activation compared with parental LNCaP cells (Hall et al. 2006). Activation of RhoC in the LNCaP collagen-binding cells is increased by binding to collagen I (Hall et al. 2008). Along with increased levels of RhoC, the cells also display increased levels of the type I collagen receptor, $\alpha_2\beta_1$. Use of a blocking antibody to $\alpha_2\beta_1$ significantly inhibits RhoC activation and invasion through type I collagen (Hall et al. 2008).

As outlined previously in this chapter, we have demonstrated that the MDA-MB-231-BO cell line – a metastatic bone variant of the original MDA-MB-231 cell line – has a very distinct pattern of RhoA and RhoC activation during IGF-I-stimulated invasion (Zhang et al. 2005). Interestingly, activation of both RhoA and RhoC increases after IGF-I stimulation with peak activation at 30 min and 3 min, respectively, suggesting coordinated yet distinct roles in the migration process. Concordantly, a decrease in RhoB activity was observed while RhoG, Rac, and Cdc42 activity remained unchanged.

In 2003–2004, there was an explosion of new reports demonstrating a correlation of RhoC expression and metastasis in a variety of human cancers (Carr et al. 2003; Marionnet et al. 2003; Horiuchi et al. 2003; Kamai et al. 2003; Shinto et al. 2003; Wang et al. 2003; Kondo et al. 2004; Shikada et al. 2003; Schwering et al. 2003). These cancers include non-Hodgkins lymphoma, melanoma, ovarian, hepatocellular, colorectal, gastric, nonsmall cell lung, bladder, and basal cell carcinomas. These studies have used in situ hybridization, microarray analysis, or immunohistochemistry to determine levels of RhoC mRNA and protein in tumor samples. Despite a lack of functional analysis of RhoC in these studies, they help to underscore the distinct and important role of RhoC in conferring a metastatic phenotype to a variety of cancers.

The most compelling evidence for the essential role of RhoC GTPase in metastasis does not come from patient studies but from a conditional knock-out mouse model (Hakem et al. 2005). Use of this model demonstrated that expression of RhoC was not need for normal embryonic development or for formation of primary, experimental breast tumors, but is absolutely essential for the formation of distant metastasis.

Since RhoC GTPase appears to be essential for metastasis, but plays a minor role in other normal processes such as development, it has become a choice target for potential antimetastatic therapies. In collaboration with George Prendergast, our laboratory has studied the potential use of farnesyl transferase inhibitors (FTIs) as an adjuvant therapy for IBC. Like other monomeric G-proteins, RhoC is modified by prenylation and inserted into the inner plasma membrane where it is thought to be activated and can come in contact with downstream effectors (Kirschmeier et al. 2001). RhoC, like other Rho proteins, is specifically geranylgeranylated (Adamson et al. 1992). Interestingly, RhoB can either be geranylgeranylated or farnesylated, with the latter modification being associated with transformation [(Prendergast et al. 1995); see Chap. 9 for more detail on RhoB and FTIs]. Treatment of cells with FTIs leads to the accumulation of the geranygeranylated form of RhoB and reversion of the transformed phenotype (Lebowitz et al. 1995, 1997; Du et al. 1999). In our study, treatment of the SUM149 IBC cell line with FTI led to the reversion of the RhoC-mediated invasive phenotype without effecting cell viability (van Golen et al. 2002a). In these treated cells, an increase in RhoB was observed; further, the

FTI effects could be phenocopied by transfection of a RhoB that could only be geranylgeranylated.

Further studies looking at the possibility of RhoC as an antimetastatic target were performed using melanoma as a model. These studies were spawned by cDNA microarray comparison of the murine B16 or human A375 nonmetastatic melanoma clones with metastatic clones of the same lines (Clark et al. 2000). This comparison demonstrated that RhoC was expressed in the metastatic clone and was required for the metastatic phenotype. Later studies by another group used artovastatin, a potent inhibitor of 3-hydroxy 3-methylglutaryl CoA (HMG-CoA) reductase, to prevent RhoC prenylation and translocation to the plasma membrane where it is activated (Hancock et al. 1991; Adamson et al. 1992). Treatment of human melanoma cells with artovastatin prevented RhoC-mediated invasion in vitro. More strikingly, at plasma levels comparable to those used to treat hypercholesterolemia, in vivo metastasis was significantly reduced. These data help to validate RhoC as an important potential anitmetastastic therapeutic target. This potential is recognized by several investigators and pharmaceutical companies alike with a push to develop small molecule inhibitors with high specificity and low toxicity.

Conclusions

Despite its apparent homology to RhoA and RhoB GTPases, RhoC uniquely confers metastatic potential to cells. It has been shown to be a potential biomarker for aggressive breast cancers with a potential to metastasize. More exciting is the potential of RhoC to be an antimetastatic target. Research on RhoC has increased significantly over the past 5 years; however, many questions remain unanswered. Continued research on this GTPase, both basic and translational, is needed to understand the physiology behind this important G-protein.

References

Adamson P, Marshall CJ, Hall A, Tilbrook PA (1992) Post-translational modifications of p21rho proteins. J Biol Chem 267:20033–20038.

Arthur WT, Burridge K (2001) RhoA inactivation by p190RhoGAP regulates cell spreading and migration by promoting membrane protrusion and polarity. Mol Biol Cell 12:2711–2720.

Avraham H (1990) Rho gene amplification and malignant transformation. Biochem Biophys Res Commun 168:114–124.

Billuart P, Winter CG, Maresh A, Zhao X, Luo L (2001) Regulating axon branch stability: the role of p190 RhoGAP in repressing a retraction signaling pathway. Cell 107:195–207.

Bishop AL, Hall A (2000) Rho GTPases and their effector proteins. Biochem J 348(Pt 2):241–255.

Brown R, Marshall CJ, Pennie SG, Hall A (1984) Mechanism of activation of an N-ras gene in the human fibrosarcoma cell line HT1080. EMBO J 3:1321–1326.
Capon DJ, Seeburg PH, McGrath JP, et al (1983) Activation of Ki-ras2 gene in human colon and lung carcinomas by two different point mutations. Nature 304:507–513.
Carr KM, Bittner M, Trent JM (2003) Gene-expression profiling in human cutaneous melanoma. Oncogene 22:3076–3080.
Clark EA, Golub TR, Lander ES, Hynes RO (2000) Genomic analysis of metastasis reveals an essential role for RhoC. Nature 406:532–535.
Crnogorac-Jurcevic T, Efthimiou E, Nielsen T, et al (2002) Expression profiling of microdissected pancreatic adenocarcinomas. Oncogene 21:4587–4594.
del Peso L, Hernandez-Alcoceba R, Embade N, et al (1997) Rho proteins induce metastatic properties in vivo. Oncogene 15:3047–3057.
Dias S, Cerione RA (2008) X-ray crystal sturctures reveal two actived states for RhoC. Biochemistry 46:6547–6558.
Du W, Lebowitz PF, Prendergast GC (1999) Cell growth inhibition by farnesyltransferase inhibitors is mediated by gain of geranylgeranylated RhoB. Mol. Cell Biol 19:1831–1840.
Feramisco JR, Gross M, Kamata T, Rosenberg M, Sweet RW (1984) Microinjection of the oncogene form of the human H-ras (T-24) protein results in rapid proliferation of quiescent cells. Cell 38:109–117.
Fritz G, Just I, Kaina B (1999) Rho GTPases are over-expressed in human tumors. Int J Cancer 81:682–687.
Groh K, Lin M, van Golen CM, van Golen KL (2006) The Rho GTPases and angiogenesis. In: Zubar R (ed) New angogenesis research. Nova Biomedical Books, New York, pp. 19–42.
Hakem A, Sanchez-Sweatman O, You-Ten A, et al (2005) RhoC is dispensable for embryogenesis and tumor initiation but essential for metastasis. Genes Dev 19:1974–1979.
Hall A (1990) The cellular functions of small GTP-binding proteins. Science 249:635–640.
Hall A (1998) Rho GTPases and the actin cytoskeleton. Science 279:509–514.
Hall A, Marshall CJ, Spurr NK, Weiss RA (1983) Identification of transforming gene in two human sarcoma cell lines as a new member of the ras gene family located on chromosome 1. Nature 303:396–400.
Hall C, Dai J, van Golen KL, Keller ET, Long M (2006) Type I colagen receptor (alpha2 beta1) signaling promotes the growth of human prostate cancer cells within the bone. Cancer Res 66:8648–8654.
Hall C, Shein D, Dubyk C, Riesenberger T, Keller ET, van Golen KL (2008) Type I Collagen receptor (alpha2beta3) signaling promotes prostate cancer cell invasion through RhoC GTPase. Neoplasia *In Press*.
Hall C, Dubyk CW, Riesenberger TA, Shein D, Keller ET, van Golen KL Neoplasia 10(8):797–803.
Hancock JF, Cadwallader K, Paterson H, Marshall CJ (1991) A CAAX or a CAAL motif and a second signal are sufficient for plasma membrane targeting of ras proteins. EMBO J 10:4033–4039.
Haskell MD, Slack JK, Parsons JT, Parsons SJ (2001) c-Src tyrosine phosphorylation of epidermal growth factor receptor, P190 RhoGAP, and focal adhesion kinase regulates diverse cellular processes. Chem Rev 101:2425–2440.
Horiuchi A, Imai T, Wang C, et al (2003) Up-regulation of small GTPases, RhoA and RhoC, is associated with tumor progression in ovarian carcinoma. Lab Invest 83:861–870.
Ihara K, Muraguchi S, Kato M, et al (1998) Crystal structure of human RhoA in a dominantly active form complexed with a GTP analoge. J Biol Chem 273:9656–9666.
Imamura F, Mukai M, Ayaki M, et al (1999) Involvement of small GTPases Rho and Rac in the invasion of rat ascites hepatoma cells. Clin Exp Metastasis 17:141–148.
Jaiyesimi IA, Buzdar AU, Hortobagyi G (1992) Inflammatory breast cancer: a review. J Clin Oncol 10:1014–1024.
Kamai T, Tsujii T, Arai K, et al (2003) Significant association of Rho/ROCK pathway with invasion and metastasis of bladder cancer. Clin Cancer Res 9:2632–2641.
Kirschmeier PT, Whyte D, Wilson O, Bishop WR, Pai JK (2001) In vivo prenylation analysis of Ras and Rho proteins. Methods Enzymol 332:115–127.

Kleer CG, van Golen KL, Merajver SD (2000) Molecular biology of breast cancer metastasis. Inflammatory breast cancer: clinical syndrome and molecular determinants. Breast Cancer Res 2:423–429.
Kleer CG, van Golen KL, Zhang Y, Wu ZF, Rubin MA, Merajver SD (2002a) Characterization of RhoC expression in benign and malignant breast disease: a potential new marker for small breast carcinomas with metastatic ability. Am J Pathol 160:579–584.
Kleer CG, Zhang Y, Pan Q, et al (2004a) WISP3 and RhoC guanosine triphosphatase cooperate in the development of inflammatory breast cancer. Breast Cancer Res Treat 6:110–115. Ref Type: Generic
Kleer CG, Zhang Y, Pan Q, Merajver SD (2004b) WISP3 (CCN6) is a secreted tumor-suppressor protein that modulates IGF signaling in inflammatory breast cancer. Neoplasia 6:179–185.
Kleer CG, Zhang Y, Pan Q, et al (2002b) WISP3 is a novel tumor suppressor gene of inflammatory breast cancer. Oncogene 21:3172–3180.
Klein WM, Hruban RH, Klein-Szanto AJ, Wilentz RE (2002) Direct correlation between proliferative activity and dysplasia in pancreatic intraepithelial neoplasia (PanIN): additional evidence for a recently proposed model of progression. Mod Pathol 15:441–447.
Kondo T, Sentani K, Oue N, Yoshida K, Nakayama H, Yasui W (2004) Expression of RHOC is associated with metastasis of gastric carcinomas. Pathobiology 71:19–25.
Lebowitz PF, Casey PJ, Prendergast GC, Thissen JA (1997) Farnesyltransferase inhibitors alter the prenylation and growth- stimulating function of RhoB. J Biol Chem 272:15591–15594.
Lebowitz PF, Davide JP, Prendergast GC (1995) Evidence that farnesyltransferase inhibitors suppress Ras transformation by interfering with Rho activity. Mol Cell Biol 15:6613–6622.
Lin M, DiVito MM, Merajver S, Boyanapalli M, van Golen KL (2005) Regulation of pancreatic cancer cell migration and invasion RhoC GTPase and caveolin-1. Mol Cancer 4:21.
Lin M, van Golen KL (2004) Rho-regulatory proteins in breast cancer cell motility and invasion. Breast Cancer Res. Treat. (2004) 84:49–60.
Logsdon CD, Simeone DM, Binkley C, et al (2003) Molecular profiling of pancreatic adenocarcinoma and chronic pancreatitis identifies multiple genes differentially regulated in pancreatic cancer. Cancer Res 63:2649–2657.
Longenecker K, Read P, Somlyo A, Nakamoto R, Derewenda Z (2003) Structure of a constitutively activated RhoA mutant (Q63L) at a 1.55 A resolution. Acta Crystalllogr D Biol Crystallogr 59:876–880.
Marionnet C, Lalou C, Mollier K, et al (2003) Differential molecular profiling between skin carcinomas reveals four newly reported genes potentially implicated in squamous cell carcinoma development. Oncogene 22:3500–3505.
McGrath JP, Capon DJ, Smith DH, et al (1983). Structure and organization of the human Ki-ras proto-oncogene and a related processed pseudogene. Nature 304:501–506.
Moorman JP, Luu D, Wickham J, Bobak DA, Hahn CS (1999). A balance of signaling by Rho family small GTPases RhoA, Rac1 and Cdc42 coordinates cytoskeletal morphology but not cell survival. Oncogene 18:47–57.
Moscow JA, He R, Gnarra JR, et al (1994) Examination of human tumors for rhoA mutations. Oncogene 9:189–194.
Pille JY, Denoyelle C, Varet J, et al (2005) Anti-RhoA and anti-RhoC siRNAs inhibit the proliferation and invasiveness of MDA-MB-231 breast cancer cells in vitro and in vivo. Mol Ther 11:267–274.
Pille JY, Li H, Bertand J, et al (2008) Intravenous delivery of anti-RhoA small interferring RNA loaded in nonoparticles of chitosan in mice: Safety and efficacy in xenografted aggressive breast cancer. Hum Mol Genet 17:1019–1026.
Prendergast GC, Khosravi-Far R, Solski PA, Kurzawa H, Lebowitz PF, Der CJ (1995) Critical role of Rho in cell transformation by oncogenic Ras. Oncogene 10:2289–2296.
Ridley AJ, Hall A (1992a) Distinct patterns of actin organization regulated by the small GTP-binding proteins Rac and Rho. Cold Spring Harb Symp Quant Biol 57:661–671.
Ridley AJ, Hall A (1992b) The small GTP-binding protein rho regulates the assembly of focal adhesions and actin stress fibers in response to growth factors. Cell 70:389–399.

Ridley AJ, Paterson HF, Johnston CL, Diekmann D, Hall A (1992) The small GTP-binding protein rac regulates growth factor-induced membrane ruffling. Cell 70:401–410.
Rochlitz CF, Scott GK, Dodson JM, et al (1989) Incidence of activating ras oncogene mutations associated with primary and metastatic human breast cancer. Cancer Res 49:357–360.
Rodenhuis S (1992) ras and human tumors. Semin Cancer Biol 3:241–247.
Rose R, Weyand M, Lammers M, Ishizaki T, Ahmadian M, Wittinghofer A (2005) Structural and mechanistic insights into the interaction between Rho and mammalian Dia. Nature 435:513–518.
Sahai E, Marshall CJ (2002) ROCK and Dia have opposing effects on adherens junctions downstream of Rho. Nat Cell Biol 4:408–415.
Sander EE, ten Klooster JP, van Delft S, van der Kammen RA, Collard JG (1999) Rac downregulates Rho activity: reciprocal balance between both GTPases determines cellular morphology and migratory behavior. J Cell Biol 147:1009–1022.
Schwering I, Brauninger A, Distler V, et al (2003) Profiling of Hodgkin's lymphoma cell line L1236 and germinal center B cells: identification of Hodgkin's lymphoma-specific genes. Mol Med 9:85–95.
Sequeira L, Dubyk C, Riesenberger T, Cooper CR, van Golen KL (2008) Rho GTPases in PC-3 prostate cancer cell morphology, invasion and tumor cell diapadesis. Clin Exp Metastasis 25(5): 569–579.
Shikada Y, Yoshino I, Okamoto T, et al (2003) Higher expression of RhoC is related to invasiveness in non-small cell lung carcinoma. Clin Cancer Res 9:5282–5286.
Shinto E, Tsuda H, Matsubara O, Mochizuki H (2003) [Significance of RhoC expression in terms of invasion and metastasis of colorectal cancer]. Nippon Rinsho 61(Suppl 7):215–219.
Simpson KJ, Dugan AS, Mercurio AM (2004) Functional analysis of the contribution of RhoA and RhoC GTPases to invasive breast carcinoma. Cancer Res 64:8694–8701.
Stacey DW, Kung HF (1984) Transformation of NIH 3T3 cells by microinjection of Ha-ras p21 protein. Nature 310:508–511.
Suwa H, Yoshimura T, Yamaguchi N, et al (1994) K-ras and p53 alterations in genomic DNA and transcripts of human pancreatic adenocarcinoma cell lines. Jpn J Cancer Res 85:1005–1014.
Takai Y, Sasaki T, Matozaki T (2001) Small GTP-binding proteins. Physiol Rev 81:153–208.
Thalmann GN, Sikes RA, Wu TT, et al (2000) LNCaP progression model of human prostate cancer: androgen-independence and osseous metastasis. Prostate 44:91–103.
van Golen KL (2003) Inflammatory breast cancer: relationship between growth factor signaling and motility in aggressive cancers. Breast Cancer Research 5:174.
van Golen KL, Bao L, DiVito MM, et al (2002a) Reversion of RhoC GTPase-induced inflammatory breast cancer phenotype by treatment with a farnesyl transferase inhibitor. Mol. Cancer Ther 1:575–583.
van Golen KL, Bao LW, Pan Q, Miller FR, Wu ZF, Merajver SD (2002b) Mitogen activated protein kinase pathway is involved in RhoC GTPase induced motility, invasion and angiogenesis in inflammatory breast cancer. Clin Exp Metastasis 19:301–311.
van Golen KL, Davies S, Wu ZF, et al (1999) A novel putative low-affinity insulin-like growth factor-binding protein, LIBC (lost in inflammatory breast cancer), and RhoC GTPase correlate with the inflammatory breast cancer phenotype. Clin Cancer Res 5:2511–2519.
van Golen KL, Wu ZF, Qiao XT, Bao L, Merajver SD (2000a) RhoC GTPase overexpression modulates induction of angiogenic factors in breast cells. Neoplasia 2:418–425.
van Golen KL, Wu ZF, Qiao XT, Bao LW, Merajver SD (2000b) RhoC GTPase, a novel transforming oncogene for human mammary epithelial cells that partially recapitulates the inflammatory breast cancer phenotype. Cancer Res 60:5832–5838.
Van Laere S, Van der Auwer I, Van den Eynden GG, Fox SB, Bianchi F, Harris AL, van Dam P, Van Marck EA, Vermuelen PB and Dirix LY Nuclear factor kappa B signature of inflammatory breast cancer by cDNA microarray validated by quantitative real-time reverse transcription-PCR, immunohistochemistry and nuclear factor kappaB DNA-binding. Breast Cancer Res Treat. (2005) 93(3):237–246.
Wang W, Yang LY, Yang ZL, Huang GW, Lu WQ (2003) Expression and significance of RhoC gene in hepatocellular carcinoma. World J Gastroenterol 9:1950–1953.

Wei Y, Zhang Y, Derewenda U, et al (1997) Crystal Structure of RhoA-GDP and its functional implications. Nat Struct Biol 4:699–703.

Wu M, Wu ZF, Kumar-Sinha C, Chinnaiyan A, Merajver SD (2004) RhoC induces differential expression of genes involved in invasion and metastasis in MCF10A breast cells. Breast Cancer Res Treat 84:3–12.

Yao H, Dashner E, van Golen CM, van Golen KL (2006) RhoC GTPase is required for PC-3 Prostate Cancer Cell Invasion but not Motility. Oncogene 25:2285–2296. Ref Type: Abstract

Zhang X, Lin M, van Golen KL, Itoh K, Yee D (2005) Multiple signaling pathways are activated during insulin-like growth factor-I (IGF-1) stimulated breast cancer cell migration. Breast Cancer Res Treat 93:159–168.

Zondag GC, Evers EE, ten Klooster JP, Janssen L, van der Kammen RA, Collard JG (2000) Oncogenic Ras downregulates Rac activity, which leads to increased Rho activity and epithelial-mesenchymal transition. J Cell Biol 149:775–782.

Chapter 9
RhoB GTPase and FTIs in Cancer

Minzhou Huang, Lisa D. Laury-Kleintop, and George C. Prendergast

Introduction

Rho GTPases constitute a subfamily of the Ras superfamily of small GTPases that integrate cytoskeletal actin organization with the control of many cellular processes, including cell adhesion, cell movement, morphogenesis, membrane dynamics, vesicle trafficking, and gene expression (Bishop and Hall 2000; Jaffe and Hall 2005). Beyond their diverse roles in normal physiology, Rho proteins also strongly influence pathological processes such as the proliferation, survival, angiogenesis, invasion, and metastasis of cancer cells (Jaffe and Hall 2002; Sahai and Marshall 2002; Ellenbroek and Collard 2007). The RhoA, RhoB, and RhoC proteins form a closely related subgroup within the Rho family. However, while these small GTPases all regulate actin stress fiber formation, they appear to have distinct physiological and pathophysiological functions in cells, probably due in part to differences in subcellular localization that allow effector interactions to be partitioned (Adamson et al. 1992b). While most Rho proteins have a positive role in proliferation and malignant transformation, the role of RhoB diverges to a significant extent. In particular, RhoB differs from RhoA and RhoC in its localization to intracellular endosomes and in its antiproliferative and proapoptotic effects in cancer cells, possibly reflecting a generalized role as a modifier of stress signaling (Prendergast 2001; Huang and Prendergast 2006).

While Ras proteins are mutated in 30% of human cancers of different origins (Bos 1989; Schubbert et al. 2007), genetic alterations in RhoB and other Rho small GTPases occur only rarely. In contrast, their altered expression or activity appears to play an important role in cancer progression. In several cancers, RhoB expression

M. Huang and L.D. Laury-Kleintop
Lankenau Institute for Medical Research, Wynnewood, PA, USA

G.C. Prendergast (✉)
Department of Pathology, Anatomy and Cell Biology and Kimmel Cancer Center, Thomas Jefferson University, Philadelphia, PA, USA
e-mail: prendergast@limr.org

K. van Golen (ed.), *The Rho GTPases in Cancer*,
DOI 10.1007/978-1-4419-1111-7_9,

levels are reduced or undetectable, especially as the cancer progresses (Fritz et al. 1999, 2002; Adnane et al. 2002; Forget et al. 2002; Mazieres et al. 2004; Sato et al. 2007), and overexpression of RhoB in cancer cells inhibits their growth, migration, invasion, and metastasis (Du and Prendergast 1999; Chen et al. 2000; Jiang et al. 2004b; Mazieres et al. 2005; Ridley 2004). Studies in genetic knockout mice establish that RhoB inhibits carcinogenesis and restricts growth factor and adhesion signaling in transformed cells (Liu et al. 2001a). Furthermore, they demonstrate that RhoB is essential for the apoptotic response of transformed cells to DNA damage or microtubule disturbing agents (Liu et al. 2001b). Taken together, these findings argue strongly that RhoB acts as a negative modifier or tumor suppressor gene in cancer.

One important line of work on RhoB has linked it to the action of farnesyltransferase inhibitors (FTIs), a class of experimental cancer therapeutics originally developed as a strategy to inhibit the function of Ras by blocking its post-translational farnesylation. FTIs are widely known for their selective effects on Ras-transformed cells, yet mechanistic studies made it clear that this cellular response was based mainly upon factors beyond Ras targeting (Lebowitz et al. 1995; Cox and Der 1997; Lebowitz and Prendergast 1998; Prendergast and Oliff 2000; Sebti and Hamilton 2000). Genetic studies performed in our laboratory developed an alternate hypothesis for drug mechanism proposing that altered prenylation of RhoB is a crucial event in mediating the cellular response to FTI treatment. First, we demonstrated that RhoB responds to FTI treatment by a gain-of-function mechanism characterized by the elevation of the geranylgeranylated isoform of RhoB (RhoB-GG) (Lebowitz et al. 1997a). Second, we showed that elevation of RhoB-GG is sufficient and essential for FTI-induced growth inhibition and apoptosis, respectively (Du et al. 1999; Liu et al. 2000). Lastly, we confirmed that RhoB-GG elevation is sufficient to mediate growth inhibition and apoptosis in human cancer cells (Du and Prendergast 1999). Another study confirmed that RhoB-GG was growth inhibitory in cancer cells but the farnesylated isoform RhoB-F was similarly growth inhibitory (Chen et al. 2000). This study confirmed that RhoB generally exerts a negative function for cancer cell growth and that geranylgeranyl transferase inhibitors (GGTIs) might also recruit this function through elevated production of RhoB-F. While these findings have been interpreted incorrectly as the evidence against the RhoB-GG hypothesis for FTI action, it is clear that while RhoB-F may be growth inhibitory in some settings obviously does not rule out a similar growth inhibitory role for RhoB-GG in mediating FTI action, nor does it refute the evidence of an essential in vivo requirement for RhoB-GG in FTI action as established by studies in RhoB knockout mice (Liu et al. 2000).

Investigations of the downstream effector pathways mediating the apoptotic and antineoplastic actions of RhoB-GG induction by FTI in transformed cells have revealed roles for three players implicated as cancer genes. Cyclin B1 was identified as a potential player through microarray analysis of the RhoB-dependent response to FTI treatment (Kamasani et al. 2003). Steady-state levels of cyclin B1 and its associated kinase Cdk1 were suppressed in a RhoB-dependent manner in cells fated to undergo FTI-induced apoptosis, but not FTI-induced growth inhibition. Moreover, enforcing cyclin B1 expression inhibited apoptosis triggered by FTI and abolished the antitumor activity of FTI in tumor graft assays (Kamasani et al. 2004). Thus, cyclin B1 downregulation was critical for the RhoB-dependent mechanism by which FTI triggers apoptosis and antitumor efficacy. Interestingly, cyclin B1 is often overexpressed in cancer (Shen et al. 2004) and it is notable that its downregulation by FTI occurs only in transformed cells (Kamasani et al. 2004), suggesting one explanation for why FTIs exhibit such specificity for transformed cells.

Other studies have also implicated the actin-microtubule organizer mDia and the BAR adapter protein Bin1 in the FTI response. mDia is a Rho effector molecule that interacts with RhoB on endosomal vesicles and promotes their trafficking by promoting actin assembly (Fernandez-Borja et al. 2005; Wallar et al. 2007). Disruption of mDia function impairs RhoB-dependent apoptosis triggered by FTI, consistent with the operation of an RhoB-mDia pathway in this response (Kamasani et al. 2007). Bin1, also known as amphiphysin II, is a BAR adapter protein that functions in membrane dynamics, endosome trafficking, and nucleocytosolic signaling. Transformed cells that lack the *Bin1* gene remain susceptible to FTI-induced actin reorganization and growth inhibition, but they are defective for FTI-induced apoptosis, specifically implicating Bin1 in this response (DuHadaway et al. 2003). While precise connections to RhoB remain undefined, it is notable that Bin1 and other BAR adapter proteins interact with Rho/ARF functions that control vesicle trafficking and that both RhoB and Bin1 have been implicated in Myc control (Elliott et al. 1999; Huang et al. 2006; Sakamuro et al. 1996). At present, evidence is lacking that cyclin B1 control may be mediated downstream of mDia or Bin1; however, given evidence of mDia localization to the mitotic spindle a connection that may permit direct communication between mDia and cyclin B1 is tempting to speculate upon (Fig. 9.2). Nevertheless, it is clear that continuing investigations of the molecular effector mechanisms downstream of RhoB will be important to gain a full understanding of its anticancer functions.

Unique Features of RhoB

As mentioned earlier, the Rho small GTPases RhoA, RhoB, and RhoC are quite similar to each other in structure (~90% identity) and, in particular, their effector domains are identical. However, several unique features distinguish RhoB from the other Rho proteins in this subgroup. First, the *RhoB* gene is smaller and contains only one exon, prompting the notion that RhoB may have evolved by reverse transcription (Karnoub et al. 2004). Second, unlike other Rho proteins, RhoB is prenylated in cells as two distinct isoforms that are either geranylgeranylated (RhoB-GG) or farnesylated (RhoB-F) (Adamson et al. 1992a; Lebowitz et al. 1997a).

This differential prenylation has consequences for localization to late endosomes versus plasma membrane or other intracellular compartments (Michaelson et al. 2001; Wherlock et al. 2004). Third, in contrast to most Ras superfamily GTPases, RhoB is short-lived and its levels fluctuate significantly during cell cycle transit and stress responses. RhoB accumulation varies throughout the cell cycle and the RhoB mRNA transcript has a half-life of ~30 min, which is substantially shorter than other Rho transcripts (Zalcman et al. 1995). A mechanism to stabilize RhoB mRNA is mediated by its interaction with the RNA-binding protein HuR (Westmark et al. 2005). RhoB protein also turns over rapidly with a half-life of ~2 h (Lebowitz et al. 1995), in contrast to most Rho GTPases that are much more stable. Levels of RhoB in cells are rapidly upregulated by a variety of stress and growth stimuli (Jahner and Hunter 1991; Fritz et al. 1995; Zalcman et al. 1995 Engel et al. 1998; Trapp et al. 2001; Chauhan et al. 2004; Gerhard et al. 2005; Huelsenbeck et al. 2007). On the other hand, levels of RhoB are reduced by Ras via the EGFR, ErbB2, and Akt/PKB pathways (Jiang et al. 2004a, b). Thus, RhoB is highly regulated at the level of expression. Lastly, in contrast to RhoA and RhoC, which promote cancer, RhoB exerts a negative effect on the growth and survival of transformed cells (Du et al. 1999; Du and Prendergast 1999; Chen et al. 2000). In summary, RhoB exhibits a number of unique features not found in its close relatives RhoA and RhoC, illustrating the hazards of predicting similar physiological functions simply on the basis of a similar primary structural relationship.

Function of RhoB in Endocytic Trafficking

Trafficking to Lysosome, Plasma Membrane, or Nucleus

The Rho family of small GTPases was originally characterized as regulators of the actin cytoskeleton, but it is now clear that a great diversity of functions for these signaling proteins arise from a root in actin organization, including prominent functions in endocytic trafficking pathways. Rho proteins can influence trafficking of endosome-containing receptor signaling complexes, leading either to lysosomes for degradation or along the recycling pathway to the plasma membrane (Ellis and Mellor 2000). Indeed, RhoB was the first member of the Rho family to be implicated in endosomal trafficking. As determined by immunofluorescence, electron microscopy, and biochemical fractionation, RhoB can localize to the both the plasma membrane and the bounding membrane of multivesicular late endosomes (MVBs), the latter of which can fuse with lysosomes or the plasma membrane (Adamson et al.1992b; Robertson et al. 1995; Mellor et al. 1998). RhoB binds the protein kinase C-related protein kinase (PRK1) and targets it to the endosomal compartment (Mellor et al. 1998). By activating PRK1, RhoB regulates the kinetics of epidermal growth factor (EGF) receptor traffic from endosomes to MVBs, which represent a prelysosomal compartment. Expression of active RhoB causes a delay in the intracellular trafficking of the EGF receptor (Gampel et al. 1999). In addition, RhoB and PRK1 cause recruitment of the PI3-kinase effector PDK1 to generate a ternary complex that depends on RhoB for the formation of endosomes (Flynn et al. 2000). The effects of RhoB at this level are important to signaling strength, since removal of receptor molecules from the cell surface by distributing them to the endocytic pathway will modify the amplitude and kinetics of signal transduction (Fernandez-Borja et al. 2005).

Studies of how RhoB affects the endocytic pathway suggest that in different settings, it may promote trafficking to the lysosome, the cell surface, or the nucleus. In one study, RhoB did not affect EGF receptor sorting into endosomes but it inhibited the terminal transfer of endosomes into the lysosome (Wherlock et al. 2004). Cell treatment with FTIs, which elicit a gain of endosomal RhoB-GG, led to a reduction in the sorting of the EGF receptor to the lysosome, thereby increasing recycling of the EGF receptor to the plasma membrane (Wherlock et al. 2004). Altering RhoB prenylation in this manner also inhibited cell proliferation, confirming the idea that RhoB-GG has an antiproliferative function. In another study, RhoB has been implicated in trafficking of the chemokine receptor CXCR2, which mediates chemotaxis (Neel et al. 2007). This study suggests that the RhoB function is essential for endosomal sorting of CXCR2 required to direct appropriate degradation and recycling.

Plasma membrane recruitment and activation of the Src tyrosine kinase has been found to rely upon RhoB (Sandilands et al. 2004). Blocking RhoB by the expression of a dominant interfering mutant not only prevented actin-dependent movement of Src to the cell surface, but also the catalytic activation of Src, illustrating a vital link between trafficking and activation. RhoB has also been found to be required for Src-mediated transport and activation of the fibroblast growth factor receptor (FGFR) (Sandilands et al. 2007). Further evidence of a connection between RhoB and Src was obtained recently in vascular smooth muscle cells, where RhoB is essential to support a nuclear trafficking pathway for Src that is stimulated by the growth factor PDGF (Huang et al. 2007). Thus, it appears that RhoB may act as a component of an "outside-in" signaling pathway that coordinates activation of Src with translocation to the plasma membrane or another site where it can function appropriately.

In vascular endothelial cells, RhoB has been implicated in supporting a nuclear trafficking pathway for Akt/PKB, where it is a pivotal regulator of cell survival (Adini et al. 2003). In these cells, RhoB colocalizes to the nuclear margin proportionally with total Akt/PKB. This colocalization is functionally relevant, because when RhoB is depleted, Akt/PKB is excluded from the nucleus resulting in degradation of total cellular Akt/PKB protein in a proteosome-dependent manner (Adini et al. 2003). Therefore, RhoB is an important determinant of Akt/PKB stability and trafficking to the nucleus in endothelial cells. Because the function of RhoB in vivo appears to be rate limiting for only endothelial cell sprouting, it appears that RhoB has a novel stage-specific function to regulate endothelial cell survival during vascular development.

Recent findings extend the paradigm of a positive function for RhoB in supporting vascular cells. Platelet-derived growth factor (PDGF) is a potent mitogen, chemoattractant, and survival factor for vascular smooth muscle cells. In response to ligand-induced receptor dimerization, autophosphorylation of the PDGF receptor occurs followed by its association with several SH2-domain containing signal transduction molecules that mediate cell growth, chemotaxis, actin reorganization, and antiapoptosis (Heldin and Westermark 1999). Although there is considerable information on PDGF receptor signaling pathways, the regulation of receptor trafficking and the mechanistic relationship between PDGF receptor trafficking and endosomal signaling are not well understood. By using a knockout mouse model system, we found that RhoB is involved in the regulation of PDGFR-β trafficking and signaling through the Akt, ERK, and Src pathways, which are known to be critical for PDGF-induced proliferation of vascular smooth muscle cells (Huang et al. 2007). RhoB was required for the association of the PDGFR-β with recycling and late endosomes that localized to a perinuclear region within the cells. By regulating endosomal trafficking of PDGFR-β, RhoB efficiently modulated the nuclear trafficking of activated Akt, ERK, and Src, all of which are involved in triggering downstream biological responses. These findings parallel those made in sprouting vascular endothelial cells, where as noted RhoB is also crucial for Akt/PKB activation, nuclear localization, and cell survival (Adini et al. 2003).

The important positive function for RhoB in survival and proliferation of vascular cells contrasts with the negative functions for growth and survival displayed in epithelial cells, oncogenically transformed cells, and cancer (Huang and Prendergast 2006). One impact of a positive role for RhoB in the vasculature is it impacts cancer, due to the implications for angiogenesis where endothelial and smooth muscle cells participate in generating the tumor vasculature. A second impact relates to how FTIs may affect vascular function, given their ability to alter RhoB, prompting the notion that FTIs might be useful to limit angiogenesis in cancer or other diseases.

Trafficking Mechanisms

How RhoB contributes to different trafficking pathways is an important question to resolve, for example, whether interactions with different effector molecules are involved, since this may also shed light on why RhoB exerts negative effects in most cell types, including transformed cells, but positive effects in vascular cells (Adini et al. 2003; Huang et al. 2007). As mentioned earlier, one mechanism that has been elucidated is mediated by the RhoB effector kinase PRK1, which restricts trafficking of the EGF receptor to MVBs (Gampel et al. 1999). This complex also recruits the Akt regulatory kinase PDK1, generating a ternary complex that has been implicated in endosome formation (Flynn et al. 2000).

More recently, the RhoB effector molecule mDia has been identified as another important effector for regulating endosome trafficking (Fernandez-Borja et al. 2005; Wallar et al. 2007). mDia1 contains an N-terminal Rho-binding domain and two C-terminally conserved forming homology (FH) domains responsible for Arp2/3-independent actin assembly and actin filament growth (Wasserman 1998; Pruyne et al. 2002; Evangelista et al. 2003). The Rho-binding domain of mDia1 interacts with RhoB as well as with RhoA and RhoC (Watanabe et al. 1997). mDia1 localizes to endosomes and it contributes significantly to Src movement (Tominaga et al. 2000), so recruitment of mDia to endosomes by RhoB seems likely to be relevant to the RhoB-dependent mechanisms of Src movement to plasma membrane and nucleus that have been described (Sandilands et al. 2004; Huang et al. 2007). In one study where the effect of RhoB on endocytosis and the actin cytoskeleton was studied extensively, it was found that activated RhoB promoted the polymerization of an actin coat around endosomes and the association of these vesicles to subcortical actin cables, effectively inhibiting further endosomal transport (Fernandez-Borja et al. 2005). Recruitment of mDia1 by activated RhoB was critical for the assembly of actin on the vesicle membrane. A second mDia family protein, mDia2, has been found to act similarly in promoting endosome and actin dynamics that are essential for RhoB to drive vesicle trafficking (Wallar et al. 2007).

Recent work has also identified a link between RhoB and microtubules, the organization of which is known to be integrated along with actin assembly by mDia (Palazzo et al. 2001). Specifically, using a two-hybrid approach one group identified LC2, the light chain associated with the microtubule-associated protein MAP1A, as a novel RhoB-binding partner that is essential for RhoB-dependent trafficking of the EGF receptor (Lajoie-Mazenc et al. 2008). The MAP1A/LC2 complex serves as a RhoB-specific effector, which helps interconnect microtubules with other cytoskeletal elements (Togel et al. 1998). Because MAP1A/LC2 can interact with several partners implicated in intracellular transport, it may have a pivotal role as an adaptor molecule that regulates cellular trafficking. Some evidence suggests that RhoB-F may prefer to localize to the plasma membrane and RhoB-GG may prefer to localize to endosomal membranes (Wherlock et al. 2004; Milia et al. 2005). Using prenyl transferase inhibitors, it was found that LC2 interaction with RhoB relied upon geranylgeranylation but not farnesylation. In support of the notion that an endosomal localization was required for LC2 interaction, treatment of cells with a GGTI inhibited the localization of RhoB to endosomes and prevented association with MAP1A/LC2 (Lajoie-Mazenc et al. 2008). Thus, MAP1A/LC2 may mediate the ability of RhoB to modify endosome-associated signaling and trafficking processes that are coordinated through interactions with the actin and microtubule networks of the cell. Together with evidence that RhoB regulates trafficking of the EGF and PDGF receptors, as well as trafficking of Akt/PKB and Src, it is apparent that RhoB acts to direct signal transduction molecules to specific subcellular locations, thereby controlling their ability to signal appropriately. A model summarizing RhoB actions in an idealized smooth muscle cell is presented in Fig. 9.1.

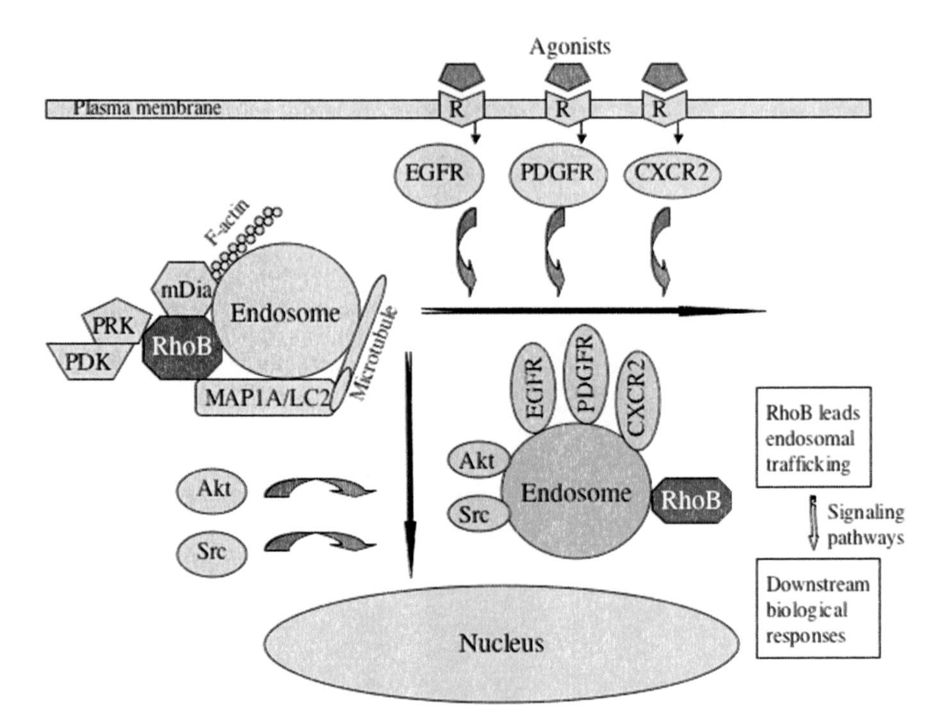

Fig. 9.1 Model for trafficking control by RhoB. RhoB localizes to endosomal compartments and mediates endosomal trafficking of a broad range of receptors and important cellular molecules such as EGFR, PDGFR, CXCR2, Akt, and Src. After agonist-stimulated receptor internalization, RhoB-regulated endosomal trafficking is an important event to deliver signal transduction pathways to trigger downstream biological responses. RhoB can recruit specific partners such as mDia proteins (controlling actin polymerization on endosome membrane), MAP1A/L2 (involved in regulation of microtubule function), resulting in control of endosomal trafficking. Akt trafficking and activity may be modulated by a ternary complex that contains RhoB, PRK (protein kinase C-related kinase), and PDK1 (Akt regulatory kinase). For simplicity, only nuclear trafficking pathways for Src and Akt controlled by RhoB in idealized vascular cells are depicted

Role of RhoB in Cancer

Evidence for the Suppressive Role of RhoB in Neoplasia

As noted above, RhoB differs from RhoA and RhoC in how it functions in cancer (Prendergast 2001). Like other Ras-like small GTPases, such as Ras, Rac1, and Cdc42, RhoA and RhoC promote oncogenesis, invasion, and metastasis (Khosravi-Far et al. 1995; Westwick et al. 1997; Whitehead et al. 1998; Pruitt and Der 2001; Ridley 2004). In contrast, RhoB has a cancer suppressive role that is manifested in various ways, including through inhibitory effects on cell proliferation, survival, invasion, and metastasis (Du et al. 1999; Du and Prendergast 1999; Chen et al. 2000; Liu et al. 2000, 2001a, b; Liu and Prendergast 2000; Delarue et al. 2001; Jiang et al. 2004a, b; Mazieres et al. 2005). Consistent with in vitro observations, RhoB null cells formed tumors more efficiently than cells expressing RhoB after injection into the intraperitoneal cavity of mice (Liu et al. 2001a). More significant support for a suppressor or negative modifier role was demonstrated by the elevated rate of papilloma formation in RhoB knockout mice subjected to a classical skin carcinogenesis assay, which involves initiation with the carcinogen 7,12-dimethylbenz[a]anthracene (DMBA) and promotion with phorbol ester (Liu et al. 2001a). Studies in a nude mouse xenograft model also showed that RhoB inhibits tumor growth, as well as metastasis (Chen et al. 2000; Jiang et al. 2004b; Mazieres et al. 2005). On the other hand, measurement of RhoB transcription in most tumor-derived cell lines showed its low expression, suggesting that low levels of RhoB expression were correlated with proliferation (Wang et al. 2003). Furthermore, many oncogenes such as the EGFR, Ras, and ErbB2 have been shown to suppress the expression of RhoB by a mechanism involving the phosphatidylinositol 3'-kinase PI3K/Akt pathway (Delarue et al. 2001; Jiang et al. 2004a, b).

Decreased Expression of RhoB in Human Cancer

Studies of patient biopsies revealed that RhoB expression levels are dramatically decreased in lung, head, and neck, as well as, brain cancer when tumors become more aggressive (Adnane et al. 2002; Forget et al. 2002; Mazieres et al. 2004). Analysis of the status of the RhoB protein in human head and neck squamous cell carcinomas has shown that RhoB expression becomes weak to undetectable as tumors become deeply invasive and poorly differentiated. In contrast, RhoA protein levels increase with tumor progression (Fritz et al. 1999; Adnane et al. 2002). Several studies reported that RhoB is downregulated in lung cancer cell lines and tumor tissues (Mazieres et al. 2004, 2007; Sato et al. 2007). Immunostaining analysis of the expression of RhoB protein in large samples of human lung cancer tissues showed that RhoB was lost in 96% of invasive tumors and reduced by 86% in poorly differentiated tumors compared with the nonneoplastic epithelium. In nonsmall cell lung cancer (NSCLC), immunohistochemical analyses showed no or weak RhoB expression in 42% of adenocarcinomas, whereas in 94% of squamous cell carcinomas. Decreased expression of RhoB was more frequently observed in poorly- or moderately differentiated adenocarcinomas than in well-differentiated ones. In the cases of brain cancer, it was demonstrated that the expression of RhoB decreased significantly in various brain tumors and this decrease was inversely related with tumor of grade II to IV malignancy (Forget et al. 2002). Together these studies illustrate a broad trend toward decreased RhoB expression in human cancer.

Mechanisms of RhoB Involvement in Cancer Suppression

Recent investigations have brought new insights into understanding the relevance of RhoB downregulation in cancer cells. Several studies have demonstrated that histone deacetylase (HDAC) inhibitors can induce a significant RhoB reexpression in lung cancer cell lines, indicating that histone modification is an important event necessary for RhoB downregulation (Wang et al. 2003; Mazieres et al. 2007; Sato et al. 2007). By using a genomic approach to identify the transcriptional consequences of HDAC1 inhibition, a link was discovered between HDAC1 inhibition and RhoB expression. Treatment of cancer cells with an HDAC inhibitor was found to induce the expression of RhoB through an inverted CCAAT box in the RhoB promoter (Wang et al. 2003). Another study also showed that RhoB expression was induced by HDAC inhibition, suggesting that RhoB downregulation may be due to histone modification (Sato et al. 2007). Additionally, loss of heterozygosity (LOH) occurs at the RhoB locus in NSCLC cell lines. Regulation of RhoB expression may occur mainly by histone deacetylation, rather than by promoter hypermethylation, perhaps modulated by specific 5' sequences within the RhoB promoter (Mazieres et al. 2007). HDAC inhibitors are currently under evaluation as potent anticancer agents because of their ability to induce growth arrest, differentiation, and apoptosis in transformed cells and tumors (Weidle and Grossmann 2000; Marks et al. 2001). Thus, the strong induction of RhoB elicited by HDAC inhibitors prompts speculation that their anticancer effects may relate in part to the recruitment of the negative functions of RhoB.

Virtually all Ras superfamily GTPases require post-translation prenylation for function, and RhoB is no exception. Site-directed mutagenesis demonstrates that RhoB prenylation is crucial for membrane localization, growth inhibition, and proapoptosis and antitumor activities (Adamson et al. 1992b; Lebowitz et al. 1995, 1997b, Wang and Sebti 2005), although prenylation is dispensable for activation of serum response factor (SRF)-dependent transcription (Lebowitz et al. 1997b). More recently, palmitoylated cysteine 192 also has been found to be crucial for tumor suppressive and proapoptotic activities, in addition to cysteine 193, which is the critical C-terminal site for prenylation (Wang and Sebti 2005). The precise manner in which prenylation and palmitoylation mediate membrane binding is unclear; nonetheless, it is clear that they are crucial in this regard (Adamson et al. 1992a). Prenylation of RhoB is essential for binding two of its interacting proteins, the Rho GDP dissociation inhibitor RhoGDI (Fukumoto et al. 1990; Miura et al. 1993), which serves as a cytosolic carrier for Rho proteins before membrane insertion, and the zinc finger transcription factor DB1/VEZF (Lebowitz and Prendergast 1998), which mediates a tissue-specific function of RhoB in vascular determination (L. Benjamin and G.C.P., unpublished observations).

RhoB has been linked to several important pathways in cancer. One study found that RhoB-GG but not RhoB-F displayed an antitransforming and antiproliferative activity in Ras-transformed NIH3T3 fibroblasts, reinforcing the concept of different functions for differently prenylated isoforms of RhoB (Mazieres et al. 2005). Significantly, neoplastic transformation, invasion, and metastasis elicited by activated forms of the EGF receptor, Ras, PI3K, and Akt/PKB each require suppression of RhoB levels that are mediated by each of these oncoproteins (Jiang et al. 2004b). The ability of RhoB to inhibit the transcription factors NF-κB and c-Myc may explain why suppression of RhoB levels by upstream-acting oncoproteins is so important. RhoB inhibits NF-kB by interfering with IkB turnover, albeit through a noncanonical mechanism that is not well understood (Fritz and Kaina 2001).

NF-κB mediates proinflammatory and protransforming signals including potent antiapoptotic signals that are likely to be important for tumor development and progression, for example, by activated Ras (Mayo et al. 1997). RhoB also facilitates turnover of the c-Myc oncoprotein by supporting nuclear accumulation of GSK-3, which phosphorylates the critical T58 site on c-Myc responsible for triggering its degradation (Huang et al. 2006). Thus, RhoB may restrict proliferation and/or promote apoptosis in part by promoting GSK-3-dependent turnover of c-Myc protein. Further, a selection for RhoB attenuation in human cancers may occur as a mechanism to promote c-Myc stabilization and thereby c-Myc-mediated cell proliferation. In summary, RhoB may suppress oncogenesis by blocking pathways required by important upstream players such as EGFR, Ras, PI3K, Akt/PKB to drive neoplastic transformation, survival, and progression.

RhoB in Apoptosis by DNA Damage

Cell suicide processes are thought to play an important role in limiting cancer progression and therapeutic response. Using a mouse knockout model, RhoB was shown to be critical for mediating the apoptotic response of oncogenically transformed cells to DNA-damaging regimens or microtubule disrupting agents (Liu et al. 2001b). Untransformed primary mouse embryonic fibroblasts (MEFs) respond to DNA damage by undergoing cell cycle arrest and chromosomal repair, whether or not RhoB is intact. In contrast, MEFs transformed by adenovirus E1A and mutant Ras, a useful model of gauging apoptotic sensitivity to cytotoxic agents (Lowe and Ruley 1993), were highly susceptible to DNA damage in a manner that relied upon RhoB integrity. After exposure to gamma irradiation or doxorubicin, which causes DNA strand breaks, E1A+Ras-transformed cells underwent apoptosis following a transient cell cycle arrest, presumably because they could not engage an appropriate arrest and repair response (Lowe and Ruley 1993). In contrast, transformed cells lacking RhoB were resistant to DNA-damage-induced apoptosis and this defect was complemented by ectopic expression of RhoB (Liu et al. 2001b).

The defective apoptotic response of transformed cells lacking RhoB improved their long-term survival in vitro after DNA damage and promoted their survival in vivo after their irradiation in the setting of a tumor xenograft assay (Liu et al. 2001b). RhoB is also required for FTIs to sensitize Ras-transformed cells to DNA-damage-induced cell death, suggesting a role for RhoB alteration in the underlying mechanism of this effect (Bernhard et al. 1996). These observations were consistent with earlier evidence that genotoxic stress stimulates transcription of RhoB (Fritz et al. 1995) and that elevating its levels can sensitize cancer cells to apoptosis by DNA alkylating agents that damage DNA (Fritz and Kaina 2000). In contrast, RhoB deletion did not affect the susceptibility of transformed MEFs to apoptosis by the broad-spectrum kinase inhibitor staurosporine (Liu et al. 2001b), indicating that RhoB is not generally required for apoptosis. However, loss of RhoB also diminished the apoptotic susceptibility to microtubule disruption by paclitaxel (Liu et al. 2001b). Consistent with previous observations

(Lowe and Ruley 1993), dose–response studies showed that DNA-damage-induced apoptosis was normally triggered after a transient arrest in the G2/M phase of the cell cycle, whereas in cells lacking RhoB there was a progressive increase in the number of cells arrested in the G2/M phase without subsequent apoptosis.

Since RhoB loss also compromised the induction of apoptosis by paclitaxel, which is thought to act by arresting the microtubule-based spindle apparatus at mitosis, the observations suggested that RhoB might participate in an apoptotic pathway that is engaged after a terminal block in mitosis. An expression microarray analysis to compare the response of transformed cells lacking RhoB to DNA damage suggested a partial defect in the p53 response, insofar as a gene module of reported p53 targets responded differently in RhoB cells (Kamasani and Prendergast 2005). In any case, it was apparent that RhoB does not participate in the activation of the DNA damage response, but rather acts downstream or in parallel to the pathway as a cell death effector or modifier, respectively. Together, these findings revealed a significant role for RhoB in facilitating apoptosis triggered by two important classes of cancer chemotherapeutic drugs that either damage DNA or disrupt microtubule integrity. Thus, RhoB might be involved in many therapeutic responses and downregulation of RhoB and its effector pathways might contribute significantly to the development of radioresistance or chemotherapeutic resistance in cancer.

Role of RhoB in FTI Anticancer Response

FTIs are a novel class of cancer therapeutics originally designed to disrupt the function of mutant Ras by blocking its post-translational farnesylation, which is essential for membrane localization and oncogenic activity. Early preclinical studies showed that FTIs were exquisitely selective antagonists of Ras-dependent neoplastic transformation, with few significant effects on untransformed cells. However, mechanistic investigations showed that Ras inhibition could not entirely explain their antitransforming properties (Cox and Der 1997; Lebowitz and Prendergast 1998). Subsequent work made it clear that although FTIs could inhibit the farnesylation of H-Ras, used in many preclinical models, this was not a sufficient event for FTI action. Furthermore, FTIs could not inhibit the prenylation of the most common Ras oncoprotein in human cancers, K-Ras, due to its ability to be cross-prenylated in FTI-treated cells by geranylgeranyl transferase I (Prendergast and Oliff 2000; Prendergast and Rane 2001). Thus, the observed biological effects of FTIs were ascribed to other mechanisms, leading to a search for alternate targets. This topic, which has engaged numerous investigators and aroused considerable debate, has been reviewed in detail elsewhere (e.g., Pan and Yeung 2005).

Among FTI targets that have been proposed, RhoB is among the better characterized in terms of those that have been both biochemically and genetically documented to be causally involved in one or more antineoplastic responses to FTI treatment. Much of the evidence justifying RhoB as a target is presented elsewhere in detail (Prendergast 2000; Prendergast and Rane 2001; Pan and Yeung 2005). Briefly, RhoB is unique among Rho proteins in existing as both farnesylated and geranylgeranylated isoforms.

Treatment of cells with FTIs causes a loss of farnesylated RhoB (RhoB-F) and a consequent increase in geranylgeranylated RhoB (RhoB-GG) as newly synthesized protein is efficiently prenylated by geranylgeranyl transferase I (Lebowitz et al. 1997a). In this manner, cell treatment with FTIs causes a rapid gain of RhoB-GG and this event is sufficient and in some cases essential for FTIs to trigger phenotypic alterations, cytoskeletal actin reorganization, growth inhibition, and apoptosis (Prendergast 2000). Briefly, one body of evidence indicates that elevating RhoB-GG levels in cells is sufficient to phenocopy the effects of FTI action (Du et al. 1999; Du and Prendergast 1999; Zeng et al. 2003), and a second body of evidence indicates that deleting RhoB from cells is sufficient to prevent features of FTI action (Liu et al. 2000). Clearly, the most powerful evidence that RhoB-GG contributes to the FTI response derives from experiments in transformed cells and mice that are genetically deficient in RhoB, where defects in apoptosis correspond with elimination of the in vivo antitumor response (Liu et al. 2000). Thus, the most intriguing feature of FTIs – their ability to selectively activate apoptosis in transformed cells – critically depends upon a gain of RhoB function in the guise of RhoB-GG.

An evaluation of the effects of RhoB-GG on oncogene-transformed epithelial cells led to a model proposing that RhoB blocks cancer by antagonizing pro-oncogenic Rho functions such as those mediated by Rac1, RhoA, or RhoC (Zeng et al. 2003). Consistent with this notion, one study demonstrated that FTI treatment could reverse the RhoC-induced phenotype of inflammatory breast cancer in a human mammary cell model through a mechanism that could be phenocopied by RhoB-GG elevation (van Golen et al. 2002). Thus, RhoB-GG might inhibit signaling by RhoC, perhaps by sequestering common downstream effectors from RhoC. Consistent with this possibility, an analysis of effector domain mutants in RhoB-GG supported a role for interactions with effector kinase PRK in mediating FTI-induced growth inhibition (Zeng et al. 2003).

Surprisingly, a recent study has argued that FT is not a critical target for inhibition by FTI, or even a critical element in tumor formation by mutant H-Ras (Mijimolle et al. 2005). However, for reasons discussed below, there is evidence that RhoB still remains relevant to the biological responses produced by FTI treatment (Pan and Yeung 2005). While FTIs clearly inhibit FT, this study offers a genetic proof that FTI inhibition can be separated completely from the desired functional inhibition of Ras. This rather shocking result, which if obtained years ago would have nullified the entire rationale for FTI development, was obtained through the use of a genetic knockout mouse for the FT beta subunit, which lacks FT activity but preserves geranylgeranyl transferase activity (Mijimolle et al. 2005). As anticipated, K-Ras could still induce tumor formation despite the absence of FT activity in the mutant mouse, due to the alternative geranylgeranylation of K-Ras, but even H-Ras could still associate with membranous structures and transform the mutant cells lacking FT activity. An inescapable conclusion was, therefore, that even H-Ras could not be a target that is relevant to the biological action of FTI. Consistent with this notion, a group studying the effect of FT siRNAs in human cancer cells also found little effect of FT knockdown on cell growth or survival, in contrast to FTIs themselves (Pan et al. 2005). Instead, this group demonstrated that the small molecule FTIs could trigger production of reactive oxygen species (ROS) that led to oxidative DNA damage and RhoB induction (Pan et al. 2005). Quenching ROS prevented induction of RhoB by FTIs, arguing that the induction of RhoB was mediated by the ROS-induced DNA damage. The notion that FTIs elevate RhoB expression at the level of transcription has been corroborated recently by others in a wide variety of human cancer cells (Delarue et al. 2007). Thus, it seems clear that the induction of high levels of RhoB-GG in FTI-treated cells likely reflects both an increase in the transcription of the RhoB gene plus a prenylation shift in the synthesis of the RhoB protein, such that all the newly synthesized protein in FTI-treated cells is expressed as RhoB-GG.

Ironically, the possibility that RhoB mediates FTI responses, even if FT does not, has been encouraged by the findings of a recent clinical study of the FTI tipifarnib, which highlights a Rho regulatory molecule as a predictor of therapeutic responses in acute myeloid leukemia (Raponi et al. 2007). Briefly, elevation of AKAP13/Lbc, a RhoGEF oncoprotein, was found to predict a negative response to FTI treatment. This RhoGEF would be expected to drive activation of pro-oncogenic Rho proteins. If RhoB-GG mediates FTI action by antagonizing the pathways activated by pro-oncogenic Rho proteins, then driving this pathway harder by elevating AKAP13/Lbc would be expected to competitively ablate the inhibitory effects of RhoB-GG, limiting the FTI response. In short, through its ability to drive oncogenic Rho signaling AKAP13/Lbc is well positioned to override the ability of RhoB-GG to inhibit Rho signals in cancer (Fig. 9.2). Of course, other interpretations

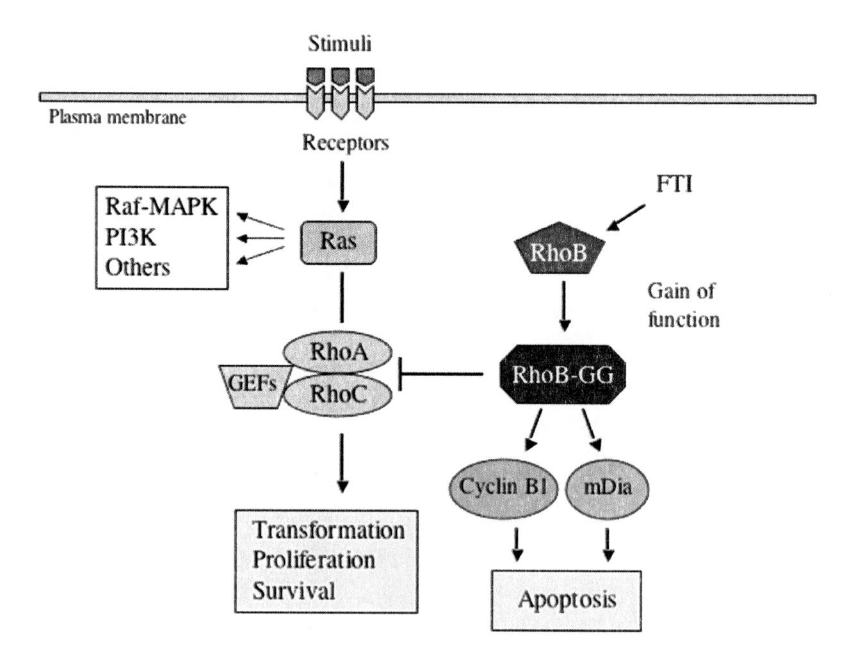

Fig. 9.2 Model for RhoB in the FTI cellular response mechanism. FTI treatment results in an increase in RhoB-GG level, which disrupts Rho oncogenic signals involved in proliferation, survival, or transformation, and which triggers the cell death program through the downstream effectors cyclin B1 and mDia

of this finding are possible and other markers may prove to be important to predicting clinical responses (Raponi et al. 2008). Nevertheless, the identification of AKAP/Lbc as a poor response marker reinforces two concepts: first, that blocking Rho signaling may be a critical element to a good FTI response in clinic, and second, that the RhoB-GG production in FTI treated cells may contribute to at least one facet of the drug mechanism in clinic.

References

Adamson P, Marshall CJ, Hall A, Tilbrook PA (1992a) Post-translational modifications of p21rho proteins. J Biol Chem 267:20033–20038.

Adamson P, Paterson HF, Hall A (1992b) Intracellular localization of the p21^{rho} proteins. J Cell Biol 119:617–627.

Adini I, Rabinovitz I., Sun JF, Prendergast GC, Benjamin LE (2003) RhoB controls Akt trafficking and stage-specific survival of endothelial cells during vascular development. Genes Dev 17:2721–2732.

Adnane J, Muro-Cacho C, Mathews L, Sebti SM, Munoz-Antonia T (2002) Suppression of rho B expression in invasive carcinoma from head and neck cancer patients. Clin Cancer Res 8:2225–2232.

Bernhard EJ, Kao G, Cox AD, Sebti SM, Hamilton AD, Muschel RJ, McKenna WG (1996) The farnesyltransferase inhibitor FTI-277 radiosensitizes H-ras-transformed rat embryo fibroblasts. Cancer Res 56:1727–1730.

Bishop AL, Hall A (2000) Rho GTPases and their effector proteins. Biochem J 348(Pt 2):241–255.

Bos JL (1989) Ras oncogenes in human cancer: a review. Cancer Res 49:4682–4689.

Chauhan S, Kunz S, Davis K et al (2004) Androgen control of cell proliferation and cytoskeletal reorganization in human fibrosarcoma cells: role of RhoB signaling. J Biol Chem 279:937–944.

Chen Z, Sun J, Pradines A, Favre G, Adnane J, Sebti SM (2000) Both farnesylated and geranylgeranylated RhoB inhibit malignant transformation and suppress human tumor growth in nude mice. J Biol Chem 275:17974–17978.

Cox AD, Der CJ (1997) Farnesyltransferase inhibitors and cancer treatment: targeting simply Ras? Biochim Biophys Acta 1333:F51–F71.

Delarue FL, Adnane J, Joshi B et al (2007) Farnesyltransferase and geranylgeranyltransferase I inhibitors upregulate RhoB expression by HDAC1 dissociation, HAT association and histone acetylation of the RhoB promoter. Oncogene 26:633–640.

Delarue FL, Taylor BS, Sebti SM (2001) Ras and RhoA suppress whereas RhoB enhances cytokine-induced transcription of nitric oxide synthase-2 in human normal liver AKN-1 cells and lung cancer A-549 cells. Oncogene 20:6531–6537.

Du W, Lebowitz PF, Prendergast GC (1999) Cell growth inhibition by farnesyltransferase inhibitors is mediated by gain of geranylgeranylated RhoB. Mol Cell Biol 19:1831–1840.

Du W, Prendergast GC (1999) Geranylgeranylated RhoB mediates inhibition of human tumor cell growth by farnesyltransferase inhibitors. Cancer Res. 59:5924–5928.

DuHadaway JB, Du W, Donover S et al (2003) Transformation-selective apoptotic program triggered by farnesyltransferase inhibitors requires Bin1. Oncogene 22:3578–3588.

Ellenbroek SI, Collard JG (2007) Rho GTPases: functions and association with cancer. Clin Exp Metastasis 24:657–672.

Elliott K, Sakamuro D, Basu A et al (1999) Bin1 functionally interacts with Myc in cells and inhibits cell proliferation by multiple mechanisms. Oncogene 18:3564–3573.

Ellis S, Mellor H (2000) Regulation of endocytic traffic by rho family GTPases. Trends Cell Biol 10:85–88.

Engel ME, Datta PK, Moses HL (1998) RhoB is stabilized by transforming growth factor beta and antagonizes transcriptional activation. J Biol Chem 273:9921–9926.
Evangelista M, Zigmond S, Boone C (2003) Formins: signaling effectors for assembly and polarization of actin filaments. J Cell Sci 116:2603–2611.
Fernandez-Borja M, Janssen L, Verwoerd D, Hordijk P, Neefjes J (2005) RhoB regulates endosome transport by promoting actin assembly on endosomal membranes through Dia1. J Cell Sci 118:2661–2670.
Flynn P, Mellor H, Casamassima A, Parker PJ (2000) Rho GTPase control of protein kinase C-related protein kinase activation by 3-phosphoinositide-dependent protein kinase. J Biol Chem 275:11064–11070.
Forget MA, Desrosiers RR, Del M et al (2002) The expression of rho proteins decreases with human brain tumor progression: potential tumor markers. Clin Exp Metastasis 19:9–15.
Fritz G, Brachetti C, Bahlmann F, Schmidt M, Kaina B (2002) Rho GTPases in human breast tumours: expression and mutation analyses and correlation with clinical parameters. Br J Cancer 87:635–644.
Fritz G, Just I, Kaina B (1999) Rho GTPases are over-expressed in human tumors. Int J Cancer 81:682–687.
Fritz G, Kaina B (2000) Ras-related GTPase RhoB forces alkylation-induced apoptotic cell death. Biochem Biophys Res Commun 268:784–789.
Fritz G, Kaina B (2001) Ras-related GTPase Rhob represses NF-kappaB signaling. J Biol Chem 276:3115–3122.
Fritz G, Kaina B, Aktories K (1995) The ras-related small GTP-binding protein RhoB is immediate-early inducible by DNA damaging treatments. J Biol Chem 270:25172–25177.
Fukumoto Y, Kaibuchi K, Hori Y, Fet al (1990) Molecular cloning and characterization of a novel type of regulatory protein (GDI) for the rho proteins, ras p21-like GTP-binding proteins. Oncogene 5:1321–1328.
Gampel A, Parker PJ, Mellor H (1999) Regulation of epidermal growth factor receptor traffic by the small GTPase rhoB. Curr Biol 9:955–958.
Gerhard R, Tatge H, Genth H et al (2005) Clostridium difficile toxin A induces expression of the stress-induced early gene product RhoB. J Biol Chem 280:1499–1505.
Heldin CH, Westermark B (1999) Mechanism of action and in vivo role of platelet-derived growth factor. Physiol Rev 79:1283–1316.
Huang M, Duhadaway JB, Prendergast GC, Laury-Kleintop LD (2007) RhoB regulates PDGFR-beta trafficking and signaling in vascular smooth muscle cells. Arterioscler Thromb Vasc Biol 27:2597–2605.
Huang M, Kamasani U, Prendergast GC (2006) RhoB facilitates c-Myc turnover by supporting efficient nuclear accumulation of GSK-3. Oncogene 25:1281–1289.
Huang M, Prendergast GC (2006) RhoB in cancer suppression. Histol Histopathol 21:213–218.
Huelsenbeck J, Dreger SC, Gerhard R, Fritz G, Just I, Genth H (2007) Upregulation of the immediate early gene product RhoB by exoenzyme C3 from Clostridium limosum and toxin B from Clostridium difficile. Biochemistry 46:4923–4931.
Jaffe AB, Hall A (2002) Rho GTPases in transformation and metastasis. Adv Cancer Res 84:57–80.
Jaffe AB, Hall A (2005) Rho GTPases: biochemistry and biology. Annu Rev Cell Dev Biol 21:247–269.
Jahner D, Hunter T (1991) The ras-related gene rhoB is an immediate-early gene inducible by v-Fps, epidermal growth factor, and platelet-derived growth factor in rat fibroblasts. Mol Cell Biol 11:3682–3690.
Jiang K, Delarue FL, Sebti SM (2004a) EGFR, ErbB2 and Ras but not Src suppress RhoB expression while ectopic expression of RhoB antagonizes oncogene-mediated transformation. Oncogene 23:1136–1145.
Jiang K, Sun J, Cheng J, Djeu JY, Wei S, Sebti S (2004b) Akt mediates Ras downregulation of RhoB, a suppressor of transformation, invasion, and metastasis. Mol Cell Biol 24:5565–5576.
Kamasani U, Duhadaway JB, Alberts AS, Prendergast GC (2007) mDia function is critical for the cell suicide program triggered by farnesyl transferase inhibition. Cancer Biol Ther 6:1422–1427.

Kamasani U, Huang M, Duhadaway JB, Prochownik EV, Donover PS, Prendergast GC (2004) Cyclin B1 is a critical target of RhoB in the cell suicide program triggered by farnesyl transferase inhibition. Cancer Res 64:8389–8396.
Kamasani U, Liu AX, Prendergast GC (2003) Genetic response to farnesyltransferase inhibitors: proapoptotic targets of RhoB. Cancer Biol Ther 2:273–280.
Kamasani U, Prendergast GC (2005) Genetic response to DNA damage: proapoptotic targets of RhoB include modules for p53 response and susceptibility to Alzheimer's disease. Cancer Biol Ther 4:282–288.
Karnoub AE, Symons M, Campbell SL, Der CJ (2004) Molecular basis for Rho GTPase signaling specificity. Breast Cancer Res Treat 84:61–71.
Khosravi-Far R, Solski PA, Clark GJ, Kinch MS, Der CJ (1995) Activation of Rac1, RhoA, and mitogen-activated protein kinases is required for Ras transformation. Mol Cell Biol 15:6443–6453.
Lajoie-Mazenc I, Tovar D, Penary M et al (2008) MAP1A light chain-2 interacts with GTP-RHOB to control EGF-dependent EGF-R signaling. J Biol Chem 283:4155–4164
Lebowitz P, Prendergast GC (1998) Functional interaction between RhoB and the transcription factor DB1. Cell Adhes Comm 4:1–11.
Lebowitz PF, Casey PJ, Prendergast GC, Thissen JA (1997a) Farnesyltransferase inhibitors alter the prenylation and growth-stimulating function of RhoB. J Biol Chem 272:15591–15594.
Lebowitz PF, Davide JP, Prendergast GC (1995) Evidence that farnesyltransferase inhibitors suppress Ras transformation by interfering with Rho activity. Mol Cell Biol 15:6613–6622.
Lebowitz PF, Du W, Prendergast GC (1997b) Prenylation of RhoB is required for its cell transforming function but not its ability to activate serum response element-dependent transcription. J Biol Chem 272:16093–16095.
Lebowitz PF, Prendergast GC (1998) Non-Ras targets of farnesyltransferase inhibitors: focus on Rho. Oncogene 17:1439–1445.
Liu A, Du W, Liu JP, Jessell TM, Prendergast GC (2000) RhoB alteration is necessary for apoptotic and antineoplastic responses to farnesyltransferase inhibitors. Mol Cell Biol 20:6105–6113.
Liu A, Prendergast GC (2000) Geranylgeranylated RhoB is sufficient to mediate tissue-specific suppression of Akt kinase activity by farnesyltransferase inhibitors. FEBS Lett 481:205–208.
Liu AX, Rane N, Liu JP, Prendergast GC (2001a) RhoB is dispensable for mouse development, but it modifies susceptibility to tumor formation as well as cell adhesion and growth factor signaling in transformed cells. Mol Cell Biol 21:6906–6912.
Liu A, Cerniglia GJ, Bernhard EJ, Prendergast GC (2001b) RhoB is required to mediate apoptosis in neoplastically transformed cells after DNA damage. Proc Natl Acad Sci U S A 98:6192–6197.
Lowe SW, Ruley HE (1993) Stabilization of the p53 tumor suppressor is induced by adenovirus 5 E1A and accompanies apoptosis. Genes Dev 7:535–545.
Marks P, Rifkind RA, Richon VM, Breslow R, Miller T, Kelly WK (2001) Histone deacetylases and cancer: causes and therapies. Nat Rev Cancer 1:194–202.
Mayo MW, Wang CY, Cogswell PC et al (1997) Requirement of NF-kappaB activation to suppress p53-independent apoptosis induced by oncogenic Ras. Science 278:1812–1815.
Mazieres J, Antonia T, Daste G et al (2004) Loss of RhoB expression in human lung cancer progression. Clin Cancer Res 10:2742–2750.
Mazieres J, Tillement V, Allal C et al (2005) Geranylgeranylated, but not farnesylated, RhoB suppresses Ras transformation of NIH-3T3 cells. Exp Cell Res 304:354–364.
Mazieres J, Tovar D, He B et al (2007) Epigenetic regulation of RhoB loss of expression in lung cancer. BMC Cancer 7:220.
Mellor H, Flynn P, Nobes CD, Hall A, Parker PJ (1998) PRK1 is targeted to endosomes by the small GTPase, RhoB. J Biol Chem 273:4811–4814.
Michaelson D, Silletti J, Murphy G, D'Eustachio P, Rush M, Philips MR (2001) Differential localization of Rho GTPases in live cells: regulation by hypervariable regions and RhoGDI binding. J Cell Biol 152:111–126.
Mijimolle N, Velasco J, Dubus P et al (2005) Protein farnesyltransferase in embryogenesis, adult homeostasis, and tumor development. Cancer Cell 7:313–324.

Milia J, Teyssier F, Dalenc F et al (2005) Farnesylated RhoB inhibits radiation-induced mitotic cell death and controls radiation-induced centrosome overduplication. Cell Death Differ 12:492–501.

Miura Y, Kikuchi A, Musha T et al (1993) Regulation of morphology by rho p21 and its inhibitory GDP/GTP exchange protein (rho GDI) in Swiss 3T3 cells. J Biol Chem 268:510–515.

Neel NF, Lapierre LA, Goldenring JR, Richmond A (2007) RhoB plays an essential role in CXCR2 sorting decisions. J Cell Sci 120:1559–1571.

Palazzo A, Cook TA, Alberts AS, Gundersen GG (2001) mDia mediates Rho-regulated formation and orientation of stable microtubules. Nature Cell Biol. 3:723–729.

Pan J, She M, Xu ZX, Sun L, Yeung SC (2005) Farnesyltransferase inhibitors induce DNA damage via reactive oxygen species in human cancer cells. Cancer Res 65:3671–3681.

Pan J, Yeung SC (2005) Recent advances in understanding the antineoplastic mechanisms of farnesyltransferase inhibitors. Cancer Res 65:9109–9112.

Prendergast GC (2000) Farnesyltransferase inhibitors: antineoplastic mechanism and clinical prospects. Curr Opin Cell Biol 12:166–173.

Prendergast GC (2001) Actin' up: RhoB in cancer and apoptosis. Nat Rev Cancer 1:162–168.

Prendergast GC, Oliff A (2000) Farnesyltransferase inhibitors: antineoplastic properties, mechanisms of action, and clinical prospects. Semin Cancer Biol 10:443–452.

Prendergast GC, Rane N (2001) Farnesyltransferase inhibitors: mechanism and applications. Expert Opin Investig Drugs 10:2105–2116.

Pruitt K, Der CJ (2001) Ras and Rho regulation of the cell cycle and oncogenesis. Cancer Lett 171:1–10.

Pruyne D, Evangelista M, Yang C et al (2002) Role of formins in actin assembly: nucleation and barbed-end association. Science 297:612–615.

Raponi M, Harousseau JL, Lancet JE et al (2007) Identification of molecular predictors of response in a study of tipifarnib treatment in relapsed and refractory acute myelogenous leukemia. Clin Cancer Res 13:2254–2260.

Raponi M, Lancet JE, Fan H et al (2008) A two-gene classifier for predicting response to the farnesyltransferase inhibitor tipifarnib in acute myeloid leukemia. Blood 111:2589–2596.

Ridley, A. J. 2004. Rho proteins and cancer. Breast Cancer Res Treat 84:13–19.

Robertson D, Paterson HF, Adamson P, Hall A, Monaghan P (1995) Ultrastructural localization of ras-related proteins using epitope-tagged plasmids. J Histochem Cytochem 43:471–480.

Sahai E, Marshall CJ (2002) Rho GTPases and cancer. Nature Rev Cancer 2:133–142.

Sakamuro D, Elliott K, Wechsler-Reya R, Prendergast GC (1996) BIN1 is a novel MYC-interacting protein with features of a tumor suppressor. Nature Genet. 14:69–77.

Sandilands E, Akbarzadeh S, Vecchione A, McEwan DG, Frame MC, Heath JK (2007) Src kinase modulates the activation, transport and signalling dynamics of fibroblast growth factor receptors. EMBO Rep 8:1162–1169.

Sandilands E, Cans C, Fincham VJ et al (2004) RhoB and actin polymerization coordinate Src activation with endosome-mediated delivery to the membrane. Dev Cell 7:855–869.

Sato N, Fukui T, Taniguchi T et al (2007) RhoB is frequently downregulated in non-small-cell lung cancer and resides in the 2p24 homozygous deletion region of a lung cancer cell line. Int J Cancer 120:543–551.

Schubbert S, Bollag G, Shannon K (2007) Deregulated Ras signaling in developmental disorders: new tricks for an old dog. Curr Opin Genet Dev 17:15–22.

Sebti SM, Hamilton AD (2000) Farnesyltransferase and geranylgeranyltransferase I inhibitors in cancer therapy: important mechanistic and bench to bedside issues. Expert Opin Investig Drugs 9:2767–2782.

Shen M, Feng Y, Gao C et al 2004. Detection of cyclin B1 expression in G(1)-phase cancer cell lines and cancer tissues by postsorting Western blot analysis. Cancer Res 64:1607–1610.

Togel M, Wiche G, Propst F (1998) Novel features of the light chain of microtubule-associated protein MAP1B: microtubule stabilization, self interaction, actin filament binding, and regulation by the heavy chain. J Cell Biol 143:695–707.

Tominaga T, Sahai E, Chardin P, McCormick F, Courtneidge SA, Alberts AS (2000) Diaphanous-related formins bridge Rho GTPase and Src tyrosine kinase signaling. Mol Cell 5:13–25.

Trapp T, Olah L, Holker I et al (2001) GTPase RhoB: an early predictor of neuronal death after transient focal ischemia in mice. Mol Cell Neurosci 17:883–894.

van Golen KL, Bao L, DiVito MM, Wu Z, Prendergast GC, Merajver SD (2002) Reversion of RhoC GTPase-induced inflammatory breast cancer phenotype by treatment with a farnesyl transferase inhibitor. Mol Cancer Ther 1:575–583.

Wallar BJ, Deward AD, Resau JH, Alberts AS (2007) RhoB and the mammalian Diaphanous-related formin mDia2 in endosome trafficking. Exp Cell Res 313:560–571.

Wang DA, Sebti SM (2005) Palmitoylated cysteine 192 is required for RhoB tumor-suppressive and apoptotic activities. J Biol Chem 280:19243–19249.

Wang S, Yan-Neale Y, Fischer D et al (2003) Histone deacetylase 1 represses the small GTPase RhoB expression in human nonsmall lung carcinoma cell line. Oncogene 22:6204–6213.

Wasserman S (1998) FH proteins as cytoskeletal organizers. Trends Cell Biol 8:111–115.

Watanabe N, Madaule P, Reid T et al (1997) p140mDia, a mammalian homolog of Drosophila diaphanous, is a target protein for Rho small GTPase and is a ligand for profilin. Embo J 16:3044–3056.

Weidle UH, Grossmann A (2000) Inhibition of histone deacetylases: a new strategy to target epigenetic modifications for anticancer treatment. Anticancer Res 20:1471–1485.

Westmark CJ, Bartleson VB, Malter JS (2005) RhoB mRNA is stabilized by HuR after UV light. Oncogene 24:502–511.

Westwick JK, Lambert QT, Clark GJ et al (1997) Rac regulation of transformation, gene expression, and actin organization by multiple, PAK-independent pathways. Mol Cell Biol 17:1324–1335.

Wherlock M, Gampel A, Futter C, Mellor H (2004) Farnesyltransferase inhibitors disrupt EGF receptor traffic through modulation of the RhoB GTPase. J Cell Sci 117:3221–3231.

Whitehead IP, Abe K, Gorski JL, Der CJ (1998) CDC42 and FGD1 cause distinct signaling and transforming activities. Mol Cell Biol 18:4689–4697.

Zalcman G, Closson V, Linares-Cruz G et al (1995) Regulation of Ras-related RhoB protein expression during the cell cycle. Oncogene 10:1935–1945.

Zeng PY, Rane N, Du W, Chintapalli J, Prendergast GC (2003) Role for RhoB and PRK in the suppression of epithelial cell transformation by farnesyltransferase inhibitors. Oncogene 22:1124–1134.

Chapter 10
Regulation of Rho GTPase Activity Through Phosphorylation Events: A Brief Overview

Heather Unger and Kenneth van Golen

Introduction

Similar to the majority of other monomeric G-proteins, the Rho GTPases transiently cycle through an inactive GDP-bound through active GTP-bound to an inactive GDP-bound state (Nobes and Hall 1995). This transient cycling is called the GTPase cycle and is essential for the proper function of the Rho proteins, leading to interactions with putative downstream effector proteins. As detailed in Part II of this book, a number of upstream regulatory proteins are involved in the activation and inactivation of the Rho proteins (Geyer and Wittinghofer 1997). Direct control of Rho GTPase activation and inactivation occurs due to the action of two main classes of proteins: guanine exchange factors (GEFs) are responsible for the exchange of GDP for GTP, while GTPase-activating proteins (GAPs) stimulate hydrolysis of GTP (Overbeck et al. 1995; Geyer and Wittinghofer 1997). Overwhelmingly, the majority of Rho GTPases exist in their inactive form in the cytoplasm. To be activated, they must be translocated to the inner plasma membrane where they are inserted via a prenylation group. The C-terminal end of Rho proteins contains a CAAX recognition sequence, which in the case of RhoA, B, and C are gernaylgeranylated (Hori, Kikuchi et al. 1991). The primary mechanism of keeping the proteins in the cytosol is through binding of a Rho guanine dissociation inhibitor (RhoGDI) (Geyer and Wittinghofer 1997; Gosser et al. 1997; Olofsson 1999). RhoGDIs contact the Rho protein via two distinct, flexible domains; an N-terminal domain that contacts and prevents GDP dissociation and hydrolysis and a C-terminal domain that binds to and stabilizes the geranylgeranyl moiety (Gosser et al. 1997).

In order to understand the canonical roles of the different Rho proteins in cells, early investigations utilized man-made dominant active or negative mutations in the Rho protein (Hall 1990; Ridley et al. 1992). These seminal experiments provided

H. Unger and K. van Golen (✉)
Center for Translational Cancer Research, Department of Biological Sciences,
Laboratory of Cytoskeletal Physiology, University of Delaware
e-mail: klvg@udel.edu

K. van Golen (ed.), *The Rho GTPases in Cancer*,
DOI 10.1007/978-1-4419-1111-7_10,

great insight into the basic roles of several Rho proteins. However, as research on Rho proteins has progressed over the years, it has become evident that simple activation or inactivation of Rho proteins cannot solely explain the variety of signals and phenotypic events initiated by the individual Rho proteins. It is becoming increasingly evident that location of the active Rho protein plays a large role in how it becomes activated and with which downstream effector it interacts.

Hand in hand with this concept is an emerging body of evidence that suggests that the phosphorylation of the Rho proteins can effect the cellular localization and activation of Rho proteins and their interaction with downstream effector molecules.

Modulation of RhoA GTPase Activity by Protein Kinases A and G

The Rap GTPases (Rap1a and Rap1b) are shown to be substrates for cAMP-dependent protein kinase A (PKA) (Lerosey et al. 1991; Quilliam et al. 1991). Introduction of cAMP elevating agents leads to phosphorylation of serine 180 on both Rap proteins (Lerosey et al. 1991). PKA positively regulates Rap activity and leads to altered interaction with Raf-1 kinase (Hu et al. 1999; Ribeiro-Neto et al. 2002). In contrast to this, evidence suggests that PKA phosphorylation of RhoA appears to negatively inhibit RhoA function both in in vitro and in vivo systems (Chen et al. 2005; Ellerbroek et al. 2003). Phosphorylation of RhoA on serine 188 or introduction of a RhoA phosphomemetic mutant (RhoAS188E) led to increased binding with RhoGDI but not RhoGEFs, GAPS, or geranylgeranyltransferase (Ellerbroek et al. 2003). Thus, this suggests that negative regulation of RhoA by phosphorylation occurs due to increased RhoGDI binding and sequestration of the Rho protein.

Treatment of cells with an activator of adenylate cyclase, such as forskolin, or direct addition of dibutyryl cAMP (Lang et al. 1996) alters cell morphology similar to what is seen with the introduction of C3-exotransferase [a general inhibitor of the RhoA, -B, and -C proteins (Aktories et al. 1989)]. From these observations, it was found that PKA and protein kinase G (PKG) phosphorylate RhoA on serine 188 (Lang et al. 1996; Sawada et al. 2001). Constitutively, active RhoA containing a S188A mutation, which cannot be phosphorylated, can effectively inhibit actin stress fiber disassembly by dibutyryl cAMP and 8-bromo-cGMP (Dong et al. 1998; Sauzeau et al. 2000). Also, introduction of a RhoAS188A can promote the formation of actin stress fibers in cells containing a constitutively active PKG (Sawada et al. 2001). Lysophosphatidic acid treatment of PC-3 prostate and SGC-7901 gastric cancer cells promotes RhoA activation, stress fiber formation, and invasion in a transwell assay that were reversed by the addition of 8-chlorophenylthio-cAMP (Chen et al. 2005). In vivo, PKA phosphorylation of RhoA is not affected by geranylgeranylation of the RhoA protein, nor does it directly affect GEF or GAP activity (Ellerbroek et al. 2003). Instead, persuasive evidence suggests that it is the addition of the negatively charged phosphate group that enhances RhoA/RhoGDI interaction. Together, these studies provide compelling evidence for the negative regulation of RhoA through phosphorylation.

Removal of RhoA from the inner plasma membrane is enhanced and linked to cAMP/PKA and cGMP/PKG signaling (O'Connor et al. 2000; Faucheux and Nagel 2002). One potential mechanism for regulating RhoA activity and removing it from the inner plasma membrane is through the promotion of formation of an RhoA-RhoGDI complex, thus translocating it into the cytosol (Lang et al. 1996; Forget et al. 2002). There has also been a suggestion that addition of cAMP analogues can lead to relocation of RhoA into the nucleus of the cell (Chen et al. 2005). Regardless, PKA phosphorylation of RhoA enhances the ability of RhoGDI to remove RhoA from the inner plasma membrane (Kwak and Uhlinger 2000).

Recent evidence also suggests that another mechanism of control may be through the interaction of RhoA with putative downstream effector proteins. Nerve growth factor (NGF) can induce cAMP-dependent PKA activation leading to neurite outgrowth in SH-SY5Y neuroblastoma cells (Sanchez et al. 2004). Introduction of a constitutively active RhoA can counter NGF-stimulated neurite outgrowth leading to neurite contraction (Sebok et al. 1999). Likewise, NGF stimulation of cells leads to RhoA phosphorylation and alteration of which effector proteins RhoA can interact with in PC12 cells (Nusser et al. 2006). Serine 188 phosphorylation interfered with interaction of active RhoA with Rho kinase but not with rhotekin, mDia1, or PKN. Introduction of a RhoAS188A mutant resulted in neurite retraction similar to what is observed with a constitutively active RhoA and introduction of a serine 188 phosphomimetic mutant led to neurite outgrowth.

Phosphorylation of Rho Proteins by Akt/Protein Kinase B

A survey of the protein sequences of Rac, Cdc42, RhoA, RhoB, RhoC, and RhoG shows a putative protein kinase B (Akt/PKB) phosphorylation site (dRIRpISYp) (Fig. 1). One of the first suggestions that Rho proteins could be phosphorylated by Akt/PKB demonstrated that Rac GTPase could be phosphorylated

RhoA	61 AGQEDYDRLRPLSYPDTDVI80
RhoC	61 AGQEDYDRLRPLSYPDTDVI80
RhoB	61 AGQEDYDRLRPLSYPDTDVI80
TC10	65 AGQEDYDRLRPLSYPMTDV83
Rac1	59 AGQEDYDRLRPLSYPQTDV77
Rac2	59 AGQEDYDRLRPLSYPQTDV77
Rac3	59 AGQEDYDRLRPLSYPQTDV77
Rac4	59 AGQEDYDRLRPLSYPQADV77
Cdc42	59 AGQEDYDRLRPLSYPQTDV 77
RhoG	59 AGQEEYDRLRTLSYPQTNV 77

Fig. 10.1 Comparison of putative Akt/PKB phosphorylation sites in various members of the Rho GTPase subfamily. Several members of the Rho subfamily contain a putative site for phosphorylation. This site lies within the GTPase switch region, potentially effecting GTPase activation or interactions with downstream effectors

on serine 71 (Kwon et al. 2000). Using recombinant Rac1 GTPase and active Akt/PKB or SK-Mel28 melanoma cell lysate (which contains active Akt/PKB), it was demonstrated that Rac1 could act as a substrate for Akt/PKB. In vitro phosphorylation of Rac1 attenuated its activation without affecting its GTPase activity. Further, treatment of the melanoma cells with wortmanin or LY29002, inhibitors of the phosphoinositide-3 kinase (PI3K) pathway, decreased Rac1 phosphorylation. Likewise, downregulation of RhoC GTPase activity in PC-3 prostate cancer cells through expression of a dominant negative RhoC led to a marked increase of Rac and Cdc42 GTPase activity, which corresponded with decreased phosphorylation at serine 71 of these two GTPases (Yao et al. 2006). The mechanism(s) by which serine 71 phosphorylation of Rac and Cdc42 prevent activation is currently unknown but may be due to changes within the activating switch region of the GTPases.

Several subsequent experiments have suggested that Akt/PKB may act as a downstream substrate of the Rac and Rho GTPases. However, studies in systems other than cancer suggest that Rac is a bona fide substrate of Akt/PKB. For example, a protective mechanism of cerebral ischemia is proposed to occur through inactivation of Rac1 GTPase through Akt/PKB-mediated phosphorylation (Zhang et al. 2006). The study suggests that the proapoptotic mixed lineage kinase 3 (MLK3)/JNK3 pathway can be inhibited through activation of the antiapoptotic PI3K-Akt/PKB pathway. The authors demonstrate that Rac1 and Akt/PKB physically interact in the hippocampal CA1 region of the brain. Ischemic stress leads to an increase in Akt/PKB activity and a concordant increase in phosphorylation of serine 71 on Rac1, thus leading to decreased Rac1 activity. Rac1 activity is required and sufficient for phosphorylation and activation of MLK2/JNK3.

Similarly, Fukuda et al. (2005) demonstrate that activation of Akt/PKB leads to phosphorylation of Rac1 on serine 71, thereby resulting in a decrease in Rac1 activity. In turn, decreased Rac1 activity leads to decreased osteoclast motility. Rac1 activity is required for osteoclast bone resorptive activity through promotion of cell survival and motility. These properties are enhanced through the introduction of a dominant active Rac1 into osteoclast-like cells and can be inhibited through the use of PI3K inhibitors (Fukuda et al. 2005).

Similarly, our laboratory group has found that Akt/PKB is active in SUM149 inflammatory breast cancer cells, probably due to the loss of phosphatase and tensin homology deleted on chromosome 10 (PTEN). Increased Akt/PKB activity results in high levels of serine phosphorylated RhoC GTPase. Like many Rho proteins, RhoC has a putative Akt/PKB phosphorylation site (Fig. 1). Inhibition of Akt/PKB activity, either by pharmacological inhibitors specific for Akt1 or through the introduction of siRNA to Akt1 or reintroduction of PTEN, leads to a significant decrease in RhoC-mediated invasion. Likewise, introduction of a RhoCS73A mutant into the inflammatory breast cancer cells significantly decreases their invasive capabilities (Fig. 2). Interestingly, downregulation of Akt2 has no effect on the SUM149 invasion but does affect MDA-MB-231 and MDA-MB-435 invasion. The mechanism through which Akt/PKB phosphorylation of RhoC affects the invasive capabilities of the cells is currently under investigation. RhoC activity does not

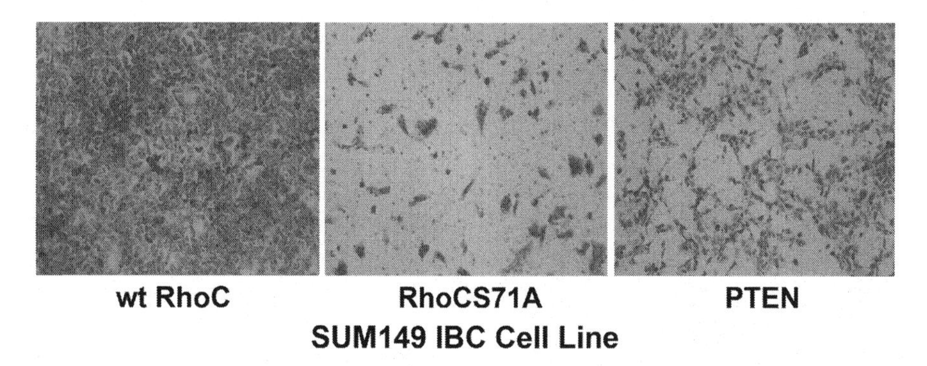

Fig. 10.2 Results of a Matrigel™ invasion assay comparing parental SUM149 inflammatory breast cancer cells with SUM149 transfected with either phosphatase and tensin homology deleted on chromosome (*PTEN*) or RhoCS71A mutant. Introduction of PTEN reduced levels of active protein kinase B (*Akt/PKB*) in the cells without effecting cell growth or survival. Invasion was significantly reduced in cells harboring PTEN or the phosphorylation mutant

appear to be affected by serine phosphorylation. Interestingly, serine 71 lies within the switch II region of RhoC. Each switch region is critical for effector binding, and X-ray crystal structures suggest that on GTP binding, RhoC undergoes two conformational changes (Dias and Cerione 2007). Phosphorylation at this site may lead to changes in RhoC conformation, thus altering its effector-binding capabilities without affecting its ability to bind GTP.

Phosphorylation of RhoA and RhoC by Protein Kinase C

Like other protein kinases, protein kinase C (PKC) is a serine/threonine kinase that has been the subject of intense research for several decades. Over the years, much has been learned about the role of active PKC isoforms in a variety of cellular activities including growth, differentiation, apoptosis, and cellular migration. Understanding the roles that each PKC isoform may play in tumor progression has been difficult due to a number of differences that exist between them. The different isoforms of PKC and their physiological differences can lead to a variety of effects in the cell, including opposing effects in some eukaryotic cells. Phorbol esters such as 12-*O*-tetradecanoylphorbol-13-acetate, which act as tumor promoters, were found to target PKC (Kikkawa et al. 1983). 12-*O*-tetradecanoylphorbol-13-acetate not only activated PKC but also stimulated its ubiquitination and degredation (Lee et al. 1996; Lu et al. 1998). It is however, not clear which isoforms of PKC are required to be activated or degraded during tumor promotion.

An example of the differences between PKC isoforms can be seen with PKCδ and PKCε. Studies demonstrate that due to structural differences, PKCε distributes to both the plasma and nuclear membrane, whereas PKCδ distributes specifically to the plasma membrane in 1,2-diacylglycerol-activated HEK293 cells (Stahelin et al. 2005). Individual isoforms of PKC, such as PKCδ, can also have opposing effects. The Kit tyrosine kinase receptor is required for normal hematopoesis, and a D816Y mutation results in an oncogenic form of the receptor. PKCδ expression inhibits growth of cells expressing wild-type Kit, while enhancing growth of cells harboring the oncogenic mutation (Jelacic and Linnekin 2005). Similarly, PKCε was shown to regulate hepatocyte growth factor/c-Met signaling in squamous cell carcinoma (SCC) (Dong et al. 2004; Kermorgant et al. 2004; Worden et al. 2005).

Aside from cell proliferation and survival, PKC has been implicated in cell motility. Insulin-like growth factor-1-stimulated multiple myleoma cells undergo tumor cell diapedesis through vascular endothelial cells (Qiang et al. 2004). Both PKC and protein kinase D, as well as RhoA GTPase were activated in a PI-3K-dependent manner in these cells. Pharmcologic inhibition of PKC led to inhibition of diapedesis. Similarly, autocrine motility factor induced Rho-dependent migration and angiogenesis and is PKC dependent (Yanagawa et al. 2004).

Recently, it was demonstrated that PKCε was a predictive marker for head and neck squamous cell carcinoma (HNSCC) and aggressive breast cancer (Pan et al. 2005, 2006). Targeted downregulation of PKCε by siRNA in the MDA-MB-231 breast cancer cell line led to a significant decrease in RhoC GTPase expression and activation (Pan et al. 2005). Using a similar approach in UMSCC11A HNSCC cells, both RhoA and RhoC expression and activation were significantly decreased (Pan et al. 2006). This correlated with a loss of serine-phosphorylated Rho proteins, suggesting that PKCε can affect expression and activation of RhoA and C via direct phosphorylation. Concordant with decreased Rho activation was a decrease in the invasive capabilities of the cells. Forced expression of either a constitutively active RhoA or RhoC in the HNSCC cells can rescue the invasive capabilities of the PKCε-deficient cells. The mechanisms involved in PKCε phosphorylation and the site at which it occurs are yet to be determined.

Conclusions

The concept that phosphorylation can affect the activity of Rho GTPases either by direct activation or through its interaction with downstream effectors is intriguing. This additional control of activation and interaction may also explain many phenotypic differences observed for different Rho GTPases in different cells types. For example, RhoA and RhoC GTPases appear to have independent functions in most cells types; however, in the MDA-MB-231 breast cancer cell line, they appear to be redundant for each other (Zhang, et al. 2005). One explanation may be how these GTPases are phosphorylated. Clearly, although this area of GTPase biology is understudied, it adds a new and intriguing level of complexity to the control of the Rho GTPase-mediated phenotype.

References

Aktories K, Braun U et al (1989) The rho gene product expressed in E. coli is a substrate of botulinum ADP-ribosyltransferase C3. Biochem Biophys Res Commun 158(1):209–213.

Chen Y, Wang Y et al (2005) The cross talk between protein kinase A- and RhoA-mediated signaling in cancer cells. Exp Biol Med (Maywood) 230(10):731–741.

Dias SM, Cerione RA (2007) X-ray crystal structures reveal two activated states for RhoC. Biochemistry 46(22):6547–6558.

Dong G, Lee TL et al (2004) Metastatic squamous cell carcinoma cells that overexpress c-Met exhibit enhanced angiogenesis factor expression, scattering and metastasis in response to hepatocyte growth factor. Oncogene 23(37):6199–6208.

Dong JM, Leung T et al (1998) cAMP-induced morphological changes are counteracted by the activated RhoA small GTPase and the Rho kinase ROKalpha. J Biol Chem 273(35):22554–22562.

Ellerbroek SM, Wennerberg K et al (2003) Serine phosphorylation negatively regulates RhoA in vivo. J Biol Chem 278(21):19023–19031.

Faucheux N and Nagel MD (2002) Cyclic AMP-dependent aggregation of Swiss 3T3 cells on a cellulose substratum (Cuprophan) and decreased cell membrane Rho A. Biomaterials 23(11):2295–2301.

Forget MA, Desrosiers RR et al (2002) Phosphorylation states of Cdc42 and RhoA regulate their interactions with Rho GDP dissociation inhibitor and their extraction from biological membranes. Biochem J 361(Pt 2):243–254.

Fukuda A, Hikita A et al (2005) Regulation of osteoclast apoptosis and motility by small GTPase binding protein Rac1. J Bone Miner Res 20(12):2245–2253.

Geyer M, Wittinghofer A (1997) GEFs, GAPs, GDIs and effectors: taking a closer (3D) look at the regulation of Ras-related GTP-binding proteins. Curr Opin Struct Biol 7(6):786–792.

Gosser YQ, Nomanbhoy TK et al (1997) C-terminal binding domain of Rho GDP-dissociation inhibitor directs N-terminal inhibitory peptide to GTPases. Nature 387(6635):814–819.

Hall A (1990) The cellular functions of small GTP-binding proteins. Science 249(4969):635–640.

Hori Y, Kikuchi A et al (1991) Post-translational modifications of the C-terminal region of the rho protein are important for its interaction with membranes and the stimulatory and inhibitory GDP/GTP exchange proteins. Oncogene 6(4):515–522.

Hu CD, Kariya K et al (1999) Effect of phosphorylation on activities of Rap1A to interact with Raf-1 and to suppress Ras-dependent Raf-1 activation. J Biol Chem 274(1):48–51.

Jelacic T, Linnekin D (2005) PKCdelta plays opposite roles in growth mediated by wild-type Kit and an oncogenic Kit mutant. Blood 105(5):1923–1929.

Kermorgant S, Zicha D et al (2004) PKC controls HGF-dependent c-Met traffic, signalling and cell migration. EMBO J 23(19):3721–3734.

Kikkawa U, Takai Y et al (1983) Protein kinase C as a possible receptor protein of tumor-promoting phorbol esters. J Biol Chem 258(19):11442–11445.

Kwak JY, Uhlinger DJ (2000) Downregulation of phospholipase D by protein kinase A in a cell-free system of human neutrophils. Biochem Biophys Res Commun 267(1):305–310.

Kwon T, Kwon DY et al. (2000). Akt protein kinase inhibits Rac1-GTP binding through phosphorylation at serine 71 of Rac1. J Biol Chem 275(1):423–428.

Lang P, Gesbert F et al (1996) Protein kinase A phosphorylation of RhoA mediates the morphological and functional effects of cyclic AMP in cytotoxic lymphocytes. EMBO J 15(3):510–519.

Lee HW, Smith L et al (1996) Ubiquitination of protein kinase C-alpha and degradation by the proteasome. J Biol Chem 271(35):20973–20976.

Lerosey I, Pizon V et al (1991) The cAMP-dependent protein kinase phosphorylates the rap1 protein in vitro as well as in intact fibroblasts, but not the closely related rap2 protein. Biochem Biophys Res Commun 175(2):430–436.

Lu Z, Liu D et al (1998) Activation of protein kinase C triggers its ubiquitination and degradation. Mol Cell Biol 18(2):839–845.

Nobes CD, Hall A (1995) Rho, rac and cdc42 GTPases: regulators of actin structures, cell adhesion and motility. Biochem Soc Trans 23(3):456–459.
Nusser N, Gosmanova E et al (2006) Serine phosphorylation differentially affects RhoA binding to effectors: implications to NGF-induced neurite outgrowth. Cell Signal 18(5):704–714.
O'Connor KL, Nguyen BK et al (2000) RhoA function in lamellae formation and migration is regulated by the alpha6beta4 integrin and cAMP metabolism. J Cell Biol 148(2):253–258.
Olofsson B (1999) Rho guanine dissociation inhibitors: pivotal molecules in cellular signalling. Cell Signal 11(8):545–554.
Overbeck AF, Brtva TR et al (1995) Guanine nucleotide exchange factors: activators of Ras superfamily proteins. Mol Reprod Dev 42(4):468–476.
Pan Q, Bao LW et al (2005) Protein kinase C epsilon is a predictive biomarker of aggressive breast cancer and a validated target for RNA interference anticancer therapy. Cancer Res 65(18):8366–8371.
Pan Q, Bao LW et al (2006) Targeted disruption of protein kinase C epsilon reduces cell invasion and motility through inactivation of RhoA and RhoC GTPases in head and neck squamous cell carcinoma. Cancer Res 66(19):9379–9384.
Qiang YW, Yao L et al (2004) Insulin-like growth factor I induces migration and invasion of human multiple myeloma cells. Blood 103(1):301–308.
Quilliam LA, Mueller H et al (1991) Rap1A is a substrate for cyclic AMP-dependent protein kinase in human neutrophils. J Immunol 147(5):1628–1635.
Ribeiro-Neto F, Urbani J et al (2002) On the mitogenic properties of Rap1b: cAMP-induced G(1)/S entry requires activated and phosphorylated Rap1b. Proc Natl Acad Sci U S A 99(8): 5418–5423.
Ridley AJ, Paterson HF et al (1992) The small GTP-binding protein rac regulates growth factor-induced membrane ruffling. Cell 70(3):401–410.
Sanchez S, Jimenez C et al (2004) A cAMP-activated pathway, including PKA and PI3K, regulates neuronal differentiation. Neurochem Int 44(4):231–242.
Sauzeau V, Le Jeune H et al (2000) Cyclic GMP-dependent protein kinase signaling pathway inhibits RhoA-induced Ca2+ sensitization of contraction in vascular smooth muscle. J Biol Chem 275(28):21722–21729.
Sawada N, Itoh H et al (2001) cGMP-dependent protein kinase phosphorylates and inactivates RhoA. Biochem Biophys Res Commun 280(3):798–805.
Sebok A, Nusser N et al (1999) Different roles for RhoA during neurite initiation, elongation, and regeneration in PC12 cells. J Neurochem 73(3):949–960.
Stahelin RV, Wang J et al (2005) The origin of C1A-C2 interdomain interactions in protein kinase Calpha. J Biol Chem 280(43):36452–36463.
Worden B, Yang XP et al (2005) Hepatocyte growth factor/scatter factor differentially regulates expression of proangiogenic factors through Egr-1 in head and neck squamous cell carcinoma. Cancer Res 65(16):7071–7080.
Yanagawa T, Funasaka T et al (2004) Novel roles of the autocrine motility factor/phosphoglucose isomerase in tumor malignancy. Endocr Relat Cancer 11(4):749–759.
Yao H, Dashner EJ et al (2006) RhoC GTPase is required for PC-3 prostate cancer cell invasion but not motility. Oncogene 25(16):2285–2296.
Zhang QG, Wang XT et al (2006) Akt inhibits MLK3/JNK3 signaling by inactivating Rac1: a protective mechanism against ischemic brain injury. J Neurochem 98(6):1886–1898.
Zhang X, Lin M et al (2005) Multiple signaling pathways are activated during insulin-like growth factor-I (IGF-I) stimulated breast cancer cell migration. Breast Cancer Res Treat 93(2):159–168.

Chapter 11
The Rho-Regulated ROCK Kinases in Cancer

Grant R Wickman, Michael S. Samuel, Pamela A Lochhead, and Michael F Olson*

Introduction

Rho-associated protein kinases (ROCK) are central and prominent downstream effectors of the RhoA, RhoB, and RhoC GTP-binding proteins. The predominant function of ROCK is the regulation and modulation of the cytoskeleton. Specifically, ROCK promotes the stabilization of actin filaments (F-actin) and the generation of actin-myosin contractility via the phosphorlyation of downstream substrates. As a consequence, ROCK is indispensable for many cellular processes dependent on the cytoskeleton including, but not limited to cell motility, adhesion, smooth muscle contraction, neurite retraction, and phagocytosis. In addition, ROCK contributes to proliferation, apoptosis, and oncogenic transformation. As a result, ROCK proteins are involved in the development and progression of human cancer in numerous ways.

ROCK Structure and Regulation

There are two isoforms of the ROCK serine/threonine kinases, ROCK I (ROKβ) and ROCK II (Rho kinase or ROKα), which share 64% overall homology with 89% homology within the kinase domain (Nakagawa et al. 1996). ROCK is most closely related to the myotonic dystrophy kinase (DMPK, 48% identity in kinase domain with ROCK I) and the DMPK-related Cdc42-binding kinases MRCKa and MRCKb (each with 53% kinase domain identity with ROCK I), which appear to share some overlapping functions in the regulation of the cytoskeleton (Wilkinson et al. 2005). The ROCK kinases contain an N-terminal kinase domain, a central coiled-coil domain (55% identity), followed by a split pleckstrin homology domain containing a C1 conserved region (80% identity) (Fig. 11.1). A Rho-binding

G.R. Wickman, M. Samuel, P.A. Lochhead and M.F. Olson (✉)
Molecular and Cellular Biology, The Beatson Institute for Cancer Research, Glasgow, United Kingdom G61 1BD
e-mail: m.olson@beatson.gla.ac.uk

K. van Golen (ed.), *The Rho GTPases in Cancer*,
DOI 10.1007/978-1-4419-1111-7_11, © Springer Science+Business Media, LLC 2010

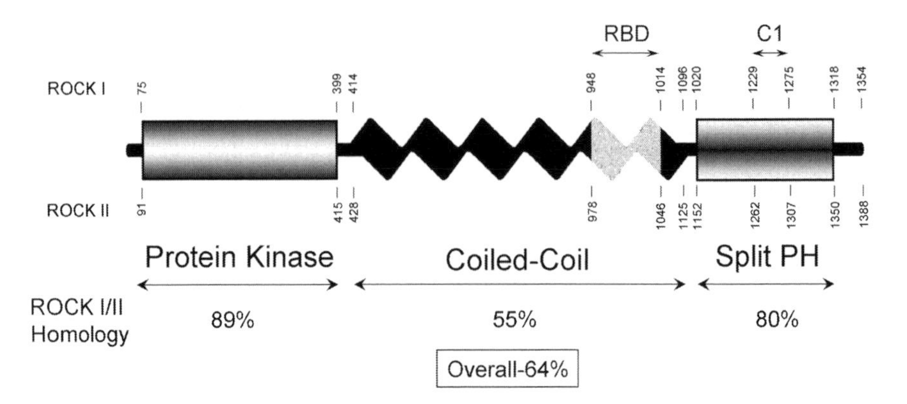

Fig. 11.1 Schematic diagram of human ROCK I/II. Protein domains, their corresponding amino acid sequence numbers, and the percentage amino acid identity of each domain as indicated diagrammatically. *RBD* Rho-binding domain, *C1* cysteine-rich region, *PH* Pleckstrin homology

domain (RBD) lies within the coiled-coil region of ROCK; structural studies have revealed that the RBD forms a coiled-coil that interacts with the switch I and II regions of GTP-bound RhoA (Shimizu et al. 2003; Dvorsky et al. 2004). Several lines of investigation have revealed that ROCK I and ROCK II form homo- and heterodimers that influence kinase activity, inhibitor sensitivity, and normal function *in vivo* (Doran et al. 2004; Jacobs et al. 2006; Yamaguchi et al. 2006). Experiments using either wild-type RhoA loaded with nonhydrolyzable GTP-γS or GTPase-deficient RhoG14V revealed that Rho-GTP can increase ROCK-specific activity in kinase assays and the formation of ROCK-dependent actin stress fibers and focal adhesions in cultured cells (Amano et al. 1997). Studies examining the mechanism of ROCK activation by Rho-GTP revealed that expression of ROCK RBD and PH fragments attenuate activation which suggests that the C-terminal domain of ROCK negatively regulates kinase activity (Amano et al. 1997; Amano et al. 1999). While it is clear from these data that Rho-GTP binding to the RBD within the C-terminus of ROCK relieves this negative regulation leading to increased kinase activity, the precise mechanistic details of this autoinhibition remain to be elucidated.

The association of additional proteins with ROCK, including other small GTP-binding proteins, appears to regulate kinase activity. RhoE, Gem, and Rad have each been shown to bind ROCK at sites distinct from the RBD (Riento et al. 2003; Ward 2002). The binding of Gem to ROCK I was found to attenuate phosphorylation of the MYPT1 regulatory subunit of the PP1M phosphatase complex but not LIM kinase, suggesting that Gem induces a shift in ROCK substrate specificity (Ward et al. 2002). In addition, the overexpression of Gem and Rad in endothelial cells result in reduced stress fibers and focal adhesions, consistent with inhibition of ROCK activity (Ward et al. 2002). Similar effects on stress fibers are reported for RhoE overexpression (Nobes et al. 1998), which was found to bind within the N-terminal kinase domain and may sterically interfere with kinase-substrate interactions (Riento et al. 2003). Alternatively, RhoE and RhoA appear to be competitive for ROCK binding, despite their binding at distinct sites, thus RhoE may antagonize ROCK function by inhibiting RhoA-mediated activation (Riento et al. 2003). In further support of a role for RhoE in ROCK regulation, the protein kinase PDK 1 has been shown to enhance ROCK I activity, not through phosphorylation, but by blocking the association of RhoE (Pinner and Sahai 2008). RhoE and Rad/Gem display discrete subcellular distributions, being located at the Golgi apparatus and the cytoskeleton, respectively, and it has been suggested that the distribution of these negative regulators is important for the modulation of ROCK activity at specific intracellular sites (Riento and Ridley 2003).

The lipid second messengers arachadonic acid and sphingosylphosphoylcholine appear to be highly efficacious ROCK activators. ROCK purified from chicken gizzard was activated five- to six fold following exposure to arachadonic acid, independent of RhoA (Feng et al. 1999). In addition, sphingosylphosphoylcholine enhanced contractility of vascular smooth muscle in a ROCK inhibitor-sensitive, but GTP-independent, manner (Shirao et al. 2002). Together these observations suggest that lipid activation is sufficient to stimulate ROCK and catalyze actin-myosin contractility. However, the relevance of these two lipid signaling pathways in the regulation of ROCK in nonsmooth muscle cell types remains to be demonstrated. A role for phosphatidyl inositides in ROCK activation has also been observed. Purified ROCK II, but not ROCK I, binds phosphatidylinositol (3,4,5)-trisphosphate (PIP_3) and phosphatidylinositol (4,5)-bisphosphate (PIP_2), which activate kinase activity independent of Rho-GTP (Yoneda et al. 2005). It has been suggested that the differential binding properties of ROCK I and ROCK II toward PIP_3 and PIP_2 may be important for subcellular regulation allowing ROCK to initiate discrete spatial functions.

Phosphorylation of ROCK II at several sites by polo-like kinase 1 appears to promote RhoA-dependent activation (Lowery et al. 2007). Although additional serine/threonine and tyrosine phosphorylation sites have been identified (http://www.phosphosite.org), which may be involved in ROCK regulation, the prevalence and function of these phosphorylations remain to be determined. Interestingly, structural studies revealed that phosphorylation within the kinase domain was not necessary for the formation of a catalytically competent conformation (Jacobs et al. 2006).

During apoptosis, the cleavage of ROCK I by caspase 3 (Coleman et al. 2001; Sebbagh et al. 2001), or ROCK II by granzyme B (Sebbagh et al. 2005), within the coiled-coil domain releases a highly active kinase fragment that is critical for membrane blebbing. In addition, the cleaved ROCK I fragment appears to be important for disruption of nuclear integrity and the packaging of fragmented DNA into membrane blebs and apoptotic bodies (Coleman et al. 2001; Croft et al. 2005). Thus, the generation of constitutively active ROCK kinase fragments appears to be a *bona fide* regulatory mechanism for the complete execution of cellular apoptosis.

ROCK Inhibitor Specificity

Many studies that have aimed to identify the biological activities of ROCK have made use of the potent small molecule inhibitors that are commercially available. The most commonly used inhibitor is Y-27632 (Uehata et al. 1997); others include Y-32885 (aka WF-536) (Uehata et al. 1997), H-89 (Chijiwa et al. 1990), HA-1077 (fasudil) (Asano et al. 1987), HA-1100 (hydroxyfasudil) (Arai et al. 1993), H-7 and H-8 (Hidaka et al. 1984), H-1152 (Ikenoya et al. 2002), Rockout (Yarrow et al. 2005), and [*N*-(4-Pyridyl)-*N*′-(2,4,6-trichlorophenyl)urea] (Takami et al. 2004). There are always questions about the specificity and selectivity of small molecule inhibitors, given that the majority act as ATP-competitors and the three-dimensional (3D) structures of protein kinases, including the ATP-binding regions, are highly related (Huse and Kuriyan 2002). Recently, a panel of ROCK inhibitors (Y-27632, H-7, H-8, HA-1077, H-89, and H-1152) was tested against a panel of 70 protein kinases (Bain et al. 2007). These experiments revealed that none of the ROCK inhibitors is absolutely specific, with the Rho-regulated PRK2 being inhibited to the same extent by all tested ROCK inhibitors (Fig. 11.2a) and both MSK1 and

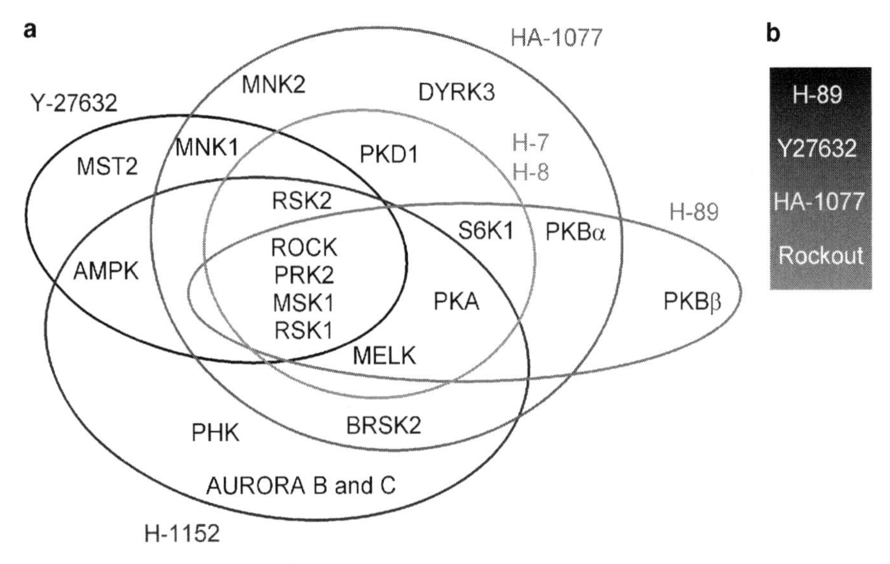

Fig. 11.2 Selectivity and potency of ROCK inhibitors. (**a**).Venn diagram shows the kinases inhibited by at least 50% at the concentrations required for >90% inhibition of ROCK II by Y-27632, HA-1077, H-89, H-1152, H-7, and H-8 (Bain et al. 2007). *AMPK* AMP-activated protein kinase, *BRSK* brain-specific kinase, *DYRK* dual specificity tyrosine phosphorylated and regulated kinase, *MELK* maternal embryonic leucine zipper kinase, *MNK*, MAPK-integrating protein kinase, *MSK* mitogen and stress-activated protein kinase, *MST* mammalian homologue Ste20-like kinase, *PHK* phosphorylase kinase, *PKA* cyclic AMP-dependent protein kinase, *PKB* protein kinase B (aka AKT), *PKD* protein kinase D, *PRK* protein kinase C-related kinase, *ROCK* Rho-dependent protein kinase, *RSK* ribosomal S6 kinase. (**b**). The potency of ROCK inhibitors is listed in order from top to bottom of IC_{50} (generated *in vitro* toward a peptide substrate at 0.1 M ATP). These are 270 nM for H-89, 800 nM for Y-27632, 1.9 μM for HA-1077 (Davies et al. 2000), and and 25 μM for Rockout (Yarrow et al. 2005)

RSK1 being at least 50% inhibited by all the tested compounds at concentrations that inhibited ROCK II by at least 90%. These studies indicate that although ROCK inhibitors are useful tools, they should not be relied upon to be conclusive.

Therefore, a greater confidence would be engendered in studies that make use of ROCK inhibitors if a number of additional conditions were satisfied including:

1. Structurally unrelated inhibitors should produce the same biological endpoints at concentrations that produce equivalent kinase inhibition. The lowest effective doses should be used to reduce off-target effects.
2. Dose-response experiments to establish rank order of potency for a set of inhibitors, that is, the most potent ROCK inhibitors, should be the most effective if a biological response is mediated by ROCK (Fig. 11.2b).
3. Examination of the relationship between ROCK inhibitor dose, substrate phosphorylation, and biological endpoint.
4. Wherever possible, additional methods should be used to inhibit ROCK function, such as RNAi-mediated knockdown.

Although care should be taken in interpreting the results of inhibitor studies, they are undoubtedly useful and convenient tools. Our understanding of the biological functions of ROCK has been vastly aided by the ready availability of these inhibitors. However, their utility is most robust in excluding the possible involvement of a particular signaling pathway in specific biological responses when adequate positive controls are in place.

Expression, Substrates, and Function

Both ROCK isoforms are ubiquitously expressed in normal tissue and appear to have enhanced expression in brain, liver, and skeletal muscle (Leung et al. 1996). Examination of intracellular localization revealed a predominantly cytosolic distribution pattern (Leung et al. 1995; Matsui et al. 1996). However, more detailed analysis has revealed ROCK localized to plasma membranes (Leung et al. 1995; Miyazaki et al. 2006), actin-myosin filaments (Leung et al. 1995; Kawabata et al. 2004), nucleus (Tanaka et al. 2006), mitotic cleavage furrow (Kosako et al. 1999; Yokoyama et al. 2005), and centrosomes (Ma et al. 2006). These observations are consistent with a role for ROCK as a key mediator of the actin cytoskeleton, and suggest that proper subcellular localization likely plays a key regulatory role.

More than 20 ROCK substrates have been identified, and while many have only been tested with one isoform, it seems likely that most substrates would be phosphorylated *in vitro* by either kinase given the 89% identity between ROCK I and ROCK II. If there were differences in substrate phosphorylation by each isoform, this would probably result from more subtle differences in subcellular localization and/or protein-protein interactions. ROCK regulates the cytoskeleton via phosphorylation of numerous downstream target proteins. ROCK phosphorylates LIM kinases-1 and -2 (LIMK1 and LIMK2) at threonine residues in their activation loops, which results in increased LIMK catalytic activity and the subsequent phosphorylation and inactivation of the actin-severing protein, cofilin. Inactivation of cofilin thereby stabilizes filamentous actin (Maekawa et al. 1999; Ohashi et al. 2000; Amano et al. 2001; Sumi et al. 2001). In addition, ROCK participates in the phosphorylation of the myosin II light chains (MLCs), a key mechanism for regulation of actin-myosin contractility (Somlyo and Somlyo 2000). MLC phosphorylation promotes the release of the myosin heavy chain tail allowing for assembly into filaments, and facilitates the association of the myosin head with F-actin. The myosin head uses ATP to "walk" toward the barbed end, when multimeric myosin is associated with more than one actin filament to provide traction; this process then allows for sliding actin of filaments in the opposite direction, thereby generating contractile force. While ROCK has been shown to phosphorylate recombinant MLC at the same site (Ser 19) as the Ca^{2+}-dependent myosin light chain kinase (MLCK) (Amano et al. 1996), experiments with GTPγS in airway smooth muscle failed to demonstrate significant MLC phosphorylation or cell contraction in Ca^{2+}-depleted and therefore MLCK-inactive conditions (Iizuka et al. 1999). These observations suggest that ROCK-mediated phosphorylation of MLC at Ser19 may not be a physiologically significant pathway to regulate cell contractility, at least in some cell types. Instead, the regulation of MLC phosphorylation, and thus cellular contraction, by ROCK may be mediated by inhibition of the MLC phosphatase PP1M. This protein complex is made up of a PP1Cδ catalytic subunit, a MLC-binding subunit and a smaller M20 subunit of unknown function (Ito et al. 2004). ROCK has been found to phosphorylate the ubiquitously expressed MYPT1 myosin-binding subunit at two sites (Thr696 and Thr853 in the human form), which inhibits MLC dephosphorylation (Feng et al. 1999; Velasco et al. 2002; Terrak et al. 2004). As a result, a net gain in MLC phosphorylation would actually require less activity by kinases such as MLCK directed toward MLC than under conditions in which PP1M was not inhibited, leading to an increase in MLC phosphorylation and greater cellular contractility (Kimura et al. 1996; Iizuka et al. 1999). In addition, many ROCK effects may be amplified by direct phosphorylation and activation of the Zipper-interacting protein kinase, which phosphorylates many of the same substrates as ROCK (Hagerty et al. 2007). Taken together, ROCK activation leads to a concerted series of events that promote actin-myosin-mediated force generation and morphological changes.

ROCK-Induced Stress Fiber Formation and Cell Adhesion

Stress fibers are bundles of contractile actin and myosin II filaments found along the length of the cell body of cultured cells (Byers et al. 1984; Pellegrin and Mellor 2007). Stress fiber formation was one of the first cellular activities identified for ROCK (Leung et al. 1996; Amano et al. 1997). Either expression of dominant negative forms of ROCK or the ROCK inhibitor Y-27632 has been found to inhibit the formation of stress fiber by active Rho (Ishizaki et al. 1997).

Enhanced cross linking and contractility of actin-myosin filaments promotes their bundling into stress fibers (Chrzanowska-Wodnicka and Burridge 1996). While ROCK is clearly involved in the generation of stress fibers, it should be noted that the expression of ROCK alone generates "stellate" stress fibers with an architecture distinctively different from those stimulated by Rho (Leung et al. 1996). Expression of ROCK with another Rho-binding protein, the Diaphanous-related forming mDia1, gave rise to stress fibers similar to those seen induced by Rho (Watanabe et al. 1999), indicating that although ROCK plays a central and critical role, Rho induction of stress fiber formation requires the input of additional signaling pathways.

Stress fibers terminate at discrete points at the membrane known as focal adhesions, which serve as attachment points for cells to anchor themselves to the extracellular matrix. While it is clear that ROCK promotes the formation of focal adhesions and thus cellular adhesion (Amano et al. 1997; Tominaga and Barber 1998), it is unclear exactly how ROCK stimulates the clustering of focal adhesion proteins. In opposition to a commonly held view that stress fibers drive focal adhesion formation, it seems as though focal adhesions act as sites of stress fiber production (Hotulainen and Lappalainen 2006; Endlich et al. 2007). Nonetheless, the formation of stress fibers and focal adhesions is critical for the generation of cellular tension and force, which in turn facilitate morphological responses and cell migration.

ROCK Functions in Embryonic Development

Targeted deletion in mice has revealed distinct roles for ROCK I and ROCK II in embryonic development and highlighted potential mechanisms for their roles in cancer. Knockout mice generated by targeted deletion of exons 3–4 of the ROCK I gene (in the C57Bl/6 strain background) were born at the expected Mendelian ratio, but were not viable and consequently were cannibalized by the mother (Shimizu et al. 2005). Developmental abnormalities of ROCK $I^{-/-}$ mice included failure in the closure of eyelids (eyelids open at birth phenotype - EOB) and of the ventral body wall. The latter phenomenon results in the protrusion of the abdominal contents into an omphalocele. Closer scrutiny revealed that polymerization of filamentous actin at the umbilical ring was reduced, resulting in incomplete closure of the abdominal wall and herniation of abdominal organs, particularly the liver and intestine. Furthermore, actin cable structures within the leading edge of cells composing the eyelid epithelial sheets were disorganized, accounting for EOB. In both cases, phosphorylation of MLC was not observed. Two further versions of ROCK I knockout mice have been generated either by targeted deletion of exon 5 (and studied in the FVB strain of mice) (Zhang et al. 2006) or exon 1 (and studied in the C57Bl/6 strain) (Rikitake et al. 2005). In neither case were EOB or omphalocele phenotypes observed, which might be explained by strain-specific genetic differences. However, in each case, underrepresentation of homozygous ROCK I knockout mice in litters and resistance of ROCK I-deficient animals to reactive cardiac fibrosis resulting from physical stress on the cardiac muscle were observed. The latter phenomenon may be mediated by reduced ROCK I-mediated expression of fibrogenic cytokines and extracellular matrix protein production by cardiac fibroblasts (Rikitake et al. 2005; Zhang et al. 2006).

Genetic deletion of ROCK II on the C57Bl/6 background also exhibited EOB and omphaloceles (Thumkeo et al. 2005), suggesting that either there is a requirement for each individual ROCK protein, or for a minimum level of total ROCK activity, in the actin-driven movement of epithelial sheets required for these developmental processes. This second possibility is supported by the observation that ROCK I/ ROCK II double heterozygous mice also exhibited EOB and omphaloceles (Thumkeo et al. 2005), although at a lower frequency than ROCK $I^{-/-}$ or ROCK $II^{-/-}$ mice. The lack of phenotype in other tissues suggests that ROCK I and ROCK II are able to functionally compensate for each other, or that the requirement for total ROCK activity is lower. Interestingly, approximately 90% of ROCK II knockout mice on the mixed C57Bl/6–129/Sv strain background died *in utero* at around 13.5 days postcoitum from coagulation of blood within the placental labyrinth layer (Thumkeo et al. 2003). Surviving ROCK $II^{-/-}$ mice were runted, suggesting nutritional deficiency resulting from placental impairment, with hemorrhages observed in hind limbs supporting the conclusion that there were defects in blood coagulation. Actin structures within the placental labyrinth layer of ROCK $II^{-/-}$ mice were apparently normal, suggesting an actin-independent role for ROCK II in the placenta. Interestingly, there may a role for ROCK I within the placenta where it robustly associates with caveolin1 within lipid rafts of syncitial trophoblasts (Rashid-Doubell et al. 2007). The strain-specific differences in the effects of ROCK deletion strongly suggest the influence of modifier genes or strain-specific differences in ROCK activity.

Contribution of ROCK Signaling to Cancer

Increased expression of Rho GTP-binding proteins has been reported for a wide variety of cancers (Benitah et al. 2004; Ellenbroek and Collard 2007). Although the mechanisms leading to elevated Rho expression have not been widely investigated, it was recently reported that metastatic breast cancer cells overexpress a microRNA that increases the expression of RhoC (Ma et al. 2007). In addition to increased levels of Rho proteins, specific examples of elevated expression or mutation of Rho activating guanine nucleotide exchange factors and downregulation or deletion of Rho inactivating GTPase-accelerating proteins have been detected (Ellenbroek and Collard 2007). These findings suggest that there may be increased ROCK activity associated with cancer. Consistent with this possibility, elevated expression of ROCK I and ROCK II was observed in bladder (Kamai et al. 2003) and testicular cancer (Kamai et al. 2002; Kamai et al. 2004). In the case of bladder cancer, elevated expression was significantly correlated with poor survival (Kamai et al. 2003). The increased expression and activity of ROCK could contribute to cancer initiation and progression in several ways. The following sections summarize evidence that supports a role for ROCK in promoting proliferation, survival, and metastasis.

Proliferation

The ready availability of potent ROCK inhibitors, especially Y-27632 (Uehata et al. 1997), has made it possible to examine the role of ROCK in mitogen-induced proliferation in a wide variety of cell types, both in tissue culture and *in vivo* (Table 11.1), albeit

Table 11.1 Antimitogenic effects of ROCK inhibitors

Cell type	ROCK inhibitor	Mitogen	References
Glioma cell lines	Y-27632	Lysophosphatidic acid	Cechin et al. 2005
	Y-27632	P2Y(12) agonists	Van Kolen and Slegers 2006
	Y-27632	TNFα	Radeff-Huang et al. 2007
Astrocytes	Y-27632	PAR-1 agonists	Nicole et al. 2005
	Y-27632	Endothelin-1	Koyama et al. 2004
	Y-27632	PAR-1 agonists, lysophosphatidic acid, sphingosine-1-phosphate	Sorensen et al. 2003
Mouse embryo fibroblasts	Y-27632	Macrophage migration inhibitory factor	Swant et al. 2005
Tenon's capsule fibroblasts	H-1152P	Wound healing	Tura et al. 2007
Oral squamous carcinoma cells	Y-27632	Sonic hedgehog	Nishimaki et al. 2004
Gastric epithelial cells	Y-27632	Glycine-extended gastrin	He et al. 2005
Umbilical vein endothelial cells	Y-27632	Oxidized LDL	Seibold et al. 2004
Vascular smooth muscle cells	Y-27632	PDGF-BB	Kamiyama et al. 2003
	Y-27632	Serotonin	Liu et al. 2004
	Y-27632	Serum	Zuckerbraun et al. 2003
	Y-27632	Urotensin-II	Sauzeau et al. 2001; Watanabe et al. 2006
	Y-27632	Thrombin	Seasholtz et al. 1999
	Y-27632	Balloon angioplasty	Iso et al. 2006

(continued)

Table 11.1 (continued)

Cell type	ROCK inhibitor	Mitogen	References
	Y-27632, Hydroxyfasudil	Stretching	Kozai et al. 2005
	Y-27632, dominant negative ROCK I	Serum, PDGF-BB, lysophosphatidic acid	Sawada et al. 2000
	Fasudil	Vein grafting	Furuyama et al. 2006
	Fasudil	Pulmonary hypertension	Abe et al. 2004
	Fasudil	Pulmonary hypertension	Li et al. 2007
	Fasudil	Angiotensin-II	Kanda et al. 2005
	Fasudil	Serum	Sase et al. 1992
	Fasudil	PDGF, serum	Shirotani et al. 1991
Airway smooth muscle cells	Y-27632	Serum	Takeda et al. 2006
	Y-27632	Endothelin-1	Yahiaoui et al. 2006
Prostatic smooth muscle cells	Y-27632		Rees et al. 2003
Cardiac myocytes	Y-27632	Serum	Zhao and Rivkees 2003
Atrial myofibroblast cells	Y-27632	Serum	Porter et al. 2004
Myoblasts	Y-27632	Serum	Dhawan and Helfman 2004
T cells	Y-27632	Concanavalin A	Mallat et al. 2003
	Y-27632 dominant negative ROCK II	Concanavalin A, anti-CD3 antibody	Tharaux et al. 2003
$CD34^+$ hematopoietic progenitor stem cells	Y-27632	CXCL12	Vichalkovski et al. 2005
	Y-27632 fasudil	Cytokine cocktail GMCSF, GCSF, IL-3, SCF	Burthem et al. 2007
Chondrocytes	Y-27632	Serum	Wang et al. 2004
Hepatic stellate cells	Y-27632	Serum	Iwamoto et al. 2000
	Y-27632 Fasudil	Serum	Fukushima et al. 2005
Pancreatic stellate cells	Y-27632 Fasudil	Serum, PDGF-BB	Masamune et al. 2003
Adrenal Zona glomerulosa cells	Y-27632	Serum	Otis and Gallo-Payet 2006

with the caveats noted above. In some instances, RNAi-mediated knockdown of ROCK expression has been found to impair proliferation (Vishnubhotla et al. 2007). The mitogenic stimuli affected by ROCK inhibitors range from agonists for G-protein-coupled serpentine receptors and tyrosine kinase growth factor receptors to mechanical stretching. The single most common cell type in which ROCK inhibition has an apparent antiproliferative effect is in vascular smooth muscle, with a number of reports indicating a similar effect in airway and prostatic smooth muscle cells. The prevalence of reports on vascular smooth muscle cells may reflect two factors; the critical role of ROCK in regulating both the contraction and proliferation of smooth muscle cells, and the high level of interest in ROCK inhibitors as agents to treat cardiovascular diseases. The lower occurrence of publications reporting an effect of ROCK inhibitors on proliferation in other tissues may reflect a role for ROCK in a limited number of cell types, or may be a function of lower interest in investigating the contribution of ROCK to proliferation in nonmuscle cells

Consistent with the antiproliferative effect of ROCK inhibitors, expression of active forms of ROCK I and ROCK II have been shown to induce proliferation of bovine endothelial cells and mouse fibroblasts, respectively (Croft and Olson 2006; Pirone et al. 2006). In addition, active ROCK I was found to cooperate with a weakly activated form of Raf1 to promote oncogenic transformation of immortalized mouse fibroblasts. Similarly, ROCK I was identified as a putative cancer gene with "driver" mutations that could contribute to oncogenesis in breast and lung cancer cells (Greenman et al. 2007). These results indicate that elevated ROCK signaling may indeed be sufficient to induce proliferation, at least in some cell types.

There are a limited number of studies reporting stimulation of proliferation by Y-27632, indicating that replication of some cell types may be restrained by ROCK activity. Treatment of rat embryonic kidney cells (Meyer et al. 2006), human primary keratinocytes (McMullan et al. 2003), and human embryonic stem cells (Watanabe et al. 2007) induced proliferation, whereas expression of a conditionally activated ROCK II in primary keratinocytes led to cell cycle arrest and terminal differentiation upon induction of catalytic activity (McMullan et al. 2003). Similarly, expression of a constitutively active ROCK II inhibited the proliferation of endothelial cells, whereas Y-27632 partially rescued cells from an inhibition of proliferation induced by TNF-α (Kishore et al. 2005). Taken together, the literature supports the conclusion of a general role for ROCK as a promoter of cell replication, with antiproliferative functions in selected cell types. What is missing from consideration are studies in which activation or inhibition of ROCK had no effect on proliferation, with a few exceptions (e.g., Sahai et al. 1999; Uchida et al. 2000; Nagatoya et al. 2002), lack-of-effect results are generally left unpublished in lab notebooks.

ROCK Regulation of Proliferation: Role for the Actin Cytoskeleton

The mechanism by which ROCK contributes to promoting the generation of actin-myosin-based contractile force has been extensively studied. However, there is not a great deal of data showing whether ROCK influences proliferation via the actin cytoskeleton or by independent signaling pathways. In a number of cases, the inhibition of mitogen-induced proliferation by Y-27632 is associated with disruption of actin cytoskeletal structures, particularly stress fibers (Iwamoto et al. 2000; Sauzeau et al. 2001; Masamune et al. 2003; Tharaux et al. 2003; Dhawan and Helfman 2004; Fukushima et al. 2005; Otis and Gallo-Payet 2006; Tura et al. 2007).

In addition, it was also determined that direct disruption of F-actin was found to inhibit mitogenesis comparably to Y-27632 (Tharaux et al. 2003; Dhawan and Helfman 2004; Koyama et al. 2004; Liu et al. 2004; Otis and Gallo-Payet 2006). The proliferation of fibroblasts induced by active ROCK could be blocked by actin-destabilizing drugs, indicating that a critical component of ROCK-induced mitogenesis is mediated via the actin cytoskeleton (Croft and Olson 2006). However, ROCK did influence the levels of key cell cycle regulatory proteins independent of the actin cytoskeleton, suggesting that in some contexts, actin-independent signaling pathways might contribute to cell cycle regulation (Croft and Olson 2006).

If the contribution of ROCK to cell proliferation is mediated by its driving actin-myosin contraction, one question that arises is whether contractility is itself sufficient to promote proliferation. By plating endothelial cells on fibronectin patches of varying sizes, it was determined that the development of cytoskeletal tension was sufficient to induce proliferation (Pirone et al. 2006). Increasing cytskeletal tension by overexpressing active RhoA or ROCK I in cells that were not fully spread, due to restricted area of extracellular matrix on which they adhered, stimulated cell cycle progression (Pirone et al. 2006). These data are consistent with a wealth of publications reporting a dependence on intact cytoskeletal structures (e.g., Bohmer et al. 1996; Huang et al. 1998; Huang and Ingber 2002) and myosin ATPase activity to generate contractile force (e.g., Huang et al. 1998; Dhawan and Helfman 2004) for mitogen-induced proliferation.

Signaling Pathways Influenced by ROCK

Mitogen-Activated Protein Kinase

The most well-characterized cellular effects of ROCK reflect its importance in regulating the organization of the actin cytoskeleton. To date however, there is limited evidence of "hardwired" signaling cascades, akin to the kinase cascade downstream of Ras that leads ultimately to the activation of gene transcription and cell cycle progression, which might be regulated ROCK. Although the lack of evidence does not exclude the possibility that there may be pathways that directly link ROCK to cell cycle regulation, many published reports indicate that ROCK has a significant impact on the activity of the mitogen-activated protein kinase (MAPK) cascade. Studies on the relationship between transient versus sustained MAPK signaling and cell cycle progression have shown that the sustained phase is required for mitogenic stimulation (Fassett et al. 2003; Villanueva et al. 2007). Inhibition of ROCK activity appears to have little effect on the rapid transient activation of MAPK but numerous studies have found a significant effect on sustained MAPK activity (Iwamoto et al. 2000; Kamiyama et al. 2003; Fukushima et al. 2005; Nicole et al. 2005; Swant et al. 2005; Otis and Gallo-Payet 2006 Yanazume, 2002), which could be mimicked by disruption of filamentous actin (Yanazume et al. 2002) and was associated in some instances with disruption of the actin cytoskeleton (Iwamoto et al. 2000; Fukushima et al. 2005; Otis and Gallo-Payet 2006). Consistent with these results, sustained ROCK signaling was sufficient to induce sustained Ras activation and increased MAPK activity, effects that were dependent on an intact actin cytoskeleton (Croft and Olson 2006). However, several studies have found that ROCK inhibitor blockade of cell proliferation was independent of effects on MAPK activation (Takeda et al. 2001; Masamune et al. 2003; Koyama et al. 2004; Cechin et al. 2005; Van Kolen and Slegers 2006; Yahiaoui et al. 2006). These discrepant findings likely highlight differences in the role played by ROCK in varying cell types and in response to different ligands. One possible pathway leads from the cell surface receptor to ROCK that via its role in regulating the actin cytoskeleton influences the sustained activation of the MAPK cascade and consequently promotes cell proliferation. An alternative pathway goes from the receptor to ROCK, which then contributes to cell cycle regulation independent of MAPK regulation. This second MAPK-independent pathway may actually coexist with the first, but would easily be obscured by the perceived consequences of decreased MAPK signaling. Further complicating these inconsistencies are the findings that ROCK activity may be required either for nuclear translocation (Liu et al. 2004; Zhao et al. 2006) or for cytosolic retention (Lai et al. 2002; Zuckerbraun et al. 2003) of active ERK, which again may reflect cell-type and/or ligand-specific functions of ROCK.

Stat3

The transcription factor Stat3 has been reported to be responsive to active RhoA (Aznar et al. 2001; Debidda et al. 2005) and to contribute to the ability of RhoA to induce proliferation and anchorage-independent growth (Aznar et al. 2001; Debidda et al. 2005). Using effector-loop RhoA mutants (Sahai et al. 1998), it was determined that ROCK played a significant role in Stat3 phosphorylation and activation (Debidda et al. 2005). In addition, the ability of active ROCK to induce anchorage-independent growth of mouse embryo fibroblasts was significantly impaired in cells genetically deleted for Stat3 (Debidda et al. 2005) However, the RhoA effector-loop mutants could not exclude the possibility that additional Rho effectors contributed to Stat3 regulation (Debidda et al. 2005). Although numerous signaling pathways have been implicated in Stat3 regulation (e.g., Liu et al. 2006), one possible connection linking ROCK to Stat3 is the JNK pathway (Aznar et al. 2001; Marinissen et al. 2004; Liu et al. 2006).

Cell Cycle Regulation

The cell cycle can be divided into four phases: the first gap phase (G_1), the DNA synthetic phase (S), the second gap phase (G_2), and mitosis (M). Cell cycle progression is propelled the cyclin-dependent kinases (CDKs), which act in concert with the regulatory cyclin subunits [reviewed in ref. Morgan (1997) and Sherr and Roberts (2004)]. During G_1 phase, the D-type cyclins (D1, D2, and D3) form active complexes with CDK4 or CDK6, whereas cyclin E1 or E2 associate with CDK2. CDK activity may also be stimulated by phosphorylation of a conserved residue within the CDK activation loop by the CDK-activating complex, which is composed of cyclin H, CDK7, and Mat1. The cyclin-CDK complexes phosphorylate numerous substrates including the retinoblastoma (Rb) "pocket protein" family (RB, p107, and p130). RB phosphorylation inhibits binding of the E2F transcription factor family, resulting in changes in gene expression that contribute to increased S-phase DNA synthesis. During S-phase, cyclin E-CDK2 activity is maintained, while A-type cyclins also form active complexes with CDK2. Once significant cyclin E-CDK2 and cyclin A-CDK2 activity has been achieved during S phase, Rb phosphorylation and cyclin D-associated CDK activity are no longer required.

Cyclin-CDK activity is regulated by the binding of two types of CDK inhibitor (CDKIs) that have different mechanisms of action. The INK4 CDKIs ($p15^{INK4b}$, $p16^{INK4a}$, $p18^{INK4c}$, and $p19^{INK4d}$) directly bind and sequester CDKs, thereby preventing them from forming active complexes with cognate cyclin proteins. The Cip/Kip-family CDKIs ($p21^{Waf1/Cip1}$ (p21), $p27^{Kip1}$ (p27), and $p57^{Kip2}$) bind to cyclin-CDK complexes. At high levels of expression, p21 and p27 potently inhibit cyclin E-CDK2 activity, resulting in cell cycle arrest. However, at moderate levels, p21 and p27 promote the assembly, stability, and nuclear retention of cyclin D-CDK4 and cyclin D-CDK6 complexes, which are not effectively inhibited by the associated Cip/Kip proteins. Cyclin D-CDK4 and cyclin D-CDK6 complexes also act as repositories for p21 and p27, thereby relieving cyclin E-CDK2 complexes from p21/p27-mediated inhibition. Therefore, the balance between cyclin, CDK, and CDKI proteins determines the net output of cyclin-CDK complexes and whether progression through the cell cycle may proceed.

Cyclins and CDKs

Active cyclin D-CDK complexes are the first to be detected after mitogenic stimulation of quiescent cells, the induction of cyclin D1 results from increased transcription, translation, and protein stability. The antimitogenic actions of ROCK inhibitors have been associated with decreased cyclin D1 levels in a number of cell types (Iwamoto et al. 2000; Nicole et al. 2005; Swant et al. 2005). The ability of ROCK inhibitors to block cyclin D1 induction was accompanied by a decrease in sustained ERK activation (Iwamoto et al. 2000; Nicole et al. 2005; Swant et al. 2005), suggesting that the requirement for ROCK to elevate cyclin D1 levels results from a role in the prolongation of MAPK signaling. Consistent with this possibility, the ability of active ROCK II to induce proliferation in mouse fibroblasts was associated with increased cyclin D1 expression and sustained ERK activity, whereas treatment with a MEK inhibitor blocked proliferation, cyclin D1 induction, and ERK activity (Croft and Olson 2006). However, given the number of reports indicating that ROCK activity is dispensable for sustained ERK signaling in some contexts, alternative pathways may also link ROCK to cyclin D1 expression. For example, the Stat3 transcription factor may be activated by ROCK (Debidda et al. 2005), which in turn contributes to increased cyclin D1 promoter activity (Debidda et al. 2005; Leslie et al. 2006).

Additional cyclins may also be regulated by ROCK signaling in some situations. Treatment with Y-27632 was found to block cyclin D3 induction by endothelin-1 in astrocytes (Koyama et al. 2004) and in cardiac myocytes in response to serum stimulation (Zhao and Rivkees 2003), whereas cyclin D2 induction by sonic hedgehog was inhibited in oral squamous carcinoma cells (Nishimaki et al. 2004). It has also been reported that cyclin E (Nishimaki et al. 2004) and cyclin A2 (Porter et al. 2004) induction were sensitive to ROCK inhibition, although whether the decreased cyclin E and cyclin A2 levels were the result of, rather than the cause of, impaired cell cycle progression was not determined. Activation of ROCK II in mouse fibroblasts elevated cyclin A2 protein levels independent of cell cycle progression, cyclin D1 expression, cyclin E-associated kinase activity, or p107 phosphorylation (Croft and Olson 2006). However, ROCK induction of cyclin A2 did require LIMK2, a kinase activated by ROCK that is most well known for a role in regulating the actin cytoskeleton (Croft and Olson 2006). Whether cyclin A2 induction by ROCK resulted from increased transcription, translation, or protein stabilization was not determined. Treatment of corneal stromal cells with ROCK inhibitor was found to result in significantly lower expression of the mitotic cyclin B1 (Harvey et al. 2004). Finally, the cell cycle arrest induced by conditionally active ROCK II in primary human keratinocytes was associated with a twofold decrease in the expression of cyclin H, a component of the CDK-activating kinase complex (McMullan et al. 2003). These data suggest that there may be several cyclins that are influenced by ROCK.

It is generally assumed that the activities of cyclin-CDK complexes are not significantly regulated by changes in CDK levels. However, inhibition of ROCK activity was correlated with decreased expression of CDK6 in mouse embryo hearts and cardiomyocytes, which was associated with an inhibition of proliferation (Zhao and Rivkees 2003). In addition, treatment with ROCK inhibitor reduced expression of the mitotic CDK1 in corneal stromal cells (Harvey et al. 2004), although it was not determined whether ROCK inhibition was sufficient to induce cell cycle arrest in this cells type.

CDK Inhibitors

The Cip/Kip type CDKI p27 is typically elevated in quiescent cells and levels fall following mitogenic stimulation. Although posttranslational modification and protein degradation contribute significantly to the regulation of p27 protein levels (Bloom and Pagano 2003), transcriptional regulation may also have some influence

(Medema et al. 2000). Treatment with ROCK inhibitors was associated with increased levels of p27 in a variety of cell types, both *in vitro* and *in vivo* (Iwamoto et al. 2000; Sawada et al. 2000; Kanda et al. 2003, 2005; Zhao and Rivkees 2003; Seibold et al. 2004; Guerin et al. 2005; Hayashi et al. 2006; Iso et al. 2006). Given that a major mechanism of p27 regulation results from phosphorylation of Thr187 by active cyclin E-CDK2 and/or cyclin A-CDK2 complexes, leading to Skp2-mediated ubiquitylation and ultimate degradation by the proteasome (Bloom and Pagano 2003), it may be difficult to determine whether the increased levels of p27 observed following treatment with ROCK inhibitors is the cause or effect of inhibited proliferation. The conditional activation of ROCK II led to lowered levels of p27 in mouse fibroblasts, even under conditions where cell cycle progression, cyclin E-associated kinase activity, and p107 phosphorylation were inhibited (Croft and Olson 2006). These data suggest that there may be additional mechanisms responsive to ROCK that regulate p27 protein levels. Phosphorylation of p27 on Tyr88 by allows for more efficient phosphorylation on Thr187 by CDK2, thereby promoting p27 degradation without the need for increased CDK2 activity (Grimmler et al. 2007). Ubiquitylation of p27 may also occur independent of Thr187 phosphorylation by the cytoplasmic E3 ubiquitin ligase KPC (Kamura et al. 2004). Ubiquitylation via this mechanism is dependent on cytoplasmic localization of p27, which may occur in response to phosphorylation on sites including Ser10 (Rodier et al. 2001; Ishida et al. 2002) and Thr157 (Liang et al. 2002; Shin et al. 2002; Viglietto et al. 2002). Although ROCK is unlikely to be a Ser10 or Thr157 kinase, and certainly is not the Tyr88 kinase, ROCK may feed into pathways that result in increased activities of the kinases responsible for one or more of these phosphorylations, with the ultimate consequence being 27-protein degradation.

The growth arrest induced by conditional activation of ROCK II in primary human keratinocytes was associated with a ninefold increase in $p57^{Kip2}$ mRNA transcripts (McMullan et al. 2003). Although it was not directly shown that the increased expression of $p57^{Kip2}$ was responsible for the observed inhibition of keratinocyte proliferation, the relatively frequent epigenetic silencing of the *CDKN1C* gene encoding $p57^{Kip2}$ in cancer (Hoffmann et al. 2005) and hyperproliferation of some specific tissues in $p57^{-/-}$ mice (Zhang et al. 1997) is consistent with the presumed role of $p57^{Kip2}$ as a critical regulator of proliferation in some contexts.

In contrast to p27 protein, p21 levels are typically low in quiescent cells and mitogens elevate p21 through increased transcription mediated principally in response to Ras and Raf-MEK-ERK/MAPK signaling (Liu et al. 1996; Lloyd et al. 1997; Pumiglia and Decker 1997; Sewing et al. 1997; Woods et al. 1997; Olson et al. 1998; Bottazzi et al. 1999). This somewhat counterintuitive pattern regulation might be explained by the role of p21 as a promoter of cyclin D1-CDK assembly, nuclear retention, and stability (LaBaer et al. 1997; Cheng et al. 1999; Alt et al. 2002), which further cell-cycle progression at low levels of expression. Although Rho has been shown to restrain p21 elevation by transcriptional and posttranscriptional mechanisms (Adnane et al. 1998; Auer et al. 1998; Olson et al. 1998; Lai et al. 2002; Zuckerbraun et al. 2003; Coleman et al. 2006), the role of ROCK has not been clearly defined. ROCK inhibition was reported to impair the ability of fibronectin to repress p21 expression (Han et al. 2005), while ROCK activation was found to promote the cytosolic retention of active ERK resulting in reduced expression of p21 (Lai et al. 2002; Zuckerbraun et al. 2003), However, several lines of evidence contradict these findings: conditional activation of ROCK II did not reduce p21 levels in mouse fibroblasts (Croft and Olson 2006) nor did ROCK inhibition elevate p21 levels in mouse mesenchymal stem cells (Pacary et al. 2007) and fibroblasts (Sahai et al. 1999; Sahai et al. 2001). In addition, it has also been reported that ROCK activity is necessary for the nuclear translocation of active ERK (Liu et al. 2004; Zhao et al. 2006), suggesting that there may be cell-specific roles of ROCK in regulating the localization of ERK. Although inhibition of Rho resulted in stabilization of p21 protein, p21 protein turnover was not impaired by ROCK inhibition nor was it accelerated by overexpression of ROCK I indicating that ROCK was unlikely to contribute to this aspect of p21 regulation (Coleman et al. 2006).

ROCK and Cell Survival

During apoptosis, cells undergo significant morphological changes including contraction, dynamic membrane blebbing, and nuclear disintegration, which are driven by ROCK-mediated actin-myosin contractile force generation (Coleman and Olson 2002). When cell death has been triggered by extrinsic factors such as TNFα, ceramide, or Fas-receptor ligation, ROCK I activation is a relatively late event (Coleman et al. 2001; Sebbagh et al. 2001) and ROCK inhibition does not halt the apoptotic process (Coleman et al. 2001). However, in some contexts chronic or high intensity ROCK activity may contribute to the initiation of apoptosis. Data from ROCK $I^{-/-}$ mice revealed that pressure overload was less effective in inducing cardiomyocyte apoptosis in relative to controls, suggesting a role for ROCK I in myocardial failure (Chang et al. 2006).

There also appears to be a role for ROCK in cell survival. Inhibition of ROCK activity induces cell death in corneal epithelial cells (Svoboda et al. 2004), airway epithelial cells (Moore et al. 2004), neointimal smooth muscle cells (Shibata et al. 2001, 2003; Matsumoto et al. 2004), vascular smooth muscle cells (Abe et al. 2004; Furuyama et al. 2006), endothelial cells (Li et al. 2002), hepatic stellate cells (Ikeda et al. 2003), spinal cord motor neurons (Kobayashi et al. 2004), rheumatoid synovial cells (Nagashima et al. 2006), and in H_2O_2-treated intestinal epithelial cells (Song et al. 2006). These effects were not limited to nontransformed cells as ROCK inhibition also induced cell death in anaplastic thyroid cancer cells (Zhong et al. 2003), glioma cells (Rattan et al. 2006), $CD34^+$ chronic myeloid leukemia progenitor cells (Burthem et al. 2007), and H_2O_2- or camptothecin-treated neuroblastoma cells (De Sarno et al. 2005).

Whether ROCK activation is proapoptotic or prosurvival is likely context specific and dependent on cell type and death stimulus. Although there are some examples of ROCK inhibitors resulting in tumor regression *in vivo* (e.g., Somlyo et al. 2003; Rattan et al. 2006), it remains to be determined whether this resulted from the inhibition of ROCK-mediated protection in the tumor cells. It is also remains to be determined whether a prosurvival activity of ROCK plays a significant role in human cancers. The precise molecular mechanisms through which ROCK might mediate its effects on cell survival remain to be elucidated.

ROCK in Tumor Cell Invasion and Metastasis

Migration

The dynamic reorganization of the cytoskeleton has long been understood as the underlying mechanism allowing cell migration and invasion (Carlier and Pantaloni 2007). Thus, given the importance of ROCK in cytoskeletal remodeling and acto-myosin contractility, it is not surprising that the enzyme has been implicated in cell migration. Expression of constitutively active ROCK in MM1 hepatoma cells significantly enhanced invasion and dissemination of tumor cells in an *in vivo* model (Itoh et al. 1999). This enhanced migration was also observed with MM1 cells expressing a constitutively active RhoAV14 and was reversed with the ROCK antagonist Y27632 (Itoh et al. 1999). These observations are supported by a significant body of evidence implicating a role for ROCK in 2D *in vitro* migration experiments (Riento and Ridley 2003). However, significant morphological differences are observed in cells grown in monolayer culture versus a more physiological 3D matrix (Yamada and Cukierman 2007). These differences call into question the relevance of 2D models of cellular migration for studying *in vivo* cell movement. Instead, 3D migration studies are required to recapitulate a more realistic environment as a model for physiological migration and invasion.

Single-cell migration within a 3D matrix can be categorized into two distinct modes, mesenchymal and amoeboid (Friedl and Wolf 2003). Mesenchymal migration is characterized by a fibroblast/spindle like morphology, dependence on focal adhesions, integrin binding, and matrix proteolysis, whereas the amoeboid form is marked by a rounded cell body, limited dependence on matrix adhesions, and no requirement for proteolytic matrix remodeling. Mesenchymal migration appears to have limited dependence on ROCK; in contrast, amoeboid migration is heavily dependent on ROCK. In evidence, Y-27632 was found to significantly inhibit the invasion of Matrigel™ by several amoeboid cell lines, while it failed to inhibit mesenchymal-like invasion (Sahai and Marshall 2003). Consistent with these observations, the amoeboid migration of DMS79 cells in collagen gels was associated with membrane blebbing, a cellular feature of ROCK activation (Coleman et al. 2001; Yamazaki et al. 2005). Membrane blebbing appears to be a general characteristic of cells undergoing amoeboid migration, and is likely important for cell movement within 3D matrices without necessitating proteolytic remodeling (Trinkaus 1973; Sahai and Marshall 2003). Furthermore, several cell lines have demonstrated plasticity in migration strategies and, in the presence of protease inhibitors, can be shifted from Y-27632-insensitive mesenchymal migration to amoeboid, whereupon cell movement is sensitive to Y-27632 treatment (Sahai and Marshall 2003). These observations clearly suggest a significant role for ROCK in cellular migration; however, a precise mechanism of control remains to be demonstrated. Nonetheless, ROCK has been proposed to regulate contraction of cortical actin rings that initiate membrane blebbing and deformation of the extra-cellular matrix (Friedl and Wolf 2003; Sahai and Marshall 2003; Wyckoff et al. 2006). In normal physiology, amoeboid migration appears to be limited to lymphocyte and neutrophil cell populations which display enhanced (10- to 30-fold) migration velocities over mesenchymal cell types (Friedl and Wolf 2003). The migration of lymphocytes is independent of integrin binding and proteolytic remodeling that effectively allows these cells to bypass the ECM (Friedl and Wolf 2003). These features of amoeboid migration are well suited to the function of immunological cells and confer a high degree of mobility within tissues as well as rapid cellular immune responses In contrast to the predominance of mesnchymal type cellular migration of nonimmune cells during normal physiological conditions, the amoeboid mechanism appears to be a significant migration strategy for tumor cells, consistent with ROCK playing a substantial role in pathological tumor invasion and metastasis (Friedl and Wolf 2003).

ROCK in In Vivo *Models of Cancer Cell Invasion and Metastasis*

A number of studies have implicated ROCK in tumor cell invasion by examining whether the invasive ability of cancer cell lines *in vivo* can be blocked by administration of ROCK inhibitors. In a recent study, HA-1077 (fasudil) reduced tumor burden and ascites resulting from peritoneally disseminated MM1 (rat hepatocellular carcinoma) cells in rats by half, decreased lung nodule formation in mice by HT1080 (human fibrosarcoma) cells by 40%, and limited breast cancer formation in mice by MDA-MB-231 (human breast carcinoma) cells threefold (Ying et al. 2006). ROCK inhibitors also were found to reduce the *in vivo* invasiveness of human hepatocellular carcinoma (Genda et al. 1999; Takamura et al. 2001), human prostate cancer (Somlyo et al. 2000), and mouse lung cancer cells (Nakajima et al. 2003). These studies are consistent with a contribution of ROCK to tumor cell invasiveness and metastasis.

A positive role for ROCK in cancer cell invasiveness *in vivo* was demonstrated with human colorectal cancer cell lines expressing a conditionally active version of ROCK II (Croft et al. 2004). These cells formed highly vascularized tumors in mouse xenografts that aggressively invaded into the surrounding stroma on ROCK activation. These results suggest that ROCK II activity is not only sufficient for tumor invasion but also sufficient for the induction of angiogenesis possibly by increasing the plasticity of tumor tissue, thereby facilitating invasion by endothelial cells, potentially aiding tumor growth and dissemination.

Conclusions

Data from a variety of sources support the hypothesis that ROCK is involved in cancer in myriad ways. Although the cell biological functions of ROCK have been relatively well studied, the precise means by which ROCK promotes cancer are less clear. Evidence suggests that ROCK may contribute by promoting proliferation, survival, and invasion. Although their use has largely been restricted because of the developmental defect limits associated with the constitutive and ubiquitous deletion of ROCK I and ROCK II, genetic mouse models should be informative in efforts to characterize how ROCK contributes to cancer. These models could be improved through the generation of conditional knockouts that would allow for tissue-selective deletion at defined time points. Alternatively, mice engineered to conditionally express shRNA, in a ubiquitous or tissue-selective manner, may be useful for these studies. In addition, the production of mice expressing conditionally active ROCK (Croft and Olson 2006) in a tissue-selective manner will make it possible to determine whether elevated ROCK activation promotes cancer initiation and progression. The refinement of these mouse models will make it possible to address significant questions about the role of ROCK in cancer and the potential value of targeting ROCK as a therapeutic strategy.

References

Abe K, Shimokawa H, Morikawa K, Uwatoku T, Oi K, Matsumoto Y, Hattori T, Nakashima Y, Kaibuchi K, Sueishi K, Takeshit A (2004) Long-term treatment with a Rho-kinase inhibitor improves monocrotaline-induced fatal pulmonary hypertension in rats. Circ Res 94:385–393.

Adnane J, Bizouarn FA, Qian Y, Hamilton AD, Sebti SM (1998) p21(WAF1/CIP1) is upregulated by the geranylgeranyltransferase I inhibitor GGTI-298 through a transforming growth factor beta- and Sp1- responsive element: involvement of the small GTPase rhoA. Mol Cell Biol 18:6962–6970.

Alt JR, Gladden AB, Diehl JA (2002) p21(Cip1) Promotes cyclin D1 nuclear accumulation via direct inhibition of nuclear export. J Biol Chem 277:8517–8523.

Amano M, Chihara K, Kimura K, Fukata Y, Nakamura N, Matsuura Y, Kaibuchi K (1997) Formation of actin stress fibers and focal adhesions enhanced by Rho-kinase. Science 275:1308–1311.

Amano M, Chihara K, Nakamura N, Kaneko T, Matsuura Y, Kaibuchi K (1999) The COOH terminus of Rho-kinase negatively regulates Rho-kinase activity. J Biol Chem 274:32418–32424.

Amano M, Ito M, Kimura K, Fukata Y, Chihara K, Nakano T, Matsuura Y, Kaibuchi K (1996) Phosphorylation and activation of myosin by Rho-associated kinase (Rho-kinase). J Biol Chem 271:20246–20249.

Amano T, Tanabe K, Eto T, Narumiya S, Mizuno K (2001) LIM-kinase 2 induces formation of stress fibres, focal adhesions and membrane blebs, dependent on its activation by Rho-associated kinase- catalysed phosphorylation at threonine-505. Biochem J 354:149–159.

Arai M, Sasaki Y, Nozawa R (1993) Inhibition by the protein kinase inhibitor HA1077 of the activation of NADPH oxidase in human neutrophils. Biochem Pharmacol 46:1487–1490.

Asano T, Ikegaki I, Satoh S, Suzuki Y, Shibuya M, Takayasu M, Hidaka H (1987) Mechanism of action of a novel antivasospasm drug, HA1077. J Pharmacol Exp Ther 241:1033–1040.

Auer KL, Park JS, Seth P, Coffey RJ, Darlington G, Abo A, McMahon M, Depinho RA, Fisher PB, Dent P (1998) Prolonged activation of the mitogen-activated protein kinase pathway promotes DNA synthesis in primary hepatocytes from p21Cip-1/WAF1-null mice, but not in hepatocytes from p16INK4a-null mice. Biochem J 336(Pt 3):551–560.

Aznar S, Valeron PF, del Rincon SV, Perez LF, Perona R, Lacal JC (2001) Simultaneous tyrosine and serine phosphorylation of STAT3 transcription factor is involved in Rho A GTPase oncogenic transformation. Mol Biol Cell 12:3282–3294.

Bain J, Plater L, Elliott M, Shpiro N, Hastie CJ, McLauchlan H, Klevernic I, Arthur JS, Alessi DR, Cohen P (2007) The selectivity of protein kinase inhibitors: a further update. Biochem J 408:297–315.

Benitah SA, Valeron PF, van Aelst L, Marshall CJ, Lacal JC (2004) Rho GTPases in human cancer: an unresolved link to upstream and downstream transcriptional regulation. Biochim Biophys Acta 1705:121–132.

Bloom J, Pagano M (2003) Deregulated degradation of the cdk inhibitor p27 and malignant transformation. Semin Cancer Biol 13:41–47.

Bohmer RM, Scharf E, Assoian RK (1996) Cytoskeletal integrity is required throughout the mitogen stimulation phase of the cell cycle and mediates the anchorage-dependent expression of cyclin D1. Mol Biol Cell 7:101–111.

Bottazzi ME, Zhu X, Bohmer RM, Assoian RK (1999) Regulation of p21(cip1) expression by growth factors and the extracellular matrix reveals a role for transient ERK activity in G1 phase. J Cell Biol 146:1255–1264.

Burthem J, Rees-Unwin K, Mottram R, Adams J, Lucas GS, Spooncer E, Whetton AD (2007) The rho-kinase inhibitors Y-27632 and fasudil act synergistically with imatinib to inhibit the expansion of ex vivo CD34(+) CML progenitor cells. Leukemia 21:1708–1714.

Byers HR, White GE, Fujiwara K (1984) Organization and function of stress fibers in cells in vitro and in situ. A review. Cell Muscle Motil 5:83–137.

Carlier MF, Pantaloni D (2007) Control of actin assembly dynamics in cell motility. J Biol Chem 282:23005–23009.

Cechin SR, Dunkley PR, Rodnight R (2005) Signal transduction mechanisms involved in the proliferation of C6 glioma cells induced by lysophosphatidic acid. Neurochem Res 30:603–611.

Chang J, Xie M, Shah VR, Schneider MD, Entman ML, Wei L, Schwartz RJ (2006) Activation of Rho-associated coiled-coil protein kinase 1 (ROCK-1) by caspase-3 cleavage plays an essential role in cardiac myocyte apoptosis. Proc Natl Acad Sci U S A 103:14495–14500.

Cheng M, Olivier P, Diehl JA, Fero M, Roussel MF, Roberts JM, Sherr CJ (1999) The p21(Cip1) and p27(Kip1) CDK 'inhibitors' are essential activators of cyclin D-dependent kinases in murine fibroblasts. Embo J 18:1571–1583.

Chijiwa T, Mishima A, Hagiwara M, Sano M, Hayashi K, Inoue T, Naito K, Toshioka T, Hidaka H (1990) Inhibition of forskolin-induced neurite outgrowth and protein phosphorylation by a newly synthesized selective inhibitor of cyclic AMP-dependent protein kinase, N-[2-(p-bromocinnamylamino)ethyl]-5-isoquinolinesulfonamide (H-89), of PC12D pheochromocytoma cells. J Biol Chem 265:5267–5272.

Chrzanowska-Wodnicka M, Burridge K (1996) Rho-stimulated contractility drives the formation of stress fibers and focal adhesions. J Cell Biol 133:1403–1415.

Coleman ML, Densham RM, Croft DR, Olson MF (2006) Stability of p21(Waf1/Cip1) CDK inhibitor protein is responsive to RhoA-mediated regulation of the actin cytoskeleton. Oncogene 25:2708–2716.

Coleman ML, Olson MF (2002) Rho GTPase signalling pathways in the morphological changes associated with apoptosis. Cell Death Differ 9:493–504.

Coleman ML, Sahai EA, Yeo M, Bosch M, Dewar A, Olson MF (2001) Membrane blebbing during apoptosis results from caspase-mediated activation of ROCK I. Nat Cell Biol 3:339–345.

Croft DR, Coleman ML, Li S, Robertson D, Sullivan T, Stewart CL, Olson MF (2005) Actin-myosin-based contraction is responsible for apoptotic nuclear disintegration. J Cell Biol 168:245–255.

Croft DR, Olson MF (2006) Conditional regulation of a ROCK-estrogen receptor fusion protein. Methods Enzymol 406:541–553.

Croft DR, Olson MF (2006) The Rho GTPase effector ROCK regulates cyclin A, cyclin D1, and p27Kip1 levels by distinct mechanisms. Mol. Cell. Biol. 26:4612–4627.

Croft DR, Sahai E, Mavria G, Li S, Tsai J, Lee WM, Marshall CJ, Olson MF (2004) Conditional ROCK activation in vivo induces tumor cell dissemination and angiogenesis. Cancer Res 64:8994–9001.

Davies SP, Reddy H, Caivano M, Cohen P (2000) Specificity and mechanism of action of some commonly used protein kinase inhibitors. Biochem J 351:95–105.

De Sarno P, Shestopal SA, Zmijewska AA, Jope RS (2005) Anti-apoptotic effects of muscarinic receptor activation are mediated by Rho kinase. Brain Research 1041:112–115.

Debidda M, Wang L, Zang H, Poli V, Zheng Y (2005) A role of STAT3 in Rho GTPase-regulated cell migration and proliferation. J Biol Chem 280:17275–17285.

Dhawan J, Helfman DM (2004) Modulation of acto-myosin contractility in skeletal muscle myoblasts uncouples growth arrest from differentiation. J Cell Sci 117:3735–3748.

Doran JD, Liu X, Taslimi P, Saadat A, Fox T (2004) New insights into the structure-function relationships of Rho-associated kinase: a thermodynamic and hydrodynamic study of the dimer-to-monomer transition and its kinetic implications. Biochem J 384:255–262.

Dvorsky R, Blumenstein L, Vetter IR, Ahmadian MR (2004) Structural insights into the interaction of ROCKI with the switch regions of RhoA. J Biol Chem 279:7098–7104.

Ellenbroek SI, Collard JG (2007) Rho GTPases: functions and association with cancer. Clin Exp Metastasis 24:657–672.

Endlich N, Otey CA, Kriz W, Endlich K (2007) Movement of stress fibers away from focal adhesions identifies focal adhesions as sites of stress fiber assembly in stationary cells. Cell Motil Cytoskeleton 64:966–976.

Fassett JT, Tobolt D, Nelsen CJ, Albrecht JH, Hansen LK (2003) The role of collagen structure in mitogen stimulation of ERK, cyclin D1 expression, and G1-S progression in rat hepatocytes. J. Biol. Chem. 278:31691–31700.

Feng J, Ito M, Ichikawa K, Isaka N, Nishikawa M, Hartshorne DJ, Nakano T (1999) Inhibitory phosphorylation site for Rho-associated kinase on smooth muscle myosin phosphatase. J Biol Chem 274:37385–37390.

Feng J, Ito M, Kureishi Y, Ichikawa K, Amano M, Isaka N, Okawa K, Iwamatsu A, Kaibuchi K, Hartshorne DJ, Nakano T (1999) Rho-associated kinase of chicken gizzard smooth muscle. J Biol Chem 274:3744–3752.

Friedl P, Wolf K (2003) Tumour-cell invasion and migration: diversity and escape mechanisms. Nat Rev Cancer 3:362–374.

Fukushima M, Nakamuta M, Kohjima M, Kotoh K, Enjoji M, Kobayashi N, Nawata H (2005) Fasudil hydrochloride hydrate, a Rho-kinase (ROCK) inhibitor, suppresses collagen production and enhances collagenase activity in hepatic stellate cells. Liver International 25:829–838.

Furuyama T, Komori K, Shimokawa H, Matsumoto Y, Uwatoku T, Hirano K, Maehara Y (2006) Long-term inhibition of Rho kinase suppresses intimal thickening in autologous vein grafts in rabbits. J Vasc Surg 43:1249–1256.

Genda T, Sakamoto M, Ichida T, Asakura H, Kojiro M, Narumiya S, Hirohashi S (1999) Cell motility mediated by rho and Rho-associated protein kinase plays a critical role in intrahepatic metastasis of human hepatocellular carcinoma. Hepatology 30:1027–1036.

Grimmler M, Wang Y, Mund T, Cilensek Z, Keidel E-M, Waddell MB, Jakel H, Kullmann M, Kriwacki RW, Hengst L (2007) Cdk-inhibitory activity and stability of p27Kip1 are directly regulated by oncogenic tyrosine kinases. Cell 128:269–280.

Guerin P, Sauzeau V, Rolli-Derkinderen M, Al Habbash O, Scalbert E, Crochet D, Pacaud P, Loirand G (2005) Stent implantation activates RhoA in human arteries: inhibitory effect of rapamycin. J Vasc Res 42:21–28.

Hagerty L, Weitzel DH, Chambers J, Fortner CN, Brush MH, Loiselle D, Hosoya H, Haystead TA (2007) ROCK1 phosphorylates and activates zipper-interacting protein kinase. J Biol Chem 282:4884–4893.

Han S, Sidell N, Roman J (2005) Fibronectin stimulates human lung carcinoma cell proliferation by suppressing p21 gene expression via signals involving Erk and Rho kinase. Cancer Lett 219:71–81.

Harvey SA, Anderson SC, SundarRaj N (2004) Downstream effects of ROCK signaling in cultured human corneal stromal cells: microarray analysis of gene expression. Invest Ophthalmol Vis Sci 45:2168–2176.

Hayashi K, Wakino S, Kanda T, Homma K, Sugano N, Saruta T (2006) Molecular mechanisms and therapeutic strategies of chronic renal injury: role of rho-kinase in the development of renal injury. J Pharmacol Sci 100:29–33.

He H, Pannequin J, Tantiongco JP, Shulkes A, Baldwin GS (2005) Glycine-extended gastrin stimulates cell proliferation and migration through a Rho- and ROCK-dependent pathway, not a Rac/Cdc42-dependent pathway. Am J Physiol Gastrointest Liver Physiol 289:G478–G488.

Hidaka H, Inagaki M, Kawamoto S, Sasaki Y (1984) Isoquinolinesulfonamides, novel and potent inhibitors of cyclic nucleotide dependent protein kinase and protein kinase C. Biochemistry 23:5036–5041.

Hoffmann MJ, Florl AR, Seifert H-H, Schulz WA (2005) Multiple mechanisms downregulate *CDKN1C* in human bladder cancer. International Journal of Cancer 114:406–413.

Hotulainen P, Lappalainen P (2006) Stress fibers are generated by two distinct actin assembly mechanisms in motile cells. J Cell Biol 173:383–394.

Huang S, Chen CS, Ingber DE (1998) Control of cyclin D1, p27(Kip1), and cell cycle progression in human capillary endothelial cells by cell shape and cytoskeletal tension. Mol Biol Cell 9:3179–3193.

Huang S, Ingber DE (2002) A discrete cell cycle checkpoint in late G(1) that is cytoskeleton-dependent and MAP kinase (Erk)-independent. Exp Cell Res 275:255–264.

Huse M, Kuriyan J (2002) The conformational plasticity of protein kinases. Cell 109:275–282.

Iizuka K, Yoshii A, Samizo K, Tsukagoshi H, Ishizuka T, Dobashi K, Nakazawa T, Mori M (1999) A major role for the rho-associated coiled coil forming protein kinase in G-protein-mediated Ca2+ sensitization through inhibition of myosin phosphatase in rabbit trachea. Br J Pharmacol 128:925–933.

Ikeda H, Nagashima K, Yanase M, Tomiya T, Arai M, Inoue Y, Tejima K, Nishikawa T, Omata M, Kimura S, Fujiwara K (2003) Involvement of Rho/Rho kinase pathway in regulation of apoptosis in rat hepatic stellate cells. Am J Physiol Gastrointest Liver Physiol 285:G880–G886.

Ikenoya M, Hidaka H, Hosoya T, Suzuki M, Yamamoto N, Sasaki Y (2002) Inhibition of rho-kinase-induced myristoylated alanine-rich C kinase substrate (MARCKS) phosphorylation in human neuronal cells by H-1152, a novel and specific Rho-kinase inhibitor. J Neurochem 81:9–16.

Ishida N, Hara T, Kamura T, Yoshida M, Nakayama K, Nakayama KI (2002) Phosphorylation of p27Kip1 on serine 10 is required for its binding to CRM1 and nuclear export. J Biol Chem 277:14355–14358.

Ishizaki T, Naito M, Fujisawa K, Maekawa M, Watanabe N, Saito Y, Narumiya S (1997) p160ROCK, a Rho-associated coiled-coil forming protein kinase, works downstream of Rho and induces focal adhesions. FEBS Lett 404:118–124.

Iso Y, Suzuki H, Sato T, Shoji M, Shimizu N, Shibata M, Koba S, Geshi E, Katagiri T (2006) Rho-kinase inhibitor suppressed restenosis in porcine coronary balloon angioplasty. Int J Cardiol 106:103–110.

Ito M, Nakano T, Erdodi F, Hartshorne DJ (2004) Myosin phosphatase: structure, regulation and function. Mol. Cell. Biochem. 259:197–209.

Itoh K, Yoshioka K, Akedo H, Uehata M, Ishizaki T, Narumiya S (1999) An essential part for Rho-associated kinase in the transcellular invasion of tumor cells. Nat Med 5:221–225.

Iwamoto H, Nakamuta M, Tada S, Sugimoto R, Enjoji M, Nawata H (2000) A p160ROCK-specific inhibitor, Y-27632, attenuates rat hepatic stellate cell growth. J Hepatol 32:762–770.

Jacobs M, Hayakawa K, Swenson L, Bellon S, Fleming M, Taslimi P, Doran J (2006) The structure of dimeric ROCK I reveals the mechanism for ligand selectivity. J. Biol. Chem. 281:260–268.

Kamai T, Arai K, Sumi S, Tsujii T, Honda M, Yamanishi T, Yoshida KI (2002) The rho/rho-kinase pathway is involved in the progression of testicular germ cell tumour. BJU Int 89:449–453.

Kamai T, Tsujii T, Arai K, Takagi K, Asami H, Ito Y, Oshima H (2003) Significant association of Rho/ROCK pathway with invasion and metastasis of bladder cancer. Clin Cancer Res 9:2632–2641.

Kamai T, Yamanishi T, Shirataki H, Takagi K, Asami H, Ito Y, Yoshida K (2004) Overexpression of RhoA, Rac1, and Cdc42 GTPases is associated with progression in testicular cancer. Clin Cancer Res 10:4799–4805.

Kamiyama M, Utsunomiya K, Taniguchi K, Yokota T, Kurata H, Tajima N, Kondo K (2003) Contribution of Rho A and Rho kinase to platelet-derived growth factor-BB-induced proliferation of vascular smooth muscle cells. J Atheroscler Thromb 10:117–123.

Kamura T, Hara T, Matsumoto M, Ishida N, Okumura F, Hatakeyama S, Yoshida M, Nakayama K, Nakayama KI (2004) Cytoplasmic ubiquitin ligase KPC regulates proteolysis of p27Kip1 at G1 phase. Nat Cell Biol 6:1229–1235.

Kanda T, Hayashi K, Wakino S, Homma K, Yoshioka K, Hasegawa K, Sugano N, Tatematsu S, Takamatsu I, Mitsuhashi T, Saruta T (2005) Role of Rho-kinase and p27 in angiotensin II-induced vascular injury. Hypertension 45:724–729.

Kanda T, Wakino S, Hayashi K, Homma K, Ozawa Y, Saruta T (2003) Effect of fasudil on Rho-kinase and nephropathy in subtotally nephrectomized spontaneously hypertensive rats. Kidney Int 64:2009–2019.

Kawabata S, Usukura J, Morone N, Ito M, Iwamatsu A, Kaibuchi K, Amano M (2004) Interaction of Rho-kinase with myosin II at stress fibres. Genes Cells 9:653–660.

Kimura K, Ito M, Amano M, Chihara K, Fukata Y, Nakafuku M, Yamamori B, Feng J, Nakano T, Okawa K, Iwamatsu A, Kaibuchi K (1996) Regulation of myosin phosphatase by Rho and Rho-associated kinase (Rho- kinase). Science 273:245–248.

Kishore R, Qin G, Luedemann C, Bord E, Hanley A, Silver M, Gavin M, Goukassian D, Losordo DW (2005) The cytoskeletal protein ezrin regulates EC proliferation and angiogenesis via TNF-alpha-induced transcriptional repression of cyclin A. J Clin Invest 115:1785–1796.

Kobayashi K, Takahashi M, Matsushita N, Miyazaki J-i, Koike M, Yaginuma H, Osumi N, Kaibuchi K, Kobayashi K (2004) Survival of developing motor neurons mediated by Rho GTPase signaling pathway through Rho-kinase. J. Neurosci. 24:3480–3488.

Kosako H, Goto H, Yanagida M, Matsuzawa K, Fujita M, Tomono Y, Okigaki T, Odai H, Kaibuchi K, Inagaki M (1999) Specific accumulation of Rho-associated kinase at the cleavage furrow during cytokinesis: cleavage furrow-specific phosphorylation of intermediate filaments. Oncogene 18:2783–2788.

Koyama Y, Yoshioka Y, Shinde M, Matsuda T, Baba A (2004) Focal adhesion kinase mediates endothelin-induced cyclin D3 expression in rat cultured astrocytes. J Neurochem 90:904–912.

Kozai T, Eto M, Yang Z, Shimokawa H, Luscher TF (2005) Statins prevent pulsatile stretch-induced proliferation of human saphenous vein smooth muscle cells via inhibition of Rho/Rho-kinase pathway. Cardiovasc Res 68:475–482.

LaBaer J, Garrett MD, Stevenson LF, Slingerland JM, Sandhu C, Chou HS, Fattaey A, Harlow E (1997) New functional activities for the p21 family of CDK inhibitors. Genes Dev 11:847–862.

Lai JM, Wu S, Huang DY, Chang, ZF (2002) Cytosolic retention of phosphorylated extracellular signal-regulated kinase and a Rho-associated kinase-mediated signal impair expression of p21(Cip1/Waf1) in phorbol 12-myristate-13- acetate-induced apoptotic cells. Mol Cell Biol 22:7581–7592.

Leslie K, Lang C, Devgan G, Azare J, Berishaj M, Gerald W, Kim YB, Paz K, Darnell JE, Albanese C, Sakamaki T, Pestell R, Bromberg J (2006) Cyclin D1 is transcriptionally regulated by and required for transformation by activated signal transducer and activator of transcription 3. Cancer Res 66:2544–2552.

Leung T, Chen XQ, Manser E, Lim L (1996) The p160 RhoA-binding kinase ROK alpha is a member of a kinase family and is involved in the reorganization of the cytoskeleton. Mol Cell Biol 16:5313–5327.

Leung T, Manser E, Tan L, Lim L (1995) A novel serine/threonine kinase binding the Ras-related RhoA GTPase which translocates the kinase to peripheral membranes. J Biol Chem 270:29051–29054.

Li FH, Xia W, Li AW, Zhao CF, Sun RP (2007) Inhibition of rho kinase attenuates high flow induced pulmonary hypertension in rats. Chin Med J (Engl) 120:22–29.

Li X, Liu L, Tupper JC, Bannerman DD, Winn RK, Sebti SM, Hamilton AD, Harlan JM (2002) Inhibition of protein geranylgeranylation and RhoA/RhoA kinase pathway induces apoptosis in human endothelial cells. J Biol Chem 277:15309–15316.

Liang J, Zubovitz J, Petrocelli T, Kotchetkov R, Connor MK, Han K, Lee JH, Ciarallo S, Catzavelos C, Beniston R, Franssen E, Slingerland JM (2002) PKB/Akt phosphorylates p27, impairs nuclear import of p27 and opposes p27-mediated G1 arrest. Nat Med 8:1153–1160.

Liu AM, Lo RK, Wong CS, Morris C, Wise H, Wong YH (2006) Activation of STAT3 by G alpha(s) distinctively requires protein kinase A, JNK, and phosphatidylinositol 3-kinase. J Biol Chem 281:35812–35825.

Liu Y, Martindale JL, Gorospe M, Holbrook NJ (1996) Regulation of p21WAF1/CIP1 expression through mitogen-activated protein kinase signaling pathway. Cancer Res 56:31–35.

Liu Y, Suzuki YJ, Day RM, Fanburg BL (2004) Rho kinase-induced nuclear translocation of ERK1/ERK2 in smooth muscle cell mitogenesis caused by serotonin. Circ Res 95:579–586.

Lloyd AC, Obermuller F, Staddon S, Barth CF, McMahon M, Land H (1997) Cooperating oncogenes converge to regulate cyclin/cdk complexes. Genes Dev 11:663–677.

Lowery DM, Clauser KR, Hjerrild M, Lim D, Alexander J, Kishi K, Ong SE, Gammeltoft S, Carr SA, Yaffe MB (2007) Proteomic screen defines the Polo-box domain interactome and identifies Rock2 as a Plk1 substrate. Embo J 26:2262–2273.

Ma L, Teruya-Feldstein J, Weinberg RA (2007) Tumour invasion and metastasis initiated by microRNA-10b in breast cancer. Nature 449:682–688.

Ma Z, Kanai M, Kawamura K, Kaibuchi K, Ye K, Fukasawa K (2006) Interaction between ROCK II and nucleophosmin/B23 in the regulation of centrosome duplication. Mol Cell Biol 26:9016–9034.

Maekawa M, Ishizaki T, Boku S, Watanabe N, Fujita A, Iwamatsu A, Obinata T, Ohashi K, Mizuno K, Narumiya S (1999) Signaling from Rho to the actin cytoskeleton through protein kinases ROCK and LIM-kinase. Science 285:895–898.

Mallat Z, Gojova A, Sauzeau V, Brun V, Silvestre JS, Esposito B, Merval R, Groux H, Loirand G, Tedgui A (2003) Rho-associated protein kinase contributes to early atherosclerotic lesion formation in mice. Circ Res 93:884–888.

Marinissen MJ, Chiariello M, Tanos T, Bernard O, Narumiya S, Gutkind JS (2004) The Small GTP-Binding Protein RhoA Regulates c-Jun by a ROCK-JNK Signaling Axis. Molecular Cell 14:29–41.

Masamune A, Kikuta K, Satoh M, Satoh K, Shimosegawa T (2003) Rho kinase inhibitors block activation of pancreatic stellate cells. Br J Pharmacol 140:1292–1302.

Matsui T, Amano M, Yamamoto T, Chihara K, Nakafuku M, Ito M, Nakano T, Okawa K, Iwamatsu A, Kaibuchi K (1996) Rho-associated kinase, a novel serine/threonine kinase, as a putative target for small GTP binding protein Rho. Embo J 15:2208–2216.

Matsumoto Y, Uwatoku T, Oi K, Abe K, Hattori T, Morishige K, Eto Y, Fukumoto Y, Nakamura K, Shibata Y, Matsuda T, Takeshita A, Shimokawa H (2004) Long-term inhibition of Rho-kinase suppresses neointimal formation after stent implantation in porcine coronary arteries: involvement of multiple mechanisms. Arterioscler Thromb Vasc Biol 24:181–186.

McMullan R, Lax S, Robertson VH, Radford DJ, Broad S, Watt FM, Rowles A, Croft DR, Olson MF, Hotchin NA (2003) Keratinocyte differentiation is regulated by the Rho and ROCK signaling pathway. Curr Biol 13:2185–2189.

Medema RH, Kops GJ, Bos JL, Burgering BM (2000) AFX-like Forkhead transcription factors mediate cell-cycle regulation by Ras and PKB through p27kip1. Nature 404:782–787.

Meyer TN, Schwesinger C, Sampogna RV, Vaughn DA, Stuart RO, Steer DL, Bush KT, Nigam SK (2006) Rho kinase acts at separate steps in ureteric bud and metanephric mesenchyme morphogenesis during kidney development. Differentiation 74:638–647.

Miyazaki K, Komatsu S, Ikebe M (2006) Dynamics of RhoA and ROKalpha translocation in single living cells. Cell Biochem Biophys 45:243–254.

Moore M, Marroquin BA, Gugliotta W, Tse R, White SR (2004) Rho kinase inhibition initiates apoptosis in human airway epithelial cells. Am J Respir Cell Mol Biol 30:379–387.

Morgan DO (1997) Cyclin-dependent kinases: engines, clocks, and microprocessors. Ann Rev Cell Develop Biol 13:261–291.
Nagashima T, Okazaki H, Yudoh K, Matsuno H, Minota S (2006) Apoptosis of rheumatoid synovial cells by statins through the blocking of protein geranylgeranylation: a potential therapeutic approach to rheumatoid arthritis. Arthritis Rheum 54:579–586.
Nagatoya K, Moriyama T, Kawada N, Takeji M, Oseto S, Murozono T, Ando A, Imai E, Hori M (2002) Y-27632 prevents tubulointerstitial fibrosis in mouse kidneys with unilateral ureteral obstruction. Kidney Int 61:1684–1695.
Nakagawa O, Fujisawa K, Ishizaki T, Saito Y, Nakao K, Narumiya S (1996) ROCK-I and ROCK-II, two isoforms of Rho-associated coiled-coil forming protein serine/threonine kinase in mice. FEBS Lett 392:189–193.
Nakajima M, Hayashi K, Katayama K, Amano Y, Egi Y, Uehata M, Goto N, Kondo T (2003) Wf-536 prevents tumor metastasis by inhibiting both tumor motility and angiogenic actions. Eur J Pharmacol 459:113–120.
Nicole O, Goldshmidt A, Hamill CE, Sorensen SD, Sastre A, Lyuboslavsky P, Hepler JR, McKeon RJ, Traynelis SF (2005) Activation of protease-activated receptor-1 triggers astrogliosis after brain injury. J Neurosci 25:4319–4329.
Nishimaki H, Kasai K, Kozaki K, Takeo T, Ikeda H, Saga S, Nitta M, Itoh G (2004) A role of activated Sonic hedgehog signaling for the cellular proliferation of oral squamous cell carcinoma cell line. Biochem Biophys Res Commun 314:313–320.
Nobes CD, Lauritzen I, Mattei MG, Paris S, Hall A, Chardin P (1998) A new member of the Rho family, Rnd1, promotes disassembly of actin filament structures and loss of cell adhesion. J Cell Biol 141:187–197.
Ohashi K, Nagata K, Maekawa M, Ishizaki T, Narumiya S, Mizuno K (2000) Rho-associated kinase ROCK activates LIM-kinase 1 by phosphorylation at threonine 508 within the activation loop. J Biol Chem 275:3577–3582.
Olson MF, Paterson HF, Marshall CJ (1998) Signals from Ras and Rho GTPases interact to regulate expression of p21Waf1/Cip1. Nature 394:295–299.
Otis M, Gallo-Payet N (2006) Differential involvement of cytoskeleton and rho-guanosine 5'-triphosphatases in growth-promoting effects of angiotensin II in rat adrenal glomerulosa cells. Endocrinology 147:5460–5469.
Pacary E, Tixier E, Coulet F, Roussel S, Petit E, Bernaudin M (2007) Crosstalk between HIF-1 and ROCK pathways in neuronal differentiation of mesenchymal stem cells, neurospheres and in PC12 neurite outgrowth. Mol Cell Neurosci 35:409–423.
Pellegrin S, Mellor H (2007) Actin stress fibres. J Cell Sci 120:3491–3499.
Pinner S, Sahai E (2008) PDK1 regulates cancer cell motility by antagonising inhibition of ROCK1 by RhoE. Nature Cell Biology 10:-.
Pirone DM, Liu WF, Ruiz SA, Gao L, Raghavan S, Lemmon CA, Romer LH, Chen CS (2006) An inhibitory role for FAK in regulating proliferation: a link between limited adhesion and RhoA-ROCK signaling. J. Cell Biol. 174:277–288.
Porter KE, Turner NA, O'Regan DJ, Balmforth AJ, Ball SG (2004) Simvastatin reduces human atrial myofibroblast proliferation independently of cholesterol lowering via inhibition of RhoA. Cardiovasc Res 61:745–755.
Pumiglia KM, Decker SJ (1997) Cell cycle arrest mediated by the MEK/mitogen-activated protein kinase pathway. Proc Natl Acad Sci U S A 94:448–452.
Radeff-Huang J, Seasholtz TM, Chang JW, Smith JM, Walsh CT, Brown JH (2007) Tumor necrosis factor-alpha-stimulated cell proliferation is mediated through sphingosine kinase-dependent Akt activation and cyclin D expression. J Biol Chem 282:863–870.
Rashid-Doubell F, Tannetta D, Redman CW, Sargent IL, Boyd CA, Linton EA (2007) Caveolin-1 and lipid rafts in confluent BeWo trophoblasts: evidence for Rock-1 association with caveolin-1. Placenta 28:139–151.
Rattan R, Giri S, Singh AK, Singh I (2006) Rho/ROCK pathway as a target of tumor therapy. J Neurosci Res 83:243–255.

Rees RW, Foxwell NA, Ralph DJ, Kell PD, Moncada S, Cellek S (2003) Y-27632, a Rho-kinase inhibitor, inhibits proliferation and adrenergic contraction of prostatic smooth muscle cells. J Urol 170:2517–2522.

Riento K, Guasch RM, Garg R, Jin B, Ridley AJ (2003) RhoE binds to ROCK I and inhibits downstream signaling. Mol Cell Biol 23:4219–4229.

Riento K, Ridley AJ (2003) Rocks: multifunctional kinases in cell behaviour. Nat Rev Mol Cell Biol 4:446–456.

Rikitake Y, Oyama N, Wang CY, Noma K, Satoh M, Kim HH, Liao JK (2005) Decreased perivascular fibrosis but not cardiac hypertrophy in ROCK1+/– haploinsufficient mice. Circulation 112:2959–2965.

Rodier G, Montagnoli A, Di Marcotullio L, Coulombe P, Draetta GF, Pagano M, Meloche S (2001) p27 cytoplasmic localization is regulated by phosphorylation on Ser10 and is not a prerequisite for its proteolysis. Embo J 20:6672–6682.

Sahai E, Alberts AS, Treisman R (1998) RhoA effector mutants reveal distinct effector pathways for cytoskeletal reorganization, SRF activation and transformation. Embo J 17:1350–1361.

Sahai E, Ishizaki T, Narumiya S, Treisman R (1999) Transformation mediated by RhoA requires activity of ROCK kinases. Curr Biol 9:136–145.

Sahai E, Marshall CJ (2003) Differing modes of tumour cell invasion have distinct requirements for Rho/ROCK signalling and extracellular proteolysis. Nat Cell Biol 5:711–719.

Sahai E, Olson MF, Marshall CJ (2001) Cross-talk between Ras and Rho signalling pathways in transformation favours proliferation and increased motility. Embo J 20:755–766.

Sase K, Yui Y, Hattori R, Shirotani M, Kawai C, Sasayama S (1992) HA-1077 suppress both proliferation of vascular smooth muscle cells and c-fos mRNA induction. Jpn Circ J 56: 1229–1233.

Sauzeau V, Le Mellionnec E, Bertoglio J, Scalbert E, Pacaud P, Loirand G (2001) Human urotensin II-induced contraction and arterial smooth muscle cell proliferation are mediated by RhoA and Rho-kinase. Circ Res 88:1102–1104.

Sawada N, Itoh H, Ueyama K, Yamashita J, Doi K, Chun TH, Inoue M, Masatsugu K, Saito T, Fukunaga Y, Sakaguchi S, Arai H, Ohno N, Komeda M, Nakao K (2000) Inhibition of rho-associated kinase results in suppression of neointimal formation of balloon-injured arteries. Circulation 101:2030–2033.

Seasholtz TM, Majumdar M, Kaplan DD, Brown JH (1999) Rho and Rho kinase mediate thrombin-stimulated vascular smooth muscle cell DNA synthesis and migration. Circ Res 84: 1186–1193.

Sebbagh M, Hamelin J, Bertoglio J, Solary E, Breard J (2005) Direct cleavage of ROCK II by granzyme B induces target cell membrane blebbing in a caspase-independent manner. J Exp Med 201:465–471.

Sebbagh M, Renvoize C, Hamelin J, Riche N, Bertoglio J, Breard J (2001) Caspase-3-mediated cleavage of ROCK I induces MLC phosphorylation and apoptotic membrane blebbing. Nat Cell Biol 3:346–352.

Seibold S, Schurle D, Heinloth A, Wolf G, Wagner M, Galle J (2004) Oxidized LDL induces proliferation and hypertrophy in human umbilical vein endothelial cells via regulation of p27Kip1 expression: role of RhoA. J Am Soc Nephrol 15:3026–3034.

Sewing A, Wiseman B, Lloyd AC, Land H (1997) High-intensity Raf signal causes cell cycle arrest mediated by p21Cip1. Mol Cell Biol 17:5588–5597.

Sherr CJ, Roberts JM (2004) Living with or without cyclins and cyclin-dependent kinases. Genes Dev 18:2699–2711.

Shibata R, Kai H, Seki Y, Kato S, Morimatsu M, Kaibuchi K, Imaizumi T (2001) Role of Rho-associated kinase in neointima formation after vascular injury. Circulation 103:284–289.

Shibata R, Kai H, Seki Y, Kusaba K, Takemiya K, Koga M, Jalalidin A, Tokuda K, Tahara N, Niiyama H, Nagata T, Kuwahara F, Imaizumi T (2003) Rho-kinase inhibition reduces neointima formation after vascular injury by enhancing Bax expression and apoptosis. J Cardiovasc Pharmacol 42(Suppl 1):S43–S47.

Shimizu T, Ihara K, Maesaki R, Amano M, Kaibuchi K, Hakoshima T (2003) Parallel coiled-coil association of the RhoA-binding domain in Rho-kinase. J Biol Chem 278:46046–46051.

Shimizu Y, Thumkeo D, Keel J, Ishizaki T, Oshima H, Oshima M, Noda Y, Matsumura F, Taketo MM, Narumiya S (2005) ROCK-I regulates closure of the eyelids and ventral body wall by inducing assembly of actomyosin bundles. J Cell Biol 168:941–953.

Shin I, Yakes FM, Rojo F, Shin NY, Bakin AV, Baselga J, Arteaga CL (2002) PKB/Akt mediates cell-cycle progression by phosphorylation of p27(Kip1) at threonine 157 and modulation of its cellular localization. Nat Med 8:1145–1152.

Shirao S, Kashiwagi S, Sato M, Miwa S, Nakao F, Kurokawa T, Todoroki-Ikeda N, Mogami K, Mizukami Y, Kuriyama S, Haze K, Suzuki M, Kobayashi S (2002) Sphingosylphosphorylcholine is a novel messenger for Rho-kinase-mediated Ca2+ sensitization in the bovine cerebral artery: unimportant role for protein kinase C. Circ Res 91:112–119.

Shirotani M, Yui Y, Hattori R, Kawai C (1991) A new type of vasodilator, HA1077, an isoquinoline derivative, inhibits proliferation of bovine vascular smooth muscle cells in culture. J Pharmacol Exp Ther 259:738–744.

Somlyo AP, Somlyo AV (2000) Signal transduction by G-proteins, rho-kinase and protein phosphatase to smooth muscle and non-muscle myosin II. J Physiol 522(Pt 2):177–185.

Somlyo AV, Bradshaw D, Ramos S, Murphy C, Myers CE, Somlyo AP (2000) Rho-kinase inhibitor retards migration and in vivo dissemination of human prostate cancer cells. Biochem Biophys Res Commun 269:652–659.

Somlyo AV, Phelps C, Dipierro C, Eto M, Read P, Barrett M, Gibson JJ, Burnitz MC, Myers C, Somlyo AP (2003) Rho kinase and matrix metalloproteinase inhibitors cooperate to inhibit angiogenesis and growth of human prostate cancer xenotransplants. Faseb J 17:223–234.

Song J, Li J, Lulla A, Evers BM, Chung DH (2006) Protein kinase D protects against oxidative stress-induced intestinal epithelial cell injury via Rho/ROK/PKC-delta pathway activation. Am J Physiol Cell Physiol 290:C1469–C1476.

Sorensen SD, Nicole O, Peavy RD, Montoya LM, Lee CJ, Murphy TJ, Traynelis SF, Hepler JR (2003) Common signaling pathways link activation of murine PAR-1, LPA, and S1P receptors to proliferation of astrocytes. Mol Pharmacol 64:1199–1209.

Sumi T, Matsumoto K, Nakamura T (2001) Specific activation of LIM kinase 2 via phosphorylation of threonine 505 by ROCK, a Rho-dependent protein kinase. J Biol Chem 276:670–676.

Svoboda KK, Moessner P, Field T, Acevedo J (2004) ROCK inhibitor (Y27632) increases apoptosis and disrupts the actin cortical mat in embryonic avian corneal epithelium. Dev Dyn 229: 579–590.

Swant JD, Rendon BE, Symons M, Mitchell RA (2005) Rho GTPase-dependent signaling is required for macrophage migration inhibitory factor-mediated expression of cyclin D1. J Biol Chem 280:23066–23072.

Takami A, Iwakubo M, Okada Y, Kawata T, Odai H, Takahashi N, Shindo K, Kimura K, Tagami Y, Miyake M, Fukushima K, Inagaki M, Amano M, Kaibuchi K and Iijima H (2004) Design and synthesis of Rho kinase inhibitors (I). Bioorg Med Chem 12:2115–2137.

Takamura M, Sakamoto M, Genda T, Ichida T, Asakura H, Hirohashi S (2001) Inhibition of intrahepatic metastasis of human hepatocellular carcinoma by Rho-associated protein kinase inhibitor Y-27632. Hepatology 33:577–581.

Takeda K, Ichiki T, Tokunou T, Iino N, Fujii S, Kitabatake A, Shimokawa H, Takeshita A (2001) Critical role of Rho-kinase and MEK/ERK pathways for angiotensin II- induced plasminogen activator inhibitor type-1 gene expression. Arterioscler Thromb Vasc Biol 21:868–873.

Takeda N, Kondo M, Ito S, Ito Y, Shimokata K, Kume H (2006) Role of RhoA inactivation in reduced cell proliferation of human airway smooth muscle by simvastatin. Am J Respir Cell Mol Biol 35:722–729.

Tanaka T, Nishimura D, Wu RC, Amano M, Iso T, Kedes L, Nishida H, Kaibuchi K, Hamamori Y (2006) Nuclear Rho kinase, ROCK2, targets p300 acetyltransferase. J Biol Chem 281:15320–15329.

Terrak M, Kerff F, Langsetmo K, Tao T, Dominguez R (2004) Structural basis of protein phosphatase 1 regulation. Nature 429:780–784.

Tharaux PL, Bukoski RC, Rocha PN, Crowley SD, Ruiz P, Nataraj C, Howell DN, Kaibuchi K, Spurney RF, Coffman TM (2003) Rho kinase promotes alloimmune responses by regulating the proliferation and structure of T cells. J Immunol 171:96–105.

Thumkeo D, Keel J, Ishizaki T, Hirose M, Nonomura K, Oshima H, Oshima M, Taketo MM, Narumiya S (2003) Targeted disruption of the mouse rho-associated kinase 2 gene results in intrauterine growth retardation and fetal death. Mol Cell Biol 23:5043–5055.

Thumkeo D, Shimizu Y, Sakamoto S, Yamada S, Narumiya S (2005) ROCK-I and ROCK-II cooperatively regulate closure of eyelid and ventral body wall in mouse embryo. Genes Cells 10:825–834.

Tominaga T, Barber DL (1998) Na-H exchange acts downstream of RhoA to regulate integrin-induced cell adhesion and spreading. Mol Biol Cell 9:2287–2303.

Trinkaus JP (1973) Surface activity and locomotion of Fundulus deep cells during blastula and gastrula stages. Dev Biol 30:69–103.

Tura A, Grisanti S, Petermeier K, Henke-Fahle S (2007) The Rho-kinase inhibitor H-1152P suppresses the wound-healing activities of human Tenon's capsule fibroblasts in vitro. Invest Ophthalmol Vis Sci 48:2152–2161.

Uchida S Watanabe G, Shimada Y, Maeda M, Kawabe A, Mori A, Arii S, Uehata M, Kishimoto T, Oikawa T, Imamura M (2000) The suppression of small GTPase rho signal transduction pathway inhibits angiogenesis in vitro and in vivo. Biochem Biophys Res Commun 269:633–640.

Uehata M, Ishizaki T, Satoh H, Ono T, Kawahara T, Morishita T, Tamakawa H, Yamagami K, Inui J, Maekawa M, Narumiya S (1997) Calcium sensitization of smooth muscle mediated by a Rho-associated protein kinase in hypertension. Nature 389:990–994.

Van Kolen K, Slegers H (2006) Atypical PKCzeta is involved in RhoA-dependent mitogenic signaling by the P2Y(12) receptor in C6 cells. Febs J 273:1843–1854.

Velasco G, Armstrong C, Morrice N, Frame S, Cohen P (2002) Phosphorylation of the regulatory subunit of smooth muscle protein phosphatase 1M at Thr850 induces its dissociation from myosin. FEBS Lett 527:101–104.

Vichalkovski A, Baltensperger K, Thomann D, Porzig H (2005) Two different pathways link G-protein-coupled receptors with tyrosine kinases for the modulation of growth and survival in human hematopoietic progenitor cells. Cell Signal 17:447–459.

Viglietto G, Motti ML, Bruni P, Melillo RM, D'Alessio A, Califano D, Vinci F, Chiappetta G, Tsichlis P, Bellacosa A, Fusco A, Santoro M (2002) Cytoplasmic relocalization and inhibition of the cyclin-dependent kinase inhibitor p27(Kip1) by PKB/Akt-mediated phosphorylation in breast cancer. Nat Med 8:1136–1144.

Villanueva J, Yung Y, Walker JL, Assoian RK (2007) ERK activity and G1 phase progression: identifying dispensable versus essential activities and primary versus secondary targets. Mol. Biol. Cell 18:1457–1463.

Vishnubhotla R, Sun S, Huq J, Bulic M, Ramesh A, Guzman G, Cho M, Glover SC (2007) ROCK-II mediates colon cancer invasion via regulation of MMP-2 and MMP-13 at the site of invadopodia as revealed by multiphoton imaging. Lab Invest 87:1149–1158.

Wang G, Woods A, Sabari S, Pagnotta L, Stanton LA, Beier F (2004) RhoA/ROCK signaling suppresses hypertrophic chondrocyte differentiation. J Biol Chem 279:13205–13214.

Ward Y, Yap SF, Ravichandran V, Matsumura F, Ito M, Spinelli B, Kelly K (2002) The GTP binding proteins Gem and Rad are negative regulators of the Rho- Rho kinase pathway. J Cell Biol 157:291–302.

Watanabe K, Ueno M, Kamiya D, Nishiyama A, Matsumura M, Wataya T, Takahashi JB, Nishikawa S, Nishikawa S, Muguruma K, Sasai Y (2007) A ROCK inhibitor permits survival of dissociated human embryonic stem cells. Nat Biotechnol 25:681–686.

Watanabe N, Kato T, Fujita A, Ishizaki T, Narumiya S (1999) Cooperation between mDia1 and ROCK in Rho-induced actin reorganization. Nat Cell Biol 1:136–143.

Watanabe T, Takahashi K, Kanome T, Hongo S, Miyazaki A, Koba S, Katagiri T, Pakara R, Benedict CR (2006) Human urotensin-II potentiates the mitogenic effect of mildly oxidized low-density lipoprotein on vascular smooth muscle cells: comparison with other vasoactive agents and hydrogen peroxide. Hypertens Res 29:821–831.

Wilkinson S, Paterson HF, Marshall CJ (2005) Cdc42-MRCK and Rho-ROCK signalling cooperate in myosin phosphorylation and cell invasion. Nat Cell Biol 7:255–261.

Woods D, Parry D, Cherwinski H, Bosch E, Lees E, McMahon M (1997) Raf-induced proliferation or cell cycle arrest is determined by the level of Raf activity with arrest mediated by p21Cip1. Mol Cell Biol 17:5598–5611.

Wyckoff JB, Pinner SE, Gschmeissner S, Condeelis JS, Sahai E (2006) ROCK- and myosin-dependent matrix deformation enables protease-independent tumor-cell invasion in vivo. Curr Biol 16:1515–1523.

Yahiaoui L, Villeneuve A, Valderrama-Carvajal H, Burke F, Fixman ED (2006) Endothelin-1 regulates proliferative responses, both alone and synergistically with PDGF, in rat tracheal smooth muscle cells. Cell Physiol Biochem 17:37–46.

Yamada KM, Cukierman E (2007) Modeling tissue morphogenesis and cancer in 3D. Cell 130:601–610.

Yamaguchi H, Kasa M, Amano M, Kaibuchi K, Hakoshima T (2006) Molecular mechanism for the regulation of rho-kinase by dimerization and its inhibition by fasudil. Structure 14:589–600.

Yamazaki D, Kurisu S, Takenawa T (2005) Regulation of cancer cell motility through actin reorganization. Cancer Sci 96:379–386.

Yanazume T, Hasegawa K, Wada H, Morimoto T, Abe M, Kawamura T, Sasayama S (2002) Rho/ROCK pathway contributes to the activation of extracellular signal-regulated kinase/GATA-4 during myocardial cell hypertrophy. J Biol Chem 277:8618–8625.

Yarrow JC, Totsukawa G, Charras GT, Mitchison TJ (2005) Screening for cell migration inhibitors via automated microscopy reveals a Rho-kinase inhibitor. Chem Biol 12:385–395.

Ying H, Biroc SL, Li WW, Alicke B, Xuan JA, Pagila R, Ohashi Y, Okada T, Kamata Y, Dinter H (2006) The Rho kinase inhibitor fasudil inhibits tumor progression in human and rat tumor models. Mol Cancer Ther 5:2158–2164.

Yokoyama T, Goto H, Izawa I, Mizutani H, Inagaki M (2005) Aurora-B and Rho-kinase/ROCK, the two cleavage furrow kinases, independently regulate the progression of cytokinesis: possible existence of a novel cleavage furrow kinase phosphorylates ezrin/radixin/moesin (ERM). Genes Cells 10:127–137.

Yoneda A, Multhaupt HA, Couchman JR (2005) The Rho kinases I and II regulate different aspects of myosin II activity. J Cell Biol 170:443–453.

Zhang P, Liegeois NJ, Wong C, Finegold M, Hou H, Thompson JC, Silverman A, Harper JW, DePinho RA, Elledge SJ (1997) Altered cell differentiation and proliferation in mice lacking p57KIP2 indicates a role in Beckwith-Wiedemann syndrome. Nature 387:151–158.

Zhang YM, Bo J, Taffet GE, Chang J, Shi J, Reddy AK, Michael LH, Schneider MD, Entman ML Schwartz, RJ, Wei L (2006) Targeted deletion of ROCK1 protects the heart against pressure overload by inhibiting reactive fibrosis. Faseb J 20:916–925.

Zhao M, Discipio RG, Wimmer AG, Schraufstatter IU (2006) Regulation of CXCR4-mediated nuclear translocation of extracellular signal-related kinases 1 and 2. Mol Pharmacol 69:66–75.

Zhao Z, Rivkees SA (2003) Rho-associated kinases play an essential role in cardiac morphogenesis and cardiomyocyte proliferation. Dev Dyn 226:24–32.

Zhong W-B, Wang C-Y, Chang T-C, Lee W-S (2003) Lovastatin induces apoptosis of anaplastic thyroid cancer cells via inhibition of protein geranylgeranylation and de novo protein synthesis. Endocrinology 144:3852–3859.

Zuckerbraun BS, Shapiro RA, Billiar TR, Tzeng E (2003) RhoA influences the nuclear localization of extracellular signal-regulated kinases to modulate p21Waf/Cip1 expression. Circulation 108:876–881.

Author Index

A
Aasheim, H.C., 70
Abe, K., 17, 172, 179
Abo, A., xvii, 16
Adams, A.E., 6, 9
Adamson, P., 129, 130, 135, 137, 138, 144
Adini, I., 139, 140
Adnane, J., 136, 143, 178
Adra, C.N., 45
Agarwal, B., 53
Aghazadeh, B., 17, 81
Ahmad, K.F., 113
Ahmadian, M.R., 31, 94
Aktories, K., 156
Allenspach, E.J., 47
Allinen, M., 119
Alt, J.R., 178
Amano, M., 164, 168, 169
Amano, T., 168
Anselmo, A.N., 45–55
Arai, M., 166
Aranda, V., xix
Aronheim, A., 6, 9, 15
Arthur, W.T., 125
Asano, T., 166
Aspenstrom, P., 3, 10, 15, 116
Auer, K.L., 178
Auld, V.J., 35
Aurandt, J., 63
Avraham, H., 59, 124
Axel, R., 4, 5
Aznar, S., 175

B
Bain, J., 166
Banach-Petrosky, W., 87
Banerjee, J., 67
Baranski, T.J., 101
Barbacid, M., 4
Barber, D.L., 169
Barber, M.A., 18
Barrett, K., 70
Bartolome, R., 78, 85
Bax, B., 94
Beder, L.B., 7, 116
Benitah, S.A., 170
Berman, D.M., 13
Bernards, A., xvi
Bernhard, E.J., 145
Betz, R., 85
Bhattacharyya, R., 66, 68
Bi, E., 18
Billadeau, D.D, 77–88
Billuart, P., 96, 97, 125
Bishop, A.L., 30, 124, 135
Bishop, A.P., 93
Blackhall, F.H., 52
Blanchoin, L., 35
Bloom, J., 177, 178
Boettner, B., 59
Bohmer, R.M., 174
Bokoch, G.M., 13, 45–55
Booden, M.A., 63, 81
Bos, J.L., 59, 135
Bottazzi, M.E., 178
Boureux, A., 10, 15, 111
Bourguignon, L.Y., 63, 78, 86
Bourmeyster, N., 48
Bourne, H.R., 30, 67
Braga, V.M., xix, 18
Brenner, J.C., 29–36
Brouns, M.R., 34, 97
Brouns, T., 97
Brown, R., 123
Brugge, J.S., xix
Brugnera, E., 14
Bruinsma, S.P., 101, 102

Brunner, T.B., 103
Burridge, K., 125, 169
Burthem, J., 172, 179
Bustelo, X.R., 17, 77
Byers, H.R., 168

C
Cagan, R.L., 101
Campostrini, N., 51
Capon, D.J., 123
Carlier, M.F., 180
Carr, K.M., 129
Casey, P.J., 11
Cechin, S.R., 171, 175
Cerione, R.A., 46, 126, 159
Chan, A.M., 14
Chang, F.K., 115
Chang, J., 179
Chapman-Shimshoni, D., 53
Chardin, P., 4, 5, 15, 115
Chauhan, S., 138
Chavrier, P., 7, 30
Chen, Y., 156, 157
Chen, Z., 66–69, 136, 138, 142, 143
Chenette, E.J., 13
Cheng, M., 178
Chiang, S.H., 97
Chiariello, M., 85
Chijiwa, T., 166
Chikumi, H., 60, 62, 63, 65, 66
Chimini, G., 30
Chrzanowska-Wodnicka, M., 169
Chuang, T.H., 48
Cichowski, K., xvi
Clark, E.A., 5, 59, 124, 130
Coleman, M.L., 165, 178–180
Collard, J.G., 135, 170
Colomba, A., 85
Corbetta, S., xx
Cordle, A., 53, 54
Coso, O.A., 16
Cote, J.F., 14
Couchman, J.R., 45
Cox, A.D., 11, 136, 146
Cramer, L.P., 36
Crespo, P., 17, 80
Crnogorac-Jurcevic, T., 127
Croft, D.R., 165, 173–175, 177–179, 181, 182
Cukierman, E., 180

D
Dallery, E., 7
Dallery-Prudhomme, E., 9
Daubon, T., 85
Davies, S.P., 166
De Sarno, P., 179
Debidda, M., 175, 177
Debnath, J., xix
Decker, S.J., 178
del Peso, L., 124
Delarue, F.L., 142, 143, 148
DeMali, K.A., 18
Der, C.J., 3–20, 63, 70, 136, 142, 146
Derewenda, U., 62, 69
DerMardirossian, C., 12, 13, 45–55
Dhanasekaran, N., 60
Dhawan, J., 172–174
Dias, S.M., 126, 159
Didsbury, J., 5
Diekmann, D., xvi, 15, 34
Dong, G., 160
Dong, J.M., 31, 156
Dong, Z., 78, 87
Doody, G.M., 80
Doran, J.D., 164
Doupnik, C.A., 64
Dovas, A., 45
Dransart, E., 45, 48
Drivas, G.T., 5, 6, 9
D'Souza-Schorey, C., 59, 93
Du, W., 129, 136, 138, 142, 147
DuHadaway, J.B., 137
Durkin, M.E., 15, 99, 100
Dvorsky, R., 31, 164

E
Eastman, A., 13, 50
Eisenhaure, T.M., 65
Ellenbroek, S.I., 135, 170
Ellerbroek, S.M., 156
Elliott, K., 137
Ellis, S., 7, 138
Endlich, N., 169
Engel, M.E., 138
Erickson, M.R., 14
Essmann, F., 13
Etienne-Manneville, S., xix, 3, 30
Eva, A., 31
Evangelista, M., 141
Evers, E.E., 30

F
Fang, G., 96
Fassett, J.T., 174
Faucheux, N., 157

Faure, J., 48
Feig, L.A., 17
Feng, J., 165, 168
Feramisco, J.R., 123
Fernandez-Borja, M., 137, 139, 141
Fernandez-Zapico, M.E., 78, 85
Finlay, B.B., 3
Flynn, P., 139, 140
Forbes, S., 116
Forget, M.A., 31, 136, 143, 157
Foster, R., 7, 9, 15, 29
Francis, S.A., 71
Frederick, M.J., 119
Freeman, S.N., 119, 120
Friedl, P., 180, 181
Fritz, G., 59, 123, 136, 138, 143–145
Fujikawa, K., 80, 85
Fukuda, A., 158
Fukuhara, S., 18, 60, 62, 63, 66
Fukumoto, Y., 13, 45, 144
Fukushima, M., 172–175
Furuyama, T., 172, 179

G

Gaggioli, C., xviii
Gallo-Payet, N., 172–175
Gamblin, S.J., 94
Gampel, A., 139, 140
Gao, G., 69, 70
Gao, Y., 88
Garrett, M.D., 14, 94
Garrett, T.A., 86
Gassama-Diagne, A., xix
Gavard, J., 87
Genda, T., 181
Gerhard, R., 138
Geyer, M., 155
Gildea, J.J., 13, 49–51, 53
Girkontaite, I., 70, 71
Gohla, A., 65
Gomez del Pulgar, T., 98
Gomez, T.S., 84
Gorvel, J.P., 45, 48, 52
Gosser, Y.Q., 13, 46, 155
Goto, T., 49, 50, 52
Govek, E.E., 3
Grabocka, E., 65, 67, 68
Greenman, 173
Grimmler, M., 178
Grizot, S., 46
Grogg, M.W., 93–182
Groh, K., 126
Grossmann, A., 144
Groysman, M., 82, 83
Gu, J.L., 71, 72
Guan, M., 99
Guerin, P., 178
Gutkind, J.S., 60, 62, 63, 66

H

Haataja, L., 5, 9
Habets, G.G., xx, 14
Hagerty, L., 168
Hajicek, N., 59–72
Hakem, A., xx, 129
Hall, A., xv–xx, 3, 13, 16–18, 30, 31, 35, 59, 93, 123–125, 135, 155
Hall, C., 128, 129
Hall, H.G., xix
Hamaguchi, M., 7, 116, 119
Hamilton, A.D., 136
Han, J., 17, 81
Han, S., 179
Hanahan, D., 49
Hancock, J.F., 130
Harhammer, R., 65
Hart, M.J., xvi, 14, 18, 31, 60, 63, 65, 70, 77
Harvey, S.A., 177
Hasegawa, H., 14
Haskell, M.D., 125
Hayashi, K., 178
He, H., 171
Heckman, B.M., 98
Heldin, C.H., 140
Helfman, D.M., 172–174
Heo, J., 82
Hidaka, H., 166
Hiley, E., 70
Hill, C.S., 16
Hirota, T., 84
Hirotani, M., 63
Hobert, O., 82, 84
Hoffman, G.R., 13, 46
Hoffmann, M.J., 178
Holinstat, M., 65
Hori, Y., 53, 155
Horii, Y., 14
Horiuchi, A., 129
Hornstein, I., 78, 85
Horvitz, Y.C., 14
Hotulainen, P., 169
Houlard, M., 82, 84
Hu, C.D., 156
Huang, M., 135–149
Huang, S., 174
Huelsenbeck, J., 138
Hunter, S.G., 86, 87
Hunter, T., 5, 138
Huse, M., 166

I
Ihara, K., 126
Iida, S., 84
Iizuka, K., 168
Ikeda, H., 179
Ikenoya, M., 166
Imamura, F., 124
Ingber, D.E., 174
Insall, R.H., 19
Ishida, N., 178
Ishizaju, 36
Ishizaki, T., 34, 168
Iso, Y., 171, 178
Ito, M., 168
Itoh, K., 180
Iwamoto, H., 172–176, 178

J
Jacks, T., xvi
Jackson, M., 62
Jacobs, M., 164, 165
Jaffe, A.B., xv–xx, 3, 18, 135
Jahner, D., 5, 138
Jaiyesimi, I.A., 128
Jardin, F., 116
Jelacic, T., 160
Jiang, K., 136, 138, 142–144
Jiang, W.G., 49, 52
Johndrow, J.E., xvii
Johnson, D.I., 6, 9
Johnstone, C.N., 102
Jones, M.B., 49, 50
Jordan, P., 5, 9
Just, I., 59

K
Kaina, B., 59, 144, 145
Kamai, T., 129, 170
Kamasani, U., 136, 137, 146
Kamiyama, M., 171, 174
Kamura, T., 178
Kanda, T., 172, 178
Kandpal, R.P., 94, 95, 98
Karnoub, A.E., 137
Katoh, H., 65
Katzav, S., 14, 17, 78, 80, 84–86
Kawabata, S., 167
Kawasaki, Y., 17, 18
Keep, N.H., 13, 46
Kermorgant, S., 160
Kerr, D.J., 102, 103
Khosravi-Far, R., 16, 59, 142
Khurana, V., 53
Kikkawa, U., 159
Kikuchi, A., 155
Kim, O., 47
Kimura, K., 19, 168
Kirchhausen, T., 18
Kirschmeier, P.T., 129
Kishore, R., 173
Kitzing, T.M., 63
Kleer, C.G., 128
Klein, W.M., 127
Knowles, M.A., 116
Kobayashi, K., 179
Kondo, T., 129
Kosako, H., 167
Kourlas, P.J., 62, 71
Koyama, Y., 171, 174, 175, 177
Kozai, T., 172
Kozasa, T., 18, 59–72
Kozma, R., 16, 97
Kranenburg, O., 65
Kreutz, B., 59–72
Krieser, R.J., 13, 50
Kristelly, R., 69, 70
Kung, H.F., 123
Kuriyan, J., 166
Kurth, I., 119
Kusama, R.P., 101
Kwak, J.Y., 157
Kwiatkowski, D.J., 9
Kwon, K.B., 50
Kwon, T., 158

L
LaBaer, J., 178
Lai, J.M., 175, 178, 179
Lai, S.Y., 78, 86
Lajoie-Mazenc, I., 141
Lamarche, N., 93
Lamarche-Vane, N., 15, 93, 94, 96
Lambert, J.M., 18
Lang, P., 156, 157
Lappalainen, P., 169
Laury-Kleintop, L.D., 135–149
Lazer, G., 84
Lebowitz, P.F., 129, 136–138, 144, 146, 147
Lee, H.W., 159
Lelias, J.M., 45
Leonard, D., 45
Leptin, M., 70
Lerosey, I., 156
Leslie, K., 177
Leung, T.H., 19, 100, 167–169

Li, F.H., 172
Li, R., 94
Li, X., 15, 29, 179
Liang, J., 178
Licht, J.D., 118
Lim, C.J., 36
Lin, M., 101, 124, 127
Linnekin, D., 160
Liu, A., 136, 142, 145, 147
Liu, A.M., 175
Liu, A.X., 136, 142, 143
Liu, H., xix
Liu, X., 81
Liu, Y., 171, 174, 175, 178, 179
Lloyd, A.C., 178
Lochhead, P.A., 163–182
Logsdon, C.D., 127
Longenecker, K.L., 46, 64, 67, 126
Lowe, S.W., 145, 146
Lowery, D.M., 165
Lu, Z., 159
Lyons, L.S., 78, 87

M
Ma, L., 170
Ma, Z., 167
Machesky, L.M., 19
Madaule, P., 4, 5
Maekawa, M., 19, 36, 168
Maesaki, R., 35
Majumdar, M., 60
Mallat, Z., 172
Malliri, A., xx
Malumbres, M., 4
Manser, E., 19
Mao, J., 60, 66
Marcoux, N., 81
Marinissen, M.J., 175
Marionnet, C., 129
Marks, P.W., 9, 144
Marshall, C.J., xviii, 3, 59, 98, 124, 135, 180, 181
Martin, C.B., 16
Martin-Belmont, F., xix
Masamune, A., 172, 173, 175
Matheson, S.F., 97
Matsui, T., 167
Matsumoto, Y., 179
Mayo, M.W., 145
Mazieres, J., 136, 142–144
McConnell, M.J., 118
McDonald, J.D., 102
McGrath, J.P., 123
McKinnon, C., 111–120
McMullan, R., 70, 173, 177, 178
Medema, R.H., 178
Mehta, D., 13, 48
Meller, N., 14
Mellor, H., 7, 111–120, 138, 168
Menna, P.L., 101
Merajver, S.D., 29–36
Meyer, T.N., 173
Michaelson, D., 31, 138
Midgley, R.S., 102, 103
Mijimolle, N., 147
Miki, H., 19
Miki, T., 14, 17
Miletic, A.V., 83
Milia, J., 141
Miller, R.T., 68
Miller, S.L., 86
Minden, A., 16
Mitchison, T.J., 36
Mitin, N., 18
Miura, Y., 144
Miyazaki, K., 167
Moers, A., 71
Moon, S.Y., 12, 94, 96, 102
Moore, M., 179
Moorman, J.P., 126
Morgan, D.O., 176
Moscow, J.A., 124
Movilla, N., 79–81, 85
Munemitsu, S., 6, 9
Munter, S., 3
Murphy, C., 7

N
Na, S., 13, 50
Nagashima, T., 179
Nagatoya, K., 173
Nagel, M.D., 157
Nakagawa, O., 163
Nakajima, M., 181
Nakamura, S., 68
Nakamura, T., 96
Nakata, Y., 49, 51
Nassar, N., 94
Nasu-Nishimura, Y., 97
Neel, N.F., 139
Nicole, O., 171, 174, 176
Nimnual, A.S., 15
Nishida, M., 83
Nishimaki, H., 171, 177
Nishiya, N., 36
Nobes, C.D., 6, 7, 9, 16, 18, 29, 35, 155, 164

Nomanbhoy, T.K., 46
Nurrish, S.J., 70
Nusser, N., 157

O
O'Brien, L.E., xix
O'Connor, K.L., 157
Offermanns, S., 68, 71
Ogura, K., 83
Ohashi, K., 168
Okabe, T., 97
Oleksy, A., 69
Oliff, A., 136, 146
Olofsson, B., 31, 93, 155
Olson, M.F., 16, 59, 163–182
Ongusaha, P.P., 7
Otis, M., 172–175
Overbeck, A.F., 155

P
Pacary, E., 179
Paduch, M., 114
Pagano, M., 177, 178
Palazzo, A., 141
Palmby, T.R., 81
Pan, J., 146, 147
Pan, Q., 160
Pantaloni, D., 180
Parks, S., 70, 71
Patel, V., 78, 86
Pedeutour, F., 70
Pellegrin, S., 111, 168
Pennica, D., 9
Perez-Torrado, R., 113, 118, 119
Perrot, V., 63
Pertz, O., xvii
Philips, A., 111
Pille, J.Y., 125
Pinner, S., 165
Pintard, L., 118
Pirone, D.M., 173, 174
Platanias, L.C., 53
Polakis, P.G., 6
Popescu, N.C., 99, 100
Porter, K.E., 172, 177
Prendergast, G.C., 129, 135–149
Prieto-Sanchez, R.M., 78, 85
Pringle, J.R., 6, 9
Pruitt, K., 142
Pruyne, D., 141
Pumiglia, K.M., 178

Q
Qian, X., 99
Qiang, Y.W., 160
Qiu, R.G., 16, 59
Quilliam, L.A., 156

R
Rabbitts, T.H., 103
Radeff-Huang, J., 171
Radhika, V., 60
Raftopoulou, M., 30
Ramos, S., 111
Rane, N., 146
Raponi, M., 148, 149
Rashid-Doubell, F., 170
Rattan, R., 179
Rees, R.W., 172
Reid, T., 19
Reimer, J., 49, 50
Repasky, G.A., 30
Reuther, G.W., 62, 63, 70
Ribeiro-Neto, F., 156
Ridley, A.J., xvii, 12, 16, 17, 19, 30, 31, 35, 125, 136, 142, 155, 165, 180
Riento, K., 164, 165, 180
Rikitake, Y., 169, 170
Rittinger, K., 94
Rivero, F., 7, 16, 111
Rivkees, S.A., 172, 177, 178
Roberts, J.M., 176
Robertson, D., 138
Rochlitz, C.F., 123
Rodenhuis, S., 123
Rodier, G., 178
Rohatgi, R., 35
Ron, D., 17
Roof, R.W., 34
Rose, R., 126
Rosen, F.S., 18
Rosenthal, D.T., 29–36
Rossman, K.L., 12, 13, 33, 70, 77, 78
Rowley, J.D., 34
Rubtsov, A., 70
Rujkijyanont, P., 72
Ruley, H.E., 145, 146
Rumenapp, U., 62
Ruppel, K.M., 71
Rushton, E., 14

S
Sahai, E., xvii, xviii, 3, 59, 98, 124, 135, 165, 173, 175, 179–181

Sahota, S.S., 116
Saitoh, O., 64
Sakamuro, D., 137
Samuel, M.S., 163–182
Sanchez, S., 157
Sander, E.E., 124
Sandilands, E., 139, 141
Sase, K., 172
Sassano, A., 53
Sato, N., 136, 143, 144
Sauzeau, V., 80, 156, 171, 173
Sawada, N., 156, 172, 178
Schapira, V., 86
Scheffzek, K., 46
Schmidt, A., 13, 18, 31
Schnelzer, A., 5, 9, 15
Schubbert, S., 135
Schuebel, K.E., 79, 80, 85
Schultz, G., 65
Schunke, D., 50, 52
Schwering, I., 129
Seabra, M.C., 31
Seasholtz, T.M., 171
Sebbagh, M., 165, 179
Sebok, A., 157
Sebti, S.M., 136, 144
Seibold, S., 171, 178
Sekimata, M., 33, 96
Sepp, K.J., 35
Sequeira, L., 125
Servitja, J.M., 85
Settleman, J., 70, 93, 97
Sewing, A., 178
Shang, X., 96, 97
Shellenberger, T.D., 119
Shen, M., 137
Sherr, C.J., 176
Shibata, R., 179
Shikada, Y., 129
Shimizu, T., 164
Shimizu, Y., 169
Shin, I., 178
Shinjo, K., 6, 9
Shinto, E., 129
Shirao, S., 165
Shirotani, M., 172
Shtivelman, E., 102
Shurin, G.V., 119
Shutes, A., 15, 88
Siderovski, D.P., 63
Simpson, K.J., 13, 125
Singer, W.D., 68
Siripurapu, V., 119
Sjoblom, T., 72
Slegers, H., 171, 175
Slep, K.C., 69
Smeland, E.B., 70
Smerdon, S.J., 94
Somlyo, A.P., 168
Somlyo, A.V., 168, 179, 181
Sondek, J., 70
Song, J., 179
Sordella, R., 98
Sorensen, S.D., 171
Stacey, D.W., 123
Stahelin, R.V., 160
Starnes, T., 119
Sternweis, P.C., 68
Stevens, E.V., 3–20
Stogios, P.J., 113, 116
Su, I.H., 82
Su, Z.L., 95
Sumi, T., 36, 168
Suwa, H., 126
Suzuki, N., 63, 64, 66, 67
Svoboda, K.K., 179
Swant, J.D., 171, 174, 176
Swiercz, J.M., 63
Symons, M., 93

T

Takahashi, K., 47
Takai, Y., 124
Takami, A., 166
Takamura, M., 181
Takeda, K., 175
Takeda, N., 172
Tanaka, T., 103, 167
Taneyhill, L., 9
Tang, K., 99
Tao, W., 6
Taya, S., 63
Taylor, J.M., 33, 96
Tcherkezian, J., 15, 93, 94, 96
Tedford, K., 80
Terrak, M., 168
Tesmer, J.J., 69, 70
Tesmer, V.M., 69
Thalmann, G.N., 128
Tharaux, P.L., 172–174
Theodorescu, D., 13, 49, 52, 53
Thompson, P.W., 18
Thumkeo, D., 170
Togashi, H., 62
Togawa, A., 47
Togel, M., 141
Tohyama, M., 47

Toksoz, D., 14
Tominaga, T., 141, 169
Trapp, T., 138
Trinkaus, J.P., 180
Tu, S., 85
Tura, A., 171, 173
Turner, M., 79
Turner, S.J., 53, 54

U
Uchida, S., 173
Ueda, T., xvi, 13
Uehata, M., 166, 171
Ugolev, Y., 48
Uhlinger, D.J., 157
Ullmannova, V., 100
Unger, H., 155–160
Upshaw, J.L., 83

V
Valencia, A., 10
Van Aelst, L., 59, 93
van den Boom, D., 99
van Golen, K., 5, 101, 123–126, 128, 129, 147, 155–160
Van Keymeulen, A., 67
Van Kolen, K., 171, 175
Van Laere, S., 128
van Leeuwen, F.N., 17
Vazquez-Prado, J., 63, 64
Vega, F.M., 12
Velasco, G., 168
Vetter, I.R., 10
Vichalkovski, A., 172
Viglietto, G., 178
Vignais, P.V., 48
Vignal, E., 6
Villanueva, J., 174
Vincent, L., 51
Vincent, S., 6, 9
Vishnubhotla, R., 172
von Tresckow, B., 102
Voyno-Yasenetskaya, T., 60
Vuori, K., 14

W
Wallar, B.J., 137, 141
Walmsley, M.J., xx
Wang, D.A., 144
Wang, G., 172
Wang, L., 98, 101, 104
Wang, S., 143, 144
Wang, W., 129
Ward, Y., 164
Wasmeier, C., 31
Wasserman, S., 141
Watanabe, G., 19
Watanabe, K., 173
Watanabe, N., 141, 169
Watanabe, T., 171
Weaver, A.M., 35
Wedegaertner, P.B., 65–68
Wei, Y., 126
Weidle, U.H., 144
Weinberg, R.A., 49, 59
Welch, H.C., 18
Wells, C.D., 62, 64–68
Wennerberg, K., 3, 10, 15, 30
Westermark, B., 140
Westmark, C.J., 138
Westwick, J.K., 142
Wherlock, M., 138, 139, 141
Whitehead, I.P., xvi, 14, 70, 142
Wickman, G.R., 163–182
Wieschaus, E., 70
Wilkins, A., 16, 116, 118
Wilkinson, S., 163
Williams, D.A., 14
Wirth, A., 71
Wittinghofer, A., 10, 155
Wolf, K., 180, 181
Wong, C.M., 99
Wong, K., 67
Woods, D., 178
Worden, B., 160
Wu, M., 128
Wu, Y.C., 14, 49
Wyckoff, J.B., 181

X
Xia, Z., 53
Xu, N., 60

Y
Yabana, N., 83
Yahiaoui, L., 172, 175
Yamada, K.M., 180
Yamada, T., 63
Yamaguchi, H., 164
Yamamoto, M., 16
Yamashita, T., 47, 48
Yamazaki, D., 180
Yanagawa, T., 160

Yanazume, T., 174
Yang, C., 35, 101
Yang, L., 96
Yao, H., 125, 128, 158
Yarrow, J.C., 166
Yau, D.M., 70
Yeramian, P., 5, 9
Yeung, S.C., 146, 147
Ying, H., 181
Yohe, M.E., 17
Yokoyama, T., 167
Yoneda, A., 165
Yoshihara, T., 120
Yuan, B.Z., 99

Z
Zalcman, G., 45, 138
Zeng, L., 85
Zeng, P.Y., 147
Zhang, B., 13, 49, 50, 52, 94
Zhang, F.L., 11
Zhang, H., 63
Zhang, P., 178
Zhang, Q.G., 158
Zhang, X., 125, 160, 219
Zhang, Y.M., 13, 49, 50, 169, 170
Zhao, M., 175, 179
Zhao, W.M., 96
Zhao, Z., 172, 177, 178
Zheng, X., 34
Zheng, Y., 12, 93, 94, 96, 97, 102
Zhong, H., 64
Zhong, W-B., 179
Zhou, Z., 83
Zigmond, S.H., 18
Zinovyeva, M., 72
Zohar, M., 35
Zohn, I.E., 16
Zondag, G.C., 124
Zrihan-Licht, S., 101
Zuckerbraun, B.S., 171, 175, 178, 179
Zugaza, J.L., 80, 82, 83

Subject Index

A

Acute myeloid leukemia, 53
Androgen receptor (AR), 87
Animal models, xx
Antiparallel coiled-coil finger (ACC finger) fold, 35

B

Biological process, upstream signaling and downstream effectors
  actin cytoskeleton regulation, 18–19
  linking extracellular signals, 17–18
  protein kinase effectors, 19
  signal transduction, 16–17
Broad-Complex, Tramtrack, and Bric-a-brac (BTB) protein. *See also* RhoBTB proteins
  BTB domain, 113
  C terminal domain, 116
  Rho domain, 113–115

C

Cell biology, xvii
Cell migration, xvii–xviii
Chimaerins, 101–102
Cyclin-dependent kinases (CDKs), 176– 178

D

Deleted in liver cancer (DLC), 99–100
Dictyostelium discoideum, 111
Downstream effectors, xvi–xvii
Drosophila melanogaster, 9, 14, 35, 71

E

Effector interaction domain, 10
Endocytic trafficking, RhoB GTPase
  Akt/PKB protein, 139
  MAP1A/LC2 complex, 141, 142
  multivesicular late endosomes (MVBs), 138
  platelet-derived growth factor (PDGF), 140
Epithelial to mesenchymal transition (EMT), 125
Evolution of Rho GTPase
  Cdc42, 9
  discovery, 4–7
  low stringency hybridization, 9
  timeline, 8
Extracellular matrix (ECM), xviii

F

Farnesyltransferase inhibitors (FTIs)
  cellular response mechanism, 148–149
  geranylgeranyl transferase I, 146–147
  reactive oxygen species (ROS), 147–148
  RhoB GTPase
    anticancer response, 146–149
    cancer suppression, 143–145
    effector molecular interactions, 140–142
    endocytic trafficking, 138–140
    G2/M phase, 146
    human cancer, 143–144
    mouse embryonic fibroblasts (MEFs), 145
    neoplasia, 142–143
    structure, 137–138
Fluorescence anisotropy, 82

G

Galpha12/13 and RGS-RhoGEFs signaling
DH/PH domains
LARG, 69–70
PDZ-RhoGEF, 69
functions
G13-mediated, 60, 61
LARG, 62–63
model organisms, 70
p115RhoGEF, 60
phenotypes, 71
RGS-RhoGEF, phenotypes, 70–71
two-hybrid screening, 62
human disease, 71–72
interaction, proteins, 63
regulatory mechanisms
Gα, 63–64
GIRK kinetics, 64
kinetic and thermodynamic analysis, 64
oligomerization, 65
phosphorylation, 65–66
subcellular localization, 66–67
structural basis
p115RhoGEF GAP activity, biochemical analysis, 68–69
RGS-like domains, 67–68
GDI displacement factors (GDFs), 46–48
GDP dissociation inhibitors (GDIs), 12
G-protein-coupled receptors (GPCRs), 16–18
GTPase-activating proteins (GAPs), xv–xvi, 12, 33–34
Guanine nucleotide dissociation inhibitors (GDIs), xv–xvi
Guanine nucleotide exchange factors (GEFs), xv–xvi
Dbl-homology (DH) domain, 31
GDP/GTP-binding motif, 33
regulators, 12

H

Head and neck small cell carcinomas (HNSCC), 116, 117, 160
Histone deacetylase (HDAC), 143–144
HMG-CoA reductase, 53
Human embryonic kidney (HEK) cells, 50
Human erythroleukemia (HEL), 54
Human microvascular endothelial cells (HMECs), 51

I

Inflammatory breast cancer (IBC), 128
Intracellular signaling proteins
biochemical activity
conformational changes, GAP, 94–95
domain features, 95–96
cellular function
negative regulators, 96–97
p85 subunits, 97
human cancers
chimaerins, 101–102
deleted in liver cancer (DLC), 99–100
other RhoGAPs, 102
p190 RhoGAP, 101
physiological study
adipogenesis/ myogenesis, 98
in vitro biochemical and cell biological studies, 97
therapeutic strategies, 102–104
Isoprenylcysteine carboxylmethyltransferase (ICMT), 10–11

L

Leukemia associated RhoGEF (LARG)
DH/PH domains, 69
fusion partner, 71
Gα regulation, 63–64
Gα13-LARG pathways, 71
glycine mutation, 68
nucleo-cytoplasmic shuttling, 67
oligomerization, 65
other protein interaction, 63
phosphorylation, 66
protooncogene products, 98
Loss of heterozygosity (LOH), 116–117
Lysophosphatic acid (LPA), 16, 18

M

Matrix metalloproteases (MMPs), xviii
Mixed lineage leukemia (MLL), 62, 78
Morphogenesis, xix
Mouse embryonic fibroblasts (MEFs), 145
Multivesicular late endosomes (MVBs), 138, 140
Myosin light chain kinase (MLCK), 168

N

Nerve growth factor (NGF), 157
Nonsmall cell lung carcinomas (NSCLCs), 52, 143, 144
Nuclear magnetic resonance (NMR), 81
Nucleo-cytoplasmic shuttling, 67

P
p21-activated kinase (PAK), 19
PDGF. *See* Platelet-derived growth factor
Phosphorylation events
  Akt/Protein Kinase B, 158–159
  protein kinase C, 159–160
  protein kinases A and G, 156–157
Platelet-derived growth factor (PDGF), 16, 140–142
Prolactin receptor (PRL), 86
Promyelocytic leukemia zinc finger protein (PLZF), 113, 114, 117–118
Prostate cancer tumorigenesis, 87
Protein kinase effectors, 19

R
Reactive oxygen species (ROS), 147–148
Receptor tyrosine kinases (RTKs), 79
Regulators of Rho GTPase
  classification, 12–13
  Dbl oncoprotein, 13–14
  DOCK proteins, 14
  and functions
    C-terminal membrane targeting domain, 10, 11
    CAAX motif, 10–11
    signaling events, 12
    switch I and switch II domains, 10
  human RhoGDIs isoforms, 13
  RhoGAP, 14–15
RhoA GTPase
  nerve growth factor (NGF), 157
  protein kinase G (PKG), 156
Rho-associated protein kinases (ROCK)
  actin cytoskeleton role, 173–174
  cancer signaling, 170–171
  cell survival, 179–180
  embryonic development
    abnormalities, 169
    genetic deletion, 170
  expression, substrates, and function, 167–168
  in vivo models, 181
  inhibitor specificity
    additional conditions, 167
    experiments, 166
  proliferation
    antiproliferative effect, 173
    tissue culture and in vivo, 171–172
  signaling pathways
    CDK inhibitors, 177–179
    cell cycle regulation, 175–176
    cyclins and CDKs, 176–177
    mitogen-activated protein kinase, 174–175
    Stat3, 175
  stress fiber formation and cell adhesion, 168–169
  structure and regulation
    apoptosis, 165
    isoforms, 163–164
    regulatory mechanism, 164–165
  tumor cell invasion and metastasis
    cellular migration, 180–181
    dynamic reorganization, 180
Rho-binding domain (RBD), 163–164
RhoB GTPase
  cancer
    decreased expression, 143
    histone deacetylase (HDAC), 143–144
    Ras superfamily, 144–145
    suppression, neoplasia, 142–143
  DNA damage
    G2/M phase, 146
    mouse embryonic fibroblasts (MEFs), 145
  farnesyltransferase inhibitors (FTIs)
    anticancer response, 146–149
    cancer suppression, 143–145
    effector molecular interactions, 140–142
    endocytic trafficking, 138–140
    G2/M phase, 146
    human cancer, 143–144
    mouse embryonic fibroblasts (MEFs), 145
    neoplasia, 142–143
    structure, 137–138
RhoBTB proteins. *See also* Broad-Complex, Tramtrack, and Bric-a-brac (BTB) protein
  BTB structure
    BTB domains, 112– 113, 114
    C-terminal, 116
    Rho domain, 114
  cell cycle, 119–120
  genetic alteration
    allelic loss/ gene silencing, 117
    loss of heterozygosity (LOH), 116–117
  signalling, 117–118
  transcriptional regulation, 118–119
  tumour suppressors, 120
RhoC GTPase
  cancer and metastasis
    collagens, 128–129

RhoC GTPase (*cont.*)
- farnesyl transferase inhibitors (FTIs), 129–130
- microarray comparison, 127–128
- vacular endotheial growth factor-C (VEGF-C), 128

vs. RhoA GTPase
- epithelial to mesenchymal transition (EMT), 125
- hypervariable region, 124
- sequence homology, 124–125
- structural analyses, 126

RhoGDP dissociation inhibitors (RhoGDIs)
- angiogenesis, 51
- drug targets, 55
- function
  - cytosol-to-membrane cycle, 45–46
  - trafficking, 47
- HMG-CoA inhibitors (statins)
  - anti-neoplastic effects, 53–54
  - protein geranylgeranylation, 53, 54
- prognostic indicators, 52–53
- regulation
  - biologically active lipids, 48
  - GDI displacement factors, 47–48
  - phosphorylation regulation, 48–49
- RhoA *vs.* Rac/Cdc42, biological effects, 52
- tumor apoptosis
  - chemotherapeutic agents, 49
  - D4GDI role, 50
- tumor invasion and metastasis, 50–51

Rho proteins, cancer
- family structure, 30–31
- malignant transformation and metastasis, 30
- regulatory
  - cell migration, 35–36
  - effector interactions, general considerations, 34–35
  - GAPs, 33–34
  - GEFs, 31–33

S

Serum response element (SRE), 60
Shwachman-Diamond syndrome, 72
Single nucleotide polymorphism (SNP), 72
Surface plasmon resonance, 64

T

Tumor cell invasion and metastasis, 180–181

V

Vacular endotheial growth factor-C (VEGF-C), 128
Vav proteins
- intracellular signaling linking
  - APS adaptor molecule, 82–83
  - CH domain, 82
  - normal *vs.* transforming potential, 83–84
  - SH3 domains, 83
  - Zyxin, 84
- role, human cancer
  - Bcr-Abl fusion protein, 85
  - intracellular signaling pathways, 85
  - neuroblastomas, 85
  - osteopontin, 86
  - prostate cancer tumorigenesis, 87
- structure, function, and regulation
  - activation mechanism, 80
  - GTPase recognition, 81
  - hematopoietic cell-specific signaling, 79
  - N-terminal calponin homology, 80
  - proto-oncogenes, 79, 82
  - Tyr^{174} phosphorylation motif, 81

VE-cadherin internalization, 88